Lecture Notes in Civil Engineering

Volume 173

Lecture Notes in Civil Engineering (LNCE) publishes the latest developments in Civil Engineering - quickly, informally and in top quality. Though original research reported in proceedings and post-proceedings represents the core of LNCE, edited volumes of exceptionally high quality and interest may also be considered for publication. Volumes published in LNCE embrace all aspects and subfields of, as well as new challenges in, Civil Engineering. Topics in the series include:

- Construction and Structural Mechanics
- Building Materials
- Concrete, Steel and Timber Structures
- Geotechnical Engineering
- Earthquake Engineering
- Coastal Engineering
- Ocean and Offshore Engineering; Ships and Floating Structures
- Hydraulics, Hydrology and Water Resources Engineering
- Environmental Engineering and Sustainability
- Structural Health and Monitoring
- Surveying and Geographical Information Systems
- Indoor Environments
- Transportation and Traffic
- Risk Analysis
- Safety and Security

To submit a proposal or request further information, please contact the appropriate Springer Editor:

- Pierpaolo Riva at pierpaolo.riva@springer.com (Europe and Americas);
- Swati Meherishi at swati.meherishi@springer.com (Asia - except China, and Australia, New Zealand);
- Wayne Hu at wayne.hu@springer.com (China).

All books in the series now indexed by Scopus and EI Compendex database!

More information about this series at https://link.springer.com/bookseries/15087

Sergey Vasil'yevich Klyuev
Editor

Digital Technologies in Construction Engineering

Selected Papers

Springer

Editor
Sergey Vasil'yevich Klyuev
Belgorod State Technological University
Belgorod, Russia

ISSN 2366-2557 ISSN 2366-2565 (electronic)
Lecture Notes in Civil Engineering
ISBN 978-3-030-81288-1 ISBN 978-3-030-81289-8 (eBook)
https://doi.org/10.1007/978-3-030-81289-8

This Springer imprint is published by the registered company Springer Nature Switzerland AG
The registered company address is: Gewerbestrasse 11, 6330 Cham, Switzerland

Organizing Conference Committee

E. I. Evtushenko, Doctor of Engineering Sciences (Advanced Doctor), Professor, Belgorod State Technological University named after V. G. Shukhov, Belgorod, Russia
T. M. Davydenko, Doctor of Pedagogical Sciences (Advanced Doctor), Professor, Belgorod State Technological University named after V. G. Shukhov, Belgorod, Russia
S. V. Klyuev, Candidate of Engineering Sciences (Ph.D.), Associate Professor, Belgorod State Technological University named after V. G. Shukhov, Belgorod, Russia
V. S. Lesovik, Doctor of Engineering Sciences (Advanced Doctor), Professor, Corresponding Member of RAASN, Belgorod State Technological University named after V. G. Shukhov, Belgorod, Russia
N. I. Vatin, Doctor of Engineering Sciences (Advanced Doctor), Professor, Peter the Great St. Petersburg Polytechnic University, St. Petersburg, Russia
R. S. Fediuk, Candidate of Technical Sciences (Ph.D.), Professor, Far Eastern, Federal University, Vladivostok, Russia

Scientific Conference Committee

Abolfazl Soltani, Ph.D., Shahid Rajaee Teacher Training University, Iran
A. Ramachandra Murthy, Ph.D., CSIR Structural Engineering Research Centre, India
Doo Yeol Yoo, Ph.D., Hanyang University, South Korea
M. A. Kovtun, Ph.D., Australia
M. I. Kozhukhova, Ph.D., University of Wisconsin-Milwaukee, USA
Mohammad Ali Asaad, Ph.D., Iraq University College, Iraq
Mohammad Ali Mosaberpanah, Ph.D., Cyprus International University, Cyprus
Mugahed Amran, Ph.D., Prince Sattam Bin Abdulaziz University, Saudi Arabia
Murali Gunasekaran, Ph.D., SASTRA Deemed University, India

Preface

The 1st Internaional Conference "Digital Technologies in Construction Engineering" was held from 8 to 9 June 2021 in Belgorod (Russian Federation), on the basis of Belgorod State Technological University named after V. G. Shukhov, Belgorod, Russia.

An important topic of the relevance of the issue of digitalization of the production process in the construction industry was discussed at the conference. Within the framework of the international conference, the participants will present reports reflecting the latest achievements of using and application of digital tools to improve delivering and operating the built environment; in the field of building materials science and building materials; technologies for their production and processing; ecological and economic efficiency of their application in various fields of modern production, as well as in the construction of building structures. As a result, there is a real opportunity among a wide range of scientists, teachers, industry representatives, and students in various fields related to material engineering, science, to exchange ideas, share knowledge, and establish close cooperation. The successful organization and holding of the conference are evidenced by the wide geography of the participants, as well as the high level of the reports presented. With its high quality, it provides an exceptional value for students, academics, and industry researchers. It also provides a premier interdisciplinary platform for researchers, practitioners, and educators to present and discuss the most recent innovations, trends, and concerns as well as practical challenges encountered and solutions adopted in the fields of construction engineering and digital technologies.

The conference has an active participation of more than 220 scientists and experts from different Russian regions (up to 40) and seven countries (Australia, Iran, South Korea, Kazakhstan, Tajikistan, the Ukraine, and Uzbekistan).

The conference covered a fairly wide range of different scientific topics, united in the following three main areas:

1. Industrial and civil engineering;
2. Special and unique structures construction;
3. Engineering structure safety, environmental engineering, and environmental protection.

A major part of theoretical information and experimental data is the results of studies, which were realized within the framework of implementation of state and national programs in the scientific field as a grant, federal targeted programs, decrees of the Government of the Russian Federation, and other programs.

Thanks to the invaluable contribution of a group of highly qualified reviewers, all articles went through high-quality peer review and several stages of editing. The organizing committee expresses its sincere gratitude to everyone who contributed to the conference. In addition, we would like to express our deep gratitude to the authors, reviewers, participants, and the entire organizing team for their support and enthusiasm, which ensured the success of the conference.

We would like to acknowledge all of those who supported 1st International Conference "Digital Technologies in Construction Engineering". Each individual and institutional help was very important for the success of this conference. Especially, we would like to thank the organizing committee for their valuable advices in the organization and helpful peer review of the papers. We sincerely hope that 1st International Conference "Digital Technologies in Construction Engineering" will be a forum for excellent discussions that will put forward new ideas and promote collaborative researches. We are sure that the proceedings will serve as an important research source of references and the knowledge, which will lead to not only scientific and engineering progress but also other new products and processes.

All conference participants express deep gratitude to the science team.

Belgorod, Russia

Sergey Vasil'yevich Klyuev
Associate Professor

Contents

Mohr–Coulomb Models for Dynamic Analysis of Granular Metamaterials in Earthquake Engineering

Sergey Kuznetsov and Ali Al Shemali

Abstract The Finite Element Analysis of Mohr–Coulomb models was used to study the effectiveness of adding a pad of granular metamaterials under the foundation to decrease influence of seismic shear waves. The FE model consists of a ten-story superstructure rested on the slab foundation, under which there is a layer of granular metamaterials. The values of five variables that affect the mechanical properties of these metamaterials have been analyzed (density—cohesion—internal friction angle—Young's modulus—Poisson's ratio) for two different pad thicknesses. The dynamic analysis performed using the software package Abaqus/CAE showed the effectiveness of the granular metamaterials in their ability to significantly reduce magnitudes of displacements, velocities and accelerations in the building compared to the same values in the absence of these metamaterials. The analysis also revealed that among the studied variables, the cohesion is the parameter most influencing the effectiveness of metamaterials in their ability to dissipate seismic waves, while no significant effect was observed for the other parameters.

Keywords Seismic waves · Mohr–Coulomb criterion · Metamaterials · Seismic pads · Finite element analysis

1 Introduction

Seismic base isolation is one of the most popular means of protecting a structure against earthquake actions [1]. However, the currently most common seismic isolation systems based on the rubber-metal seismic isolators have some disadvantages, the most important of which are: the underground parts of structures remain unprotected from vibrations, which are caused by the diffracted seismic waves transmitted directly from soil; insufficient durability, for example the rubber-metal

S. Kuznetsov
Institute for Problems in Mechanics, Russian Academy of Sciences, Moscow, Russia

A. Al Shemali (✉)
Moscow State University of Civil Engineering, Moscow, Russia

S. V. Klyuev (ed.), *Digital Technologies in Construction Engineering*,
Lecture Notes in Civil Engineering 173, https://doi.org/10.1007/978-3-030-81289-8_1

seismic isolator service life is usually limited to 20–25 years [2]. Consequently, nowadays there is a strong demand in developing more robust and durable seismic protection systems, including seismic barriers and seismic pads. This paper is intended to investigate the effectiveness of adding a seismic pad under the slab foundation of a multi-storey building exposed to the impact of seismic shear waves.

Granular metamaterials used for fillings of seismic barriers and seismic pads are of extensive research at different laboratories, institutes, and companies across the globe [3]. These materials are designed to handle earthquake vibrations. They are characterized by mechanical properties, including: the density $\rho = 1200 - 1700\,\mathrm{kg/m^3}$, the angle of internal friction $\phi \leq 1°$ and the cohesion $C = 1 - 100\,\mathrm{kPa}$.

In this study, the Finite Element code was used to create and analyze the Mohr–Coulomb models for these metamaterials. The Finite Element Method (FEM) is an effective way for treating wave diffraction problems in various media. Applications FE of modeling to note account of energy dissipation in structures, to imposing non-reflecting boundary conditions, to modeling seismic sources and to seismic protection are considered in [4–8].

2 Methods and Materials

To study the effect of granular metamaterials on the dispersion of seismic shear waves, a dynamic analysis of a finite element model was performed which consisted of a ten-story building plus a pad under the mat. The properties of the construction material (concrete) are shown in the Table 1. An 8-node linear hexahedral solid element with reduced integration (C3D8R) was used for mesh generation to solve the FE model. Figures 1 and 2 show respectively a floor plan of this building and the 3D model with pad generated in engineering software Abaqus/CAE 6.14.

In order to simulate the movement of seismic shear waves, an acceleration was applied under the pad in two directions (x, z), which changes with time according to the sinusoidal periodic law [9], and so that the maximum amplitude is equal to 5 m/s^2, considering that the construction is located in a zone of seismic intensity 9 according to the modified Mercalli scale [10].

To obtain the desired results, the effect of a set of parameters on the values of displacement, velocity and acceleration was studied for two different thicknesses of the pad (d = 0.1 m, 0.5 m) according to the following cases, when only one parameter was varying, while other were remained unchanged:

1. Varying density values (ρ);
2. Varying cohesion (C);
3. Varying angle of friction (ϕ);

Table 1 The material properties of concrete used in software Abaqus

Modulus of elasticity	Poisson's ratio	Density
30×10^9 N/m^2	0.25	2200 kg/m^3

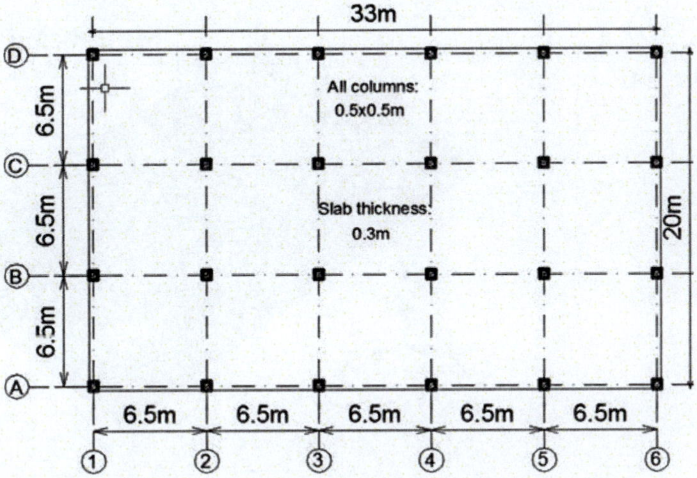

Fig. 1 Floor plan of the building

Fig. 2 3D Finite element model of the building with pad

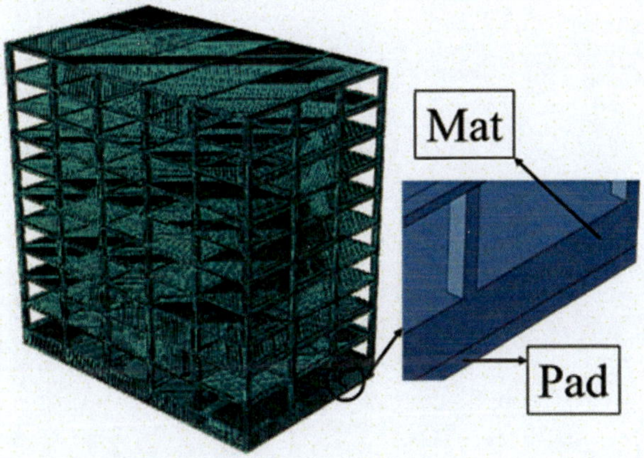

4. Varying Young's modulus (E), at thickness of the pad 10 cm;
5. Varying Poisson's ratio (ν), at thickness of the pad 10 cm.

Abaqus allows changing the values of these parameters by defining material behaviors (density—elasticity—Mohr Coulomb plasticity).

3 Results and Discussion

The explicit dynamic analysis was performed for the seismic frequency f = 2 Hz over a time period of 5 s for all the above-mentioned cases. After the analysis, the values of displacements (U), velocities (V), and accelerations (A) in the x-direction

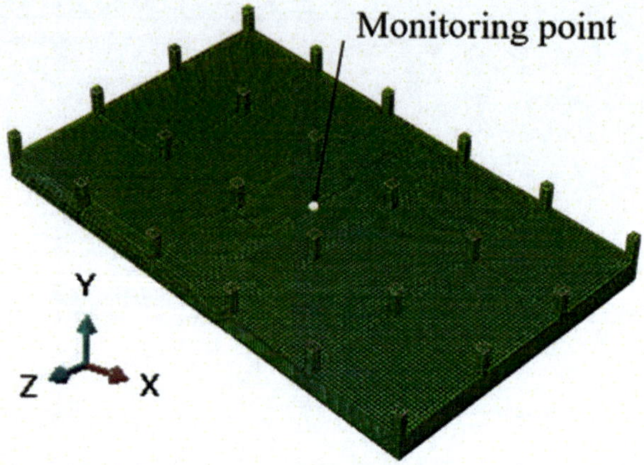

Fig. 3 Location of monitoring point on the upper surface of the mat

were considered at a point located on the upper surface of the mat, as shown in the Fig. 3.

The values obtained were organized in the following tables for each case of the study:

Case 1:

$$E = 30\,\text{MPa},\ \nu = 0.25,\ \phi = 0.1°,\ C = 1\,\text{kPa}.$$

See Table 2.

Case 2:

$$E = 30\,\text{MPa},\ \nu = 0.25,\ \rho = 1700\,\text{kg/m}^3,\ \phi = 0.1°.$$

See Table 3.

Case 3:

$$E = 30\,\text{MPa},\ \nu = 0.25,\ \rho = 1700\,\text{kg/m}^3,\ C = 1\,\text{kPa}.$$

See Table 4.

Table 2 Values of displacements, velocities, and accelerations when changing the density of metamaterial

	$\rho = 1200\,\text{kg/m}^3$		$\rho = 1500\,\text{kg/m}^3$		$\rho = 1700\,\text{kg/m}^3$	
	d = 0.1 m	d = 0.5 m	d = 0.1 m	d = 0.5 m	d = 0.1 m	d = 0.5 m
U1 (m)	0.7894	0.7480	0.7885	0.7410	0.7877	0.7372
V1 (m/s)	0.2721	0.2596	0.2720	0.2595	0.2721	0.2589
A1 (m/s^2)	−0.2425	−0.2166	−0.2414	−0.2107	−0.2407	−0.2070

Table 3 Values of displacements, velocities, and accelerations when changing the cohesion of granular metamaterial

	C = 1 kPa		C = 5 kPa		C = 10 kPa	
	d = 0.1 m	d = 0.5 m	d = 0.1 m	d = 0.5 m	d = 0.1 m	d = 0.5 m
U1 (m)	0.7877	0.7372	1.6076	1.5335	1.7923	1.7518
V1 (m/s)	0.2721	0.2589	0.4764	0.4585	0.5793	0.5523
A1 (m/s^2)	−0.2407	−0.2070	−1.2264	−1.0771	−2.4464	−2.0861

Table 4 Values of displacements, velocities, and accelerations when changing the angle of friction of granular metamaterial

	$\phi = 0.1°$		$\phi = 0.5°$		$\phi = 1°$	
	d = 0.1 m	d = 0.5 m	d = 0.1 m	d = 0.5 m	d = 0.1 m	d = 0.5 m
U1 (m)	0.7877	0.7372	0.7877	0.7369	0.7874	0.7366
V1 (m/s)	0.2721	0.2589	0.2722	0.2589	0.2721	0.2588
A1 (m/s^2)	−0.2407	−0.2070	−0.2407	−0.2070	−0.2407	−0.2069

Case 4:

$$\nu = 0.25,\ \rho = 1700\,\mathrm{kg/m^3},\ \phi = 0.1°,\ C = 1\,\mathrm{kPa}.$$

See Table 5.

Case 5:

$$E = 30\,\mathrm{MPa},\ \rho = 1700\,\mathrm{kg/m^3},\ \phi = 0.1°,\ C = 1\,\mathrm{kPa}.$$

See Table 6.

Table 5 Values of displacements, velocities, and accelerations when changing Young's modulus, d = 0.1 m

	E = 30 MPa	E = 50 MPa
U1 (m)	0.7877	0.7881
V1 (m/s)	0.2721	0.2722
A1 (m/s^2)	−0.2407	−0.2409

Table 6 Values of displacements, velocities, and accelerations when changing Poisson's ratio, d = 0.1 m

	$\nu = 0.25$	$\nu = 0.35$
U1 (m)	0.7877	0.7878
V1 (m/s)	0.2721	0.2721
A1 (m/s^2)	−0.2407	−0.2407

Table 7 Values of displacements, velocities, and accelerations without seismic pad

U1 (m)	2.0095
V1 (m/s)	0.8061
A1 (m/s^2)	4.5873

All values in the above tables were compared with each other and with the values of displacements, velocities and acceleration in the absence of a seismic pad under the foundation of the building, which are shown in the Table 7.

By looking at the values shown in the tables, a set of results were drawn, which are listed below.

- Placing a pad of granular metamaterials under the slab foundation of the building reduced the displacement values, velocities by more than 60% and reduced the acceleration values by more than 90% in the case of the smallest value for the cohesion of metamaterials.
- Among the mechanical properties of metamaterials, cohesion has the greatest influence, as it has been found that the effectiveness of the pad is significantly reduced with a small increase in cohesion.
- The influence of the density, angle of friction, Young's modulus and Poisson's ratio of the granular metamaterials on the decrease in the values of displacements, velocities and accelerations is almost imperceptible.
- An increase in the seismic pad thickness does not have a noticeable effect in reducing the values of displacements, velocities and accelerations.

According to these results, the thickness of the seismic pad 10 cm can be adopted for the building under study, as increasing the thickness of the pad will not be economically feasible.

4 Conclusion

The effectiveness of placing a pad of granular metamaterials is proved in significantly reducing the vibration of a multi-storey building under the influence of seismic shear waves according to the results of dynamic analysis using the Mohr–Coulomb failure criterion. Moreover, to obtain more objective results, it is recommended to perform the analysis for different seismic frequencies. It is also recommended to study the effect of the number of building floors, in addition to its geometry, on the thickness of the seismic pad.

References

1. Datta TK (2010) Seismic analysis of structures. John Wiley & Sons
2. Vershinin V, Javkhlan S, Saidmukaram S (2019) Seismic pads to protect buildings and structures from bulk seismic waves. E3S Web Conf. 97:04047
3. Goldstein RV et al (2016) The modified Cam-Clay (MCC) model: cyclic kinematic deviatoric loading. Arch Appl Mech 86:2021–2031
4. Jones RM (2009) Deformation theory of plasticity. Bull Ridge Corporation
5. Li S et al (2019) Hybrid asynchronous absorbing layers based on Kosloff damping for seismic wave propagation in unbounded domains. Comput Geotech 109:69–81
6. Kuznetsov SV, Terentjeva EO (2015) Planar internal Lamb problem: waves in the epicentral zone of a vertical power source. Acoust Phys 61:356–367
7. Kuznetsov SV (2011) Seismic waves and seismic barriers. Acoust Phys 57(3):420–426
8. Dudchenko A (2018) Numerical analysis of surface Rayleigh wave interaction with seismic barriers and pile fields accounting elastic-plastic soil behaviour. PhD thesis, UJF: Grenoble
9. Al Shemali A (2021) Diffraction of harmonic S-waves into frame buildings. In: Proceedings of the international conference industrial and civil construction 2021, pp 85–92. Springer International Publishing
10. Jain AK. Performance based seismic design of tall buildings: risks and responsibilities. https://cecr.in/CurrentIssue/pages/20147

Distribution of the Mixed Bulk Components After Impact on the Baffle Plane

A. B. Kapranova, D. V. Stenko, and N. M. Lozovoy

Abstract To obtain a high-quality mixture of granular media for various purposes, it is required to establish the appropriate rational limits for changing the parameters of this technological process. In the case of mixing bulk materials in a rotary way, the authors previously identified the following most significant parameters: the angular speed of rotation of the mixing drum with elastic non-radial blades and the rate of their deformation. The purpose of this work is to study the influence of the main operating and design parameters during the formation of scattered flares of bulk components during the operation of the specified mixing drum (stage 1) on their distribution in reflected flares after hitting the baffle plane (stage 2). A reflection of the application of the energy method is the obtained relationship between the distribution laws of granular media at the listed stages 1 and 2. In particular, the results showed that an increase in the angular speed of rotation of the mixing drum by a factor of 1.27 leads to a (1.11–1.25) -fold decrease in the values of the differential distribution function of granular components for the first extremum of the angle of reflection from the bump plane. The specified limit is characteristic for each row of elastic blades fixed on the surface of the mixing drum with a fixed angle of displacement.

Keywords Rotary mixing · Bulk materials · Elastic blades · Baffle plane · Model

1 Introduction

The design features of mixing equipment are directly related to its functional tasks within the framework of the implementation of a specific technological chain. The need to improve the mixing process is observed in almost all areas of processing bulk materials for many reasons. These factors include, for example: expanding the

A. B. Kapranova · D. V. Stenko
Yaroslavl State Technical University, Yaroslavl, Russia

N. M. Lozovoy (✉)
Belgorod State Technological University Named After V.G. Shukhov, Belgorod, Russia

S. V. Klyuev (ed.), *Digital Technologies in Construction Engineering*,
Lecture Notes in Civil Engineering 173, https://doi.org/10.1007/978-3-030-81289-8_2

list of required ready-made bulk products; constantly changing conditions for obtaining the constituent working components; increasing quality requirements [1]; updating of technological operations, etc. In particular, the production of general-purpose building materials [2] or preliminary compositions for obtaining glass requires compliance with the regulatory mixing proportions of granular components, including 1:10 or more.

The accepted standards for the composition of the final bulk product are reflected in the constructive ways of achieving them [3], taking into account cohesive and segregation effects [4].

The purpose of this research is to study the effect of the main operating and design parameters of the designed rotary mixer [5], which correspond to different stages of its operation, on the mixing process of bulk components. In particular, the relationship of the process parameters is analyzed at stage 1 (when scattered flares of bulk components are formed during the operation of the mixing drum) and stage 2 (when particles are distributed in reflected flares after hitting the baffle plane).

The difficulties of designing a new apparatus for the successful mixing of bulk materials cannot be overcome without the development of the theoretical foundations of the technological process itself according to the system-structural approach [6, 7]. The use of a combination of several mixing methods within one apparatus is justified in previous studies [8–11] related to the development of equipment based on the principle of penetration of scattered flares of one bulk component into another. For example, the effectiveness of this approach has been proven when implementing the previously mentioned stages 1 and 2 on a moving belt [8, 9], on a chute of a gravitational device [9–11]. The main objectives of this work are as follows:

(a) establishing the factors of the consistent influence of the design and operating parameters of mixing stage 1 on the corresponding parameters of stage 2 using the previously proposed principles for the formation of sets of information variables;
(b) predicting the behavior of particles of the components of a free-flowing mixture at stage 2 when changing the parameters of stage 1 within the selected range of variation from the standpoint of stochastic modeling by the energy method.

Note that this method can be considered as an alternative to other well-known stochastic methods for describing the kinetics of the behavior of the mixed components, for example, in the framework of the application of Markov A.A. chains, these are.

2 Materials and Methods

The use of the proposed rotary method for mixing granular materials (stage 1) in combination with a blow on the bumping plane (stage 2) suggests various options for obtaining a free-flowing mixture in the working volume of the new apparatus.

For example, this process can be implemented on a moving belt with additional pressing of the finished mixture by a compaction roller or in the working chamber of the apparatus with a system of baffle elements while removing the mixture by gravity. However, to form the theoretical foundations for the design of new mixers with the indicated stages 1 and 2, it is necessary to determine the distribution of particles of each component in the scattered flows formed at these stages, in particular, for the latter, depending on the angle of reflection γ_{2i} from the baffle plane.

At stage 1, the main unit of the apparatus is a mixing drum with elastic non-radial blades [5]. The location of these projections of the blades coincides with the direction of rotation of the mixing drum with an angular velocity ω. These rectangular blades are fixed on the cylindrical surface of the drum in rows so that the projections of the blades on the transverse plane of the section to the axis of rotation of the drum are perpendicular to its radius. The notation be adopted: r_b drum radius, m; β is offset angle for installed rows of blades, rad; k is number of their ranks; n_0 is the number of blades in each row; l_b, q_b, h_b are length, width and height of a rectangular blade, m; h_0 is the distance between the drum and the surface from which the blades of particles of bulk components are captured, m. Note that the specified bottom surface can have a different profile, for example, (a) coincide with the horizontal plane of the movable belt [5]; (b) correspond to a cylindrical surface located coaxially [6] or eccentrically relative to the drum axis. The loading of the mixed components onto this lower surface occurs vertically from the metering hoppers. The results of the operation of the apparatus at stage 1 are scattered torches of particles of each component, which are pulled out from the described gap at a scattering angle λ_i to the tangent bottom surface.

At stage 2, the previously formed scattered torches of particles of the mixed materials interact in an impact manner with an inclined baffle plane. The result of the operation of the designed device at stage 2 corresponds to the mixture obtained after the reflection of the primary flares from the specified baffle element. In this case, the designed bumper can have both a corrugated surface [5] and coincide with the inclined walls of the apparatus body [6]. The finished granular mixture consist of two components (i = 1, 2). the particles of each working granular material have a spherical surface, their diameters $D_i = N_v^{-1}\sum_{v=1}^{N_v} D_{iv}$ coincide with the diameters obtained by averaging the sizes D_{iv} by selected fractions N_v.

As already noted in the introduction, the construction of a stochastic description of the behavior of macrosystems of particles in the case under consideration presupposes the use of the energy method Yu L Klimontovich. Assumptions are made, similar to work [2, 3] for stage 1 about the absence of: energy exchange of these macrosystems; macroscale fluctuations during their transition from one state to another; homogeneity, continuity and stationarity of a random process Markov A.A.

Therefore, the construction of differential distribution functions of the number of particles N_{iv} for each component $u_{iv}(\gamma_{2i}) = N_{iv}^{-1}dN_{iv}/d\gamma_{2i}$; $\gamma_{2i} = k_0^{-1}\sum_{v=1}^{k_0}\gamma_{2iv}$ at stage 2, proposed in this work, is performed taking into account the previously established laws of distribution of components $g_{iv}(\varphi_{iv}) = N_{iv}^{-1}dN_{iv}/d\varphi_{iv}$ at stage 1, as stationary solutions of the kinetic Fokker–Planck equation with a given element of the phase volume $d\Sigma_{iv}$. For the convenience of the geometric interpretation of the problem, a transition was made from some phase variables (components of the velocity of motion of the center of mass of particles $d\Sigma_{iv} = dV_{xiv}dV_{yiv}$) to others ($d\Sigma_{iv} = -\omega^2 r_{iv}^2 dr_{iv} d\varphi_{iv}$, где ($r_{iv}$, φ_{iv}) are polar coordinates. In this case, the center of the introduced polar coordinate system coincides with the center of the Archimedes spiral—a curve that is described by those points of the projection of the blades that have finished sliding along the lower surface of the apparatus after their clearance with the drum.

Taking the position of a stationary solution of the Fokker–Planck equation

$$dN_{iv} = \alpha_{iv} \exp(-E_{iv}/E_{0iv}) d\Sigma_{iv} \tag{1}$$

apply a linear expansion of the relationship for the relationship between the angle of dispersion of particles φ_{iv} in stage 1 and the angle of reflection γ_{2iv} for stage 2

$$\varphi_{iv}(\psi, \delta, \gamma_{2i}) = c_2(\psi, \delta)\gamma_{2iv} + c_1(\psi, \delta). \tag{2}$$

Here it is indicated: α_{iv} is normalization parameter; E_{iv} is energy of stochastic motion of component particles i = 1, 2 when working a deformed blade in a row of the total k, J; E_{0iv} is energy parameter determined according to the energy balance of the particles of the mixed materials at the moment of separation from the blade, J; ψ is bump angle, rad; $\delta = h_0/l_b$ is blade deformation index; $c_2(\psi, \delta)$ и $c_1(\psi, \delta)$ are functional dependencies between the geometric parameters of the working volume of the apparatus and the kinematic characteristic of the movement of particles upon impact on the bump, for example, the recovery factor k_{Vi} for the direction-averaged particle velocities. Proposed calculation method $U_{iv}(\gamma_{2i})$ allows to determine the complete distribution functions of particles of each component

$$U_i(\gamma_{2i}) = \prod\nolimits_{v=1}^{k_0} u_{iv}(\gamma_{2i}) \tag{3}$$

in scattered flares, reflected from an inclined bump surface, necessary for further assessment of the quality of the mixture.

3 Results and Discussion

The performed calculation examples for the functions $U_{iv}(\gamma_{2i})$ and $U_i(\gamma_{2i})$ according to expression (3) are based on the following input data for the design and operating parameters of the apparatus (Figs. 1, 2 and 3): $r_b = 3.0 \times 10^{-2}$ m, $\beta = 0.3490$ rad, $h_0 = 3.0 \times 10^{-2}$ m, $r_b = 3.0 \times 10^{-2}$ m, $k = 3$, $\omega = (40 - 58)$ s^{-1}, $\delta = (0.66 - 0.70)$, $\psi = (0.6 - 0.7)$ rad. The work assumes a mixing of two granular components, the physical and mechanical properties of which are similar to those of bulk materials: natural sand GOST 8736-93 (i = 1, $D_1 = 1.5 \times 10^{-4}$ m, substance density $\rho_1 = 1.525 \times 10^3$ kg/m^3, $k_{V1} = 0.1853$) and semolina GOST 7022-97 ($i = 2$, $D_2 = 4.0 \times 10^{-4}$ m, substance density $\rho_2 = 1.440 \times 10^3$ kg/m^3, $k_{V2} = 0.1587$). Within the framework of the tasks set, the influence of the previously established significant parameters of the mixing process of granular components at stage 1 on the nature of the distribution of particles in scattered flows at stage 2 during the operation of the rotary apparatus was investigated. In particular, the analysis of the surfaces shown in Fig. 1 for the functions $U_{i\nu}(\gamma_{2i})$ at stage 2 shows a connection with the nature of the distribution of particles in scattered flares after the completion of stage 1 of the process of their mixing. At the same time, the pronounced first burst (graphs 1–3, Fig. 1a; graphs 1–3, Fig. 1b) corresponds to the number of particles that is (1.5–2) times greater than the number of particles for the second burst of functions $U_{i\nu}(\gamma_{2i})$. This reflects the behavior of particles during repeated self-cleaning of the blades, when the drum is rotated through an angle within (0.3–0.7) rad.

The observed coincidence of the ranges of variation of the extrema for the angles of reflection $\gamma_{2i} \leq 0.02$ rad for both components in the case of a function $U_i(\gamma_{2i})$ becomes a prerequisite for their effective mixing, for example, graph 1, Fig. 2a; graph 1, Fig. 2b and graph 1, Fig. 3a; graph 1, Fig. 3b. Comparison of the behavior of graphs 1–3, Fig. 3a; graphs 1–3, Fig. 3b allows us to conclude that even a slight increase of 0.01 units. the degree of deformation of the blades δ (stage 1) leads to a significant decrease by almost 2 times in the values of the functions $U_i(\gamma_{2i})$ (stage 2). Similarly, an increase in the values of the operating parameter from the range $\omega = (40-58)$ s^{-1} by 1.27 times is accompanied by a drop in the values for functions by (1.11–1.25) times for the first extremum of the angle of reflection from the bump plane (graphs 1–3 Fig. 2a; graphs 1–3, Fig. 2b).

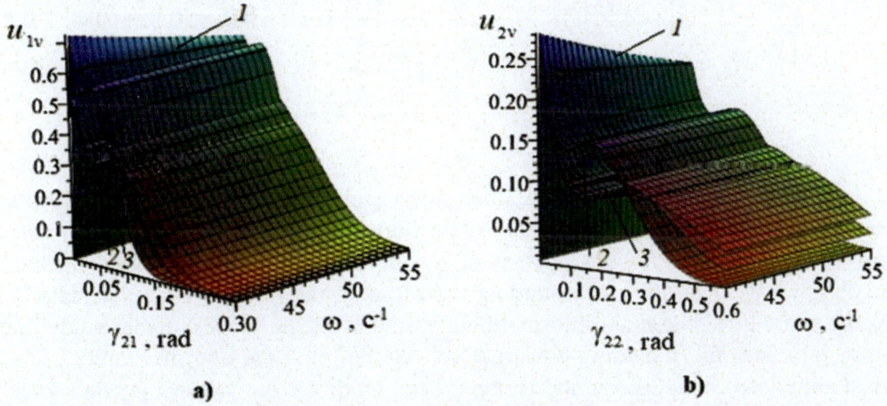

Fig. 1 Relationship between the differential distribution functions $U_i(\gamma_{2i}, \omega)$ by the angle of their reflection γ_{2i} and the rotation speed of the drum ω: **a** natural sand GOST 8736-93 ($i = 1$); **b** semolina GOST 7022-97 ($i = 2$); $\delta = 0.67$; $\psi = 0.6109$ rad; *1*—$\nu = 1$; 2—$\nu = 2$; *3*—$\nu = 3$

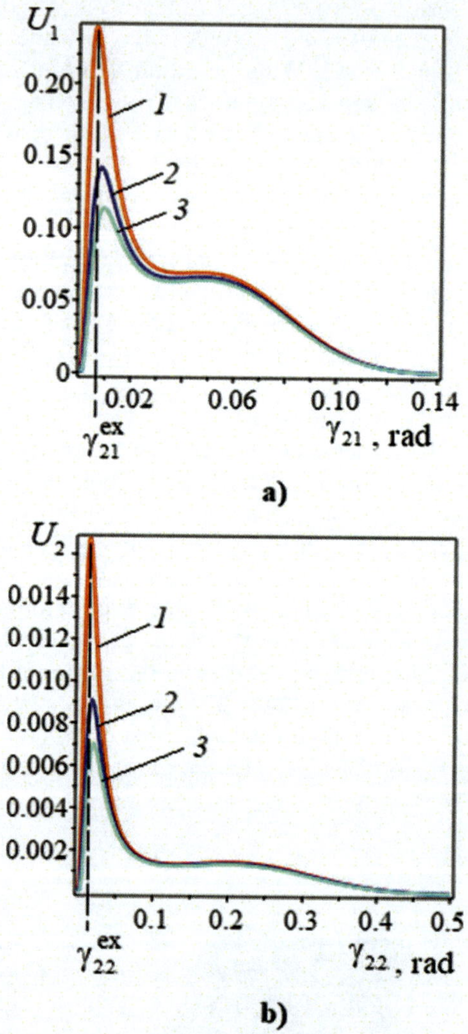

Fig. 2 Relationship between the differential distribution functions $Ui(\gamma_{2i})$ by the angle of their reflection γ_{2i} and the rotation speed of the drum ω: **a** natural sand GOST 8736-93 (i = 1); **b** semolina GOST 7022-97 (i = 2); δ = 0.67; ψ = 0.6109 rad; *1* —ω = 41.9 s^{-1}; *2* —ω = 47.1 s^{-1}; *3* —ω = 52.4 s^{-1}

In this work, according to the stochastic description of the behavior of bulk components proposed by the authors upon impact on the baffle plane of a rotary mixer, the analysis of the influence of the design and operating parameters of mixing at stage 1 on the corresponding indicators of the mixing process at stage 2. At the same time, the main informational variables of the process under study are taken into account, including constructive, regime, physical and mechanical. The most significant conclusions and results of the work are the following provisions.

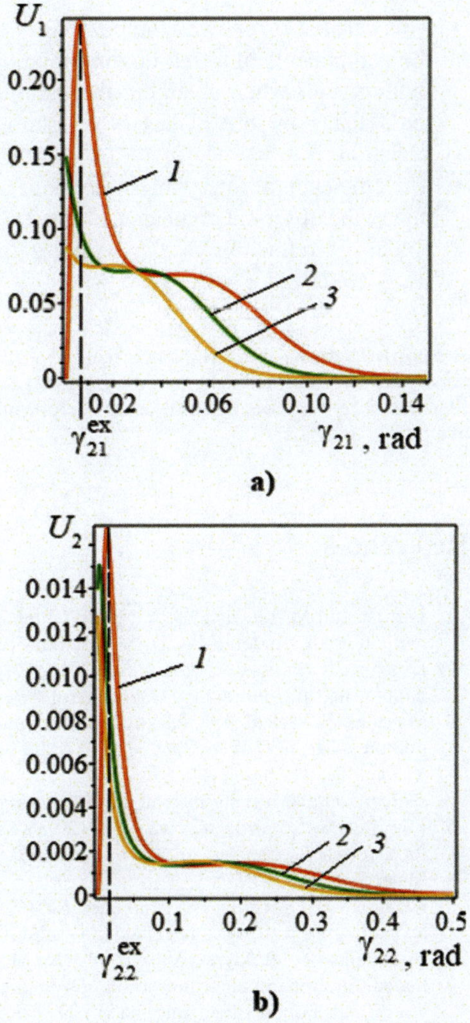

Fig. 3 Relationship between the differential distribution functions $Ui(\gamma_{2i})$ by the angle of their reflection γ_{2i} and the degree of deformation of the elastic blades δ: **a** natural sand GOST 8736-93 ($i = 1$); **b** semolina GOST 7022-97 ($i = 2$); $\omega = 41.9\ s^{-1}$; $\psi = 0.6109$ rad; *1*—$\delta = 0.67$; *2*—$\delta = 0.68$; *3*—$\delta = 0.69$

4 Conclusion

The generalization of the tasks of the design of mixing when dividing this process of processing bulk materials into two functionally related stages has been carried out.

- For the model components of a granular mixture, the calculation of indicators of a random mixing process in scattered flares at impact interaction with the baffle plane is performed, as stage 2 of the operation of the rotary apparatus. The results obtained will allow in the future to carry out a theoretical assessment of the quality of the finished mixture.

- This calculation showed the possibility of predicting the behavior of particles of the components of a free-flowing mixture at stage 2 (impact interaction with the baffle plane) when changing the parameters of stage 1 (the formation of scattered flares by the blades of a rotating drum) within the selected range of variation.
- The efficiency of rotary mixing of bulk components is theoretically substantiated due to the observed coincidence of the regions of variation of the extrema for the angles of reflection of the particles of the mixed components at values of $\gamma_{2i} \leq 0.02$ rad.

Acknowledgements This work was realized in the framework of the Program of flagship university development on the base of the Belgorod State Technological University named after V G Shukhov. The work was realized using equipment of High Technology Center at BSTU named after V. G. Shukhov.

References

1. Wen Y, Liu M, Liu B, Shao Y (2015) Comparative study on the characterization method of particle mixing index using DEM method. Procedia Eng 102:1630–1642
2. Dyomin OV, Pershin VF, Smolin D (2012) Preparation of mixture of free-flowing materials in double-shaft paddle mixers. Chem Petrol Eng 48(3–4):213–216
3. Mansour W, Melekhin D, Pasko A (2020) Continuous production of multicomponent powder mixtures. MATEC Web Conf 315(1):04003
4. Yazdani E, Hashemabadi SH (2019) The influence of cohesiveness on particulate bed segregation and mixing in rotating drum using DEM Phys A Stat Mech Appl 525
5. Lozovaya S, Lozovoy N, Uvarov V (2019) Computer simulating of strength characteristics for deformable chambers and most loaded mixer elements with a vertical shaft. IOP Conf Ser: Mater Sci Eng 560:1–6
6. Kafarov VV, Dorokhov IN, Arutyunov SY (1985) Sistemnyy analiz khimicheskikh protsessov tekhnolo-gii. Protsessy izmel'cheniya i smeshivaniya sypuchikh materialov. Moscow, Nauka
7. Bogomolov B, Boldyrev V, Zubarev A, Meshalkin V, Men'shikov V (2019) Intelligent logical information algorithm for choosing energy- and resource-efficient chemical technologies. Theor Found Chem Eng 53(5):709–718
8. Kapranova AB, Bakin MN, Verloka II (2018) Simulation of the quality criterion of a mixture in a drum-belt apparatus. Chem Petrol Eng 54(5–6):287–297
9. Kapranova AB, Verloka II (2018) Stochastic description of the formation of flows of particulate components in apparatuses with brush elements. Theor Found Chem Eng 52(6): 1004–1018
10. Verloka I, Kapranova A, Tarshis M, Cherpitsky S (2018) Stochastic modeling of bulk components batch mixing process in gravity apparatus. Int J Mech Eng Technol 9(2):438–444
11. Kapranova AB, Verloka II, Bahaeva DD, Tarshis MYu, Cherpitsky SN (2020) To the calculation of the average value of the volume fraction of the key bulk component at the intermediate stage of mixing with an inclined bump. Front Energy Res Process Energy Syst 8(135):1–11

Impact of PMB Preparation Technology on the Properties of Road Composites

M. Vysotskaya, D. Kuznetsov, A. Burgonutdinov, and K. Istomina

Abstract The paper deals with the problem of the quality of bitumen binders for road construction. It is noted that by basing the development of a raw bitumen base for the preparation of modified bitumen binders on a purposeful approach to the group composition of bitumen dispersions, rather than on strict compliance with current regulations, through multi-stage compounding, it is possible to solve the problem of developing effective, time-stable and reproducible modified bitumen compositions for road composites. On the example of stone-mastic asphalt concrete SMA-20, the influence of a bituminous binder made on the basis of a bituminous compound, the preparation of which is based on the sequential unification of oil refining products, is considered. It is established that such a composite is characterized by higher compressive strength at test temperatures of 20 °C and 50 °C relative to the composite on the base polymer-bitumen binder and is 5% and 12%, respectively. There is an increase in the water resistance index of the studied composition of the SMA. It was found that the SMA prepared on the basis of PMB is characterized by greater resistance to thermal-oxidative aging processes, which indicates a greater compatibility of the residual compounded bitumen with the polymer and, as a result, the formation of a structure resistant to destructive factors of the modified binder. The conducted research suggests that the developed composition is effective, and the proposed approach to its design is justified.

Keywords Bituminous binder · Stone-Mastic asphalt concrete

M. Vysotskaya · D. Kuznetsov (✉)
Belgorod State Technological University Named After V.G. Shukhov,
Kostyukov St., 46, Belgorod 308012, Russia
e-mail: roads-bstu@yandex.ru

A. Burgonutdinov · K. Istomina
Perm National Research Polytechnic University, Perm 614990, Russia

S. V. Klyuev (ed.), *Digital Technologies in Construction Engineering*,
Lecture Notes in Civil Engineering 173, https://doi.org/10.1007/978-3-030-81289-8_3

1 Introduction

One of the main reasons for the development of various types of defects on asphalt pavement, such as: plastic track, cracks, patching, as well as a reduction in the length of the cycle of asphalt concrete performance, include the use of bituminous binders with unsatisfactory characteristics.

At present, a considerable amount of knowledge was accumulated on technological methods of bitumen production and their modification [1–9]. However, bitumen is a complex colloidal system, which is characterized by a significant variation in the properties of the indicators due to the difference in the group composition of oil and the technological factors of its production and processing, and unstable initial characteristics of bitumen, which are manifested in the variability of the group composition of bitumen dispersions. This does not always make it possible to obtain stable bituminous binders with a set of required rheological and technological characteristics using known modification techniques.

It can be assumed that by combining the components of oil refining that have the maximum affinity and can act as a dispersion medium and stabilizers of the asphaltene phase, it is possible to obtain stable bituminous colloidal systems that can be effectively used in the technology of producing modified bituminous binders and road composites based on them.

2 Materials and Methods

In the present study, two types of bitumen binders were considered, obtained on the basis of bitumen of the BND 50/70 brand (No. 1) and bitumen compound (No. 2), corresponding to the penetration of the 50/70 brand. Each of the binders was subsequently modified with a polymer. Oil refining products were considered as objects for compounding: tar; fuel oil; de-asphalting asphalt. The polymer of the SBS L 30-01A brand was used as a modifier.

In general, the plan of the study was to develop an effective compound by combining an unoxidized bitumen base with a penetration of 50–70 mm^{-1} and a plasticizing composition, followed by polymer modification of the resulting bitumen raw material base and standard BND 50/70 bitumen (together with a plasticizer). Further, the selection of the composition and preparation of samples of SMA-20 and the assessment of the influence of the type of binder on the reliability of the composite.

The preparation of the modified bituminous binder compositions included the following steps:

1) heating of the components and their mixing for the preparation of the raw material base;

2) heating the resulting mixtures to the preparation temperature and mixing on a paddle mixer. Technological parameters: speed 200–400 rpm; temperature 165–175 °C; time 30 min;
3) insertion of a polymer modifier into the prepared bitumen system and mixing for 60 min.

Standard indicators of the properties of bitumen and bituminous binders, including PMB, were performed according to the methods in accordance with the current GOST, and the stability of the modified bitumen during storage was additionally evaluated in accordance with GOST EN 13399-2013.

Additionally, the aging rate of the composite was determined. Modeling of aging processes was carried out in a furnace at a temperature of 160 °C for 2 and 5 h. Asphalt concrete samples were subsequently formed from the mixtures subjected to thermal-oxidative destruction.

3 Results and Discussion

At the first stage of the study, several series of non-oxidized bitumen binders were developed, which are close in properties to BND 50/70 bitumen. From the developed line of compositions, the most promising (No. 2) was selected, the properties of which, as well as the control bitumen of the BND 50/70 brand (No. 1) are presented in Table 1.

Compositions of modified bitumen binders PMB60 were prepared on the basis of bitumen. The formulation is presented in Table 2. The composition of the modified bitumen binder based on the bitumen compound was conditionally called c.

Indicators of the properties of PMB compositions are presented in Table 3.

Table 1 Test results of bituminous binders with a penetration of 50–70 mm^1

Name of indicators	GOST requirements	Type of binders (composition №)	
		BND (№ 1)	Compound (№ 2)
Needle penetration depth, mm^{-1} at 25 °C	51–70	56	55
Ring and ball softening temperature, °C	51	53	55
Extensibility, cm, at temperature 0 °C	3.5	4.1	3.5
Fraas brittleness temperature, °C	−16	−17	−16
Flash point, °C	230	247	284
Change in sample mass after aging, %	0.6	0.6	0.4
Change in softening temperature after aging, °C	7	6	5

Table 2 The formula for preparing samples of PMB-60

Name of component	Content of the component in the composition of PMB 60, %
BND-based formula	
Bitumen BND 50/70	92.6
Plasticizer	7.4
Polymer	3.7
Compound-based formula (PMBc)	
Compound composition No. 2 includes: tar 20%; fuel oil 30%; asphalt deasphalting 20%	70
Plasticizing composition: tar and vacuum distillate	30
Polymer	3.7

Table 3 Indicators of the properties of PMB 60 samples

Name of indicators	RD requirements	PMB 60 on a binder of series	
		№ 1	№ 2
Uniformity	Homogeneous	Hom	Hom
Needle penetration depth at 25 °C, 0.1 mm	No more than 60	72	80
Needle penetration depth at 0 °C, 0.1 mm	No more than 32	29	33
Softening temperature by the ring and ball method, °C	No less than 58	83	83
Fraas brittleness temperature, °C, not higher	No more than −20	−25	−26
Extensibility at 25 °C, cm	No less than 25	72	79
Elasticity at 25 °C, %	No less than 80	94	96
Extensibility at 0 °C, cm	No less than 11	22	28
Elasticity at 0 °C, %	No less than 70	70	75
Change in the softening temperature after warming up, °C	No more than 5	5	1
Flash point, °C	No less than 230	282	275
Grip with marble or sand	Control sample № 2	№ 2	№ 2

The study of the properties of road composites was carried out on SMA-20 using PMB-60 series No. 1 and No. 2 as a binder.

From the selected mixture of SMA-20, cylinder samples were molded and tested. The properties of the prepared SMA formulas are shown in Table 4.

Asphalt concretes made on the basis of the investigated bituminous binders meet the requirements of the standards. However, the compositions of SMA with the use of PMBc have higher water resistance indicators.

The operational reliability and durability of road surfaces depends on the speed of the aging process of the binder. The ability to resist thermal-oxidative processes was evaluated by changing the strength properties of asphalt concrete samples at

Table 4 Properties of SMA-20 depending on the type of binder and the time of temperature control

Name of indicator	Requirements of GOST 31015-2002	Property indicators of SMA-20					
		type of bitumen binder					
		PMB			PMBc		
		temperature control time of the mixture, h					
		0	2	5	0	2	5
Porosity of the mineral part, %	From 15 to 19	15.42	15.6	15.4	15.31	18.7	18.7
Residual porosity, %	From 2.0 to 4.5	2.26	3.1	4.2	2.24	2.7	3.0
Run-off of bituminous binder, %		0.04	–	–	0.02	–	–
Water saturation, %	From 1.5 to 4.0	2.1	2.9	4.3	1.80	2.1	3.3
Compressive strength, MPa,							
– at temperature 20 °C	2.5	3.8	4.3	5.7	4.0	4.4	5.0
– at temperature 50 °C	0.70	1.25	2.3	3.1	1.40	2.1	2.6
Shift resistance:							
– internal friction coefficient,	0.94	0.95	–	–	0.97	–	–
– shear strength at 50 °C, MPa	No less than 0.20	0.45			0.46		
Crack resistance to the ultimate tensile strength at 0 °C, MPa	3.0–6.5	5.4	6.1	7.5	4.7	4.9	5.8
Water resistance under long-term water saturation	No less than 0.75	0.80	0.76	0.63	0.89	0.85	0.77

different test temperatures, as well as changes in water resistance, shear resistance and crack resistance (Table 4).

Comparison of the data shows that the SMA on PMBc is characterized by a less intensive dynamics of the increase in strength indicators after the mixture is thermostated. This is true for composites both after 2 h and after 5 h of aging. The change in the water resistance of composites during the aging of the mixture is also interesting: during 5 h of temperature control, the water resistance decreased for the control composition by 21%, while the studied composition showed a change in water resistance by 13%.

Analysis of the data after the aging of the SMA mixtures showed that after 5 h of temperature control, asphalt concrete on the control PMB becomes ineffective. Taking into account that the difference in the compositions was only in the binders, it is obvious that the key role was played by the bituminous binder.

4 Conclusion

In this paper, on the example of SMA-20, the influence of a binder made on the basis of a bituminous compound obtained by sequential compounding of oil refining products is considered. It is found that such a composite is characterized by higher strength indicators relative to the composite on the base PMB, and an increase in the water resistance index is observed. The crack resistance of the SMA prepared on the control composition of PMB-60 exceeds by almost 15% the same index of the composite obtained on PMBc, which indicates an increase in the rigidity of the composite at low temperatures.

It was found that the SMA made on the basis of PMBc is characterized by greater resistance to thermal-oxidative aging processes, which indicates a greater compatibility of the residual compounded bitumen with the polymer and, as a result, the formation of a structure resistant to destructive factors of the modified binder.

The carried out tests allow asserting that the developed composition is effective, and the proposed approach to its design is justified.

Acknowledgements This work was realized under the support of the framework of the Program of flagship university development on the base of the Belgorod State Technological University named after V.G. Shukhov, using equipment of High Technology Center at BSTU named after V. G. Shukhov.

References

1. Gokhman LM (2019) Association of researchers of asphalt concrete. Automobile and Road State Technical University, Moscow, pp 39–46
2. Soenen H, Lu X, Redelius P. SBS modified bitumen's: Does their morphology and storage stability influence asphalt mix performance? https://www.nynas.com/en/product-areas/bitumen-for-paving/bitumen-for-pavingapplications/knowledge-tank/hej
3. Holleran G, Holleran I, Wilson DJ (2017) Building up efficient and sustainable transport infrastructure. Conf Ser: Mater Sci Eng 236(1):012009
4. Kindeev ON, Vysotskaya MA, Shekhovtsova SY (2016) Bulletin of BSTU named V.G. Shukhov 1, pp 26–31
5. Domínguez FJN (2011) The use of waste polymers to modify bitumen. In: Polymer Modified Bitumen, pp 98–135
6. Recasene R, Martinez A, Jimenez F, Bianchetto H (2005) Effect of filler on the aging potential of asphalt mixtures. J Transp Res Board 1901:10–17
7. Bahia HU (1993) Glass transition behavior and physical hardening of asphalt binders. J Assoc Asphalt Paving Technol 62
8. Micaelo R, Santos A, Duarte C (2012) Mixing and compaction temperatures of asphalt mixtures with modified bitumen. In: Proceedings of 5th eurasphalt and eurobitume congress, Istanbul, pp. 7
9. Kuznetsov DA, Kurlykina AV, Shiryaev AO, Litovchenko DP (2020) Influence of modified bituminous binders on the properties of stone mastic asphalt. In: Lecture notes in civil engineering, vol 95, pp 287–293

Physical–Chemical Aspects of Modifying Cement Systems with Individual and Complex Plasticizing Admixtures

M. M. Kosukhin and A. M. Kosukhin

Abstract The research findings of rheological properties of cement systems, modified with plasticizing mono-admixtures, containing adsorption-active groups of various natures, and with complexes, based on them, are presented. It has been demonstrated that the cement paste's rheology can be efficiently regulated not only by changing the chemical composition and structure of modifiers, but also by combining the types and amounts of active admixtures of polar organic substances with various molecular weight. It has been determined that the alteration of hydrophilic groups' nature by replacing sulfo-groups with hydroxyl groups increases the admixture's plasticizing activity, which is both due to the higher hydrophilic ability of phenolic hydroxyl groups, and to their lower molecular weight. The inverse proportion between the cement paste mini-cone flow and its yield point has been identified, characterized with the minor alteration of d at the considerable initial alteration of τ_0 and indicating that after the equilibrium is reached, the gravity force is balanced with cement suspension's particles interaction force. Experimental data indicate that the nature of the plasticizing admixture's hydrophilic groups influences such its properties, as plasticizing ability, plasticizing activity and selectivity; with that, hydroxyl groups provide the better properties of the admixture, in comparison with sulfo-groups.

Keywords Polyfunctional modifiers · Plasticizing mono-admixtures · Adsorption-active groups · Rheological properties · Yield point · Plastic viscosity · Plasticizing activity · Mechanism of action

1 Introduction

The modern achievements in the science of concretes and their production practice are based on using the integrated action of physical, mechanical and chemical methods on raw components and are the basis for the theory of producing materials

M. M. Kosukhin (✉) · A. M. Kosukhin
Belgorod State Technological University Named After V.G. Shukhov, Belgorod, Russia

S. V. Klyuev (ed.), *Digital Technologies in Construction Engineering*,
Lecture Notes in Civil Engineering 173, https://doi.org/10.1007/978-3-030-81289-8_4

with tailor-made properties. This rightfully allows considering the modern concrete technology among advanced innovative technologies.

The crucial role in solving this problem is assigned to using efficient, available and low-cost chemical admixtures. The achieved results and the opportunities of improving the properties and technologies of concrete demonstrate that in this area there are still considerable resources and prospects for the further improvement of concrete, as a material with unique properties. One of the most promising trends of the further research in creating efficient modifiers for concrete, determined in the work [1], is the targeted regulation of properties of concretes and concrete mixes by adding modifiers in form of combined polyfunctional water-soluble or water-reducible products on the basis of organic compounds and electrolytes of various natures and mechanisms of action. These can be mixture compositions of conventional additives in new commodity forms or specially synthesized organic products.

Identifying the relations between molecular structure of organic compounds, properties of adsorption layers and behavior of cement systems laid the foundation for understanding complicated colloid-chemical and physical processes, occurring in cement systems and changing the conditions of their proceeding and the resulting properties of the hardened cement. These achievements became the starting point for creating complex chemical new-generation multipurpose modifiers and up-to-date concretes on their basis.

The high efficiency of individual and complex admixtures of various natures is achieved by increasing their functionality—that is, their ability to perform their functions. From this perspective, the functionality of admixtures for concrete is predetermined by their chemical composition and determines their performance efficiency. The concept of functionality is made a basis of creating high performance concretes (High Performance Concrete—HPC). A new strategic concept of defined performance concrete technology (Defined Performance Concrete—DPC) is being formed [2, 3].

In this regard, the purpose of this work is substantiating the necessity to develop high-efficient and low-cost modifiers on the basis of chemical production waste for mass-produced general-purpose concretes of the transitional period.

2 Materials and Methods

During the experiment, different mono-admixtures, varying in their nature and structure, were used, which contained various adsorption-active groups: fluidizing agent S-3, LST—hydrophilic sulfo-groups—SO_3^-, superplasticizer (SP) SB-3—hydroxyl groups and their mixes.

To determine the composition and structure of oligomeric molecules such methods as the gas–liquid and liquid chromatography, ultraviolet and infrared spectroscopy, NMR-spectroscopy, conductometry and potentiometry were used.

The molecular weight of the synthesized oligomers was determined by cryoscopy method.

The rheological properties of cement paste and mortars were studied and the cement concrete's composition was proportioned in accordance with the methodology, developed in CRCRI, concerning the application of various types of admixtures in the pre-cast and cast-in-situ concrete technologies [4].

The fluidity of the modified cement suspensions was determined by means of mini-cone method, according with the methodology of CRCRI (Concrete and Reinforced Concrete Research Institute) of Gosstroy [5], which consists in determining the diameter of cement suspension flow under the action of gravity.

The rheological parameters of suspensions were studied by means of a rotary visco-simeter «Reotest-2.1». The concentration of admixtures (Cm) was calculated in dry wt% of the dispersed phase amount. In the course of the research the dependence between the values of shearing stress and shear rate were determined. According to the obtained data, the rheological curves were plotted, from which the yield point τ_0 and the plastic viscosity η_{pl} were found.

3 Results and Discussions

Elaborating the mechanism of action and the compatibility of complex modifiers' components, the authors of the work [6] at the 7th International Congress on Cement Chemistry have demonstrated that at the combined usage of two or more same-type admixtures with the similar mechanism of action, the effect is weaker than the additive effect. At using complex admixtures, containing components of various natures and with various mechanisms of action, a synergetic effect is implemented, i.e. the effect of mutual strengthening of each admixture's plasticizing activity.

At studying the efficiency of complex admixtures, consisting of various surface-active agents (SAA), in the work [7] it is demonstrated that the maximum increase of the mortar mix flowability is achieved at adding an admixture, consisting of two SAA with different surface activity. The mechanism of action of such a combination of surface-active agents is confirmed by the authors of [8] and explained by independent or competitive adsorption at the phase boundary. The complicated nature of structure-regulating function of electrolytes in conjunction with superplasticizer in cement systems' rheology is also noted in the work [9]. The alteration of physical–chemical properties of SAA under the action of electrolytes, caused by the changes of aggregative stability and macromolecular shape of SAA, results in the change of its efficiency, which is also one of the causes of synergistic effect or antagonistic effect.

As for the superplasticizer's mechanism of action, in spite of a large amount of experimental data, available for its explanation, there is still some controversy in various academic circles about the prevailing role of adsorption-solvation and electrostatic factors and about the role of spatial or linear structure of SP molecular chains. Though, for example, in our research [10, 11] it has been determined that the superplasticizer's mechanism of action is conditioned only by the combined action of electrostatic and adsorption-solvation factors.

Taking into account the fact, that all phenomena and processes, caused by adding admixtures, take place at phase boundaries, it is important that these complexes contained components, which would influence the kinetics and character of these processes, and were compatible with each other. Of special interest are the complexes, the components of which at their combined use influence the simultaneous alteration of aggregative stability, rheology and other properties of dispersed systems, which result in the alteration of physical–mechanical properties of concretes.

The rheological properties of cement pastes with admixtures were assessed by mini-cone flow method and by change of the system's rheological parameters, assessed by means of rotational viscometry. The first method allows visualizing the influence of admixtures on a standard cone flow of the concrete mix, and the second method is necessary at studying the mechanism of admixtures' plasticizing action.

In Fig. 1 the influence of S-3, SB-3 and LST on the mini-cone flow is shown. As we can see in the diagram, all curves are of S-shape. In the low concentrations range the plasticizing is manifested only marginally, and then, with the increase of the admixtures' concentration, the flow rises up sharply, and with the further increase of the admixtures' concentration, becomes practically constant. The same S-shape is characteristic for other admixtures and manifests itself in the same way at the large cone test method. As we can see in Fig. 1, of two admixtures (S-3 and LST), which contain sulfo-groups $-SO_3$ as hydrophilic groups, the higher fluidizing ability is possessed by superplasticizer S-3, which is confirmed by a large amount of literature data [12]. The plasticizing activity, which is determined [13] as the admixture concentration, necessary for achieving the maximum fluidizing effect, makes up 0.3; 0.45 and 0.4% of cement weight, for SB-3, S-3 and LST, respectively (Fig. 1). The alteration of hydrophilic groups' nature by replacing sulfo-groups with hydroxyl groups increases the admixture's plasticizing activity.

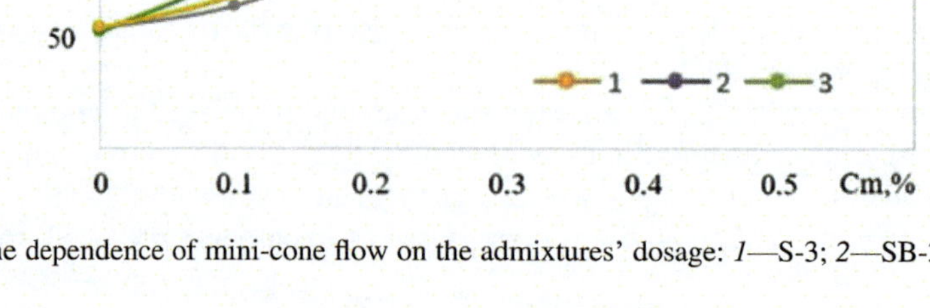

Fig. 1 The dependence of mini-cone flow on the admixtures' dosage: *1*—S-3; *2*—SB-3; *3*—LST

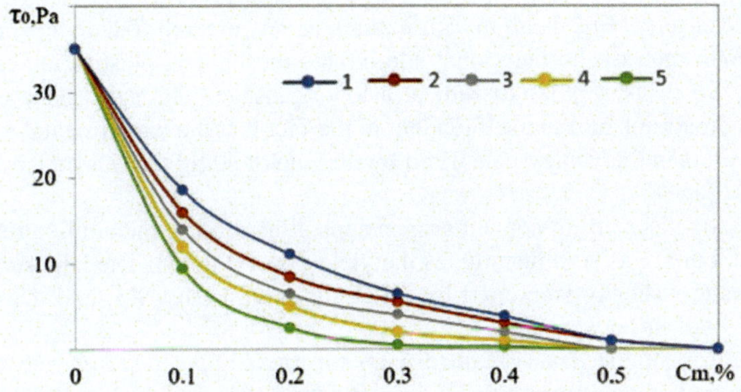

Fig. 2 The dependence of cement paste yield point on the admixtures' dosage: *1*—S-3; *2*—LST; *3*—SB-3; *4*—LST + SB-3; *5*—S-3 + SB-3

As previously noted [14], this is due to the higher hydrophilic ability of phenolic hydroxyl groups, and to their lower molecular weight. Replacement of a sulfo-group with molecular weight 80 with a hydroxyl group with molecular weight 16 results, ceteris paribus, in the decrease of the plasticizer's molecular weight by 30–35%, which corresponds to about the same alteration of the plasticizing activity of SB-3 in comparison with S-3. LST, the molecule of which contains various hydrophilic groups (SO_3^-, O^-, COO^-), falls in between SB-3 and S-3 in its plasticizing activity.

According to the results of shearing stress alteration depending on shear rate, the rheological curves were plotted, from which the yield point and the plastic viscosity were found.

The influence of dosage of SB-3, S-3, LST, and complex admixtures on their basis (at optimal concentration of SB-3, equal to 70%) on rheological parameters of cement paste at W/C = 0.3 is shown in Figs. 2, 3.

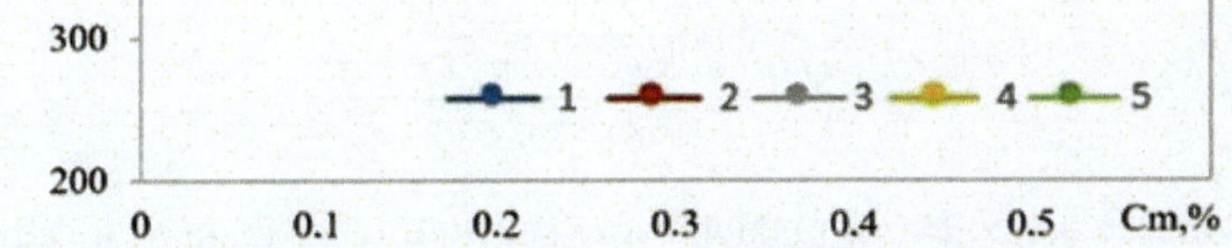

Fig. 3 The dependence of cement paste plastic viscosity on the admixtures' dosage: *1*—S-3; *2*—LST; *3*—SB-3; *4*—LST + SB-3; *5*—S-3 + SB-3

As we can see in Fig. 2, all the admixtures under research reduce the yield point to zero. With that, of the individual admixtures the highest plasticizing activity is shown by SB-3, as τ_0 goes to zero at a lower dosage. The synergistic effect for complex admixtures makes itself evident in the fact that the experimental values for τ_0 are lower, than the values, calculated by the rule of additivity, and τ_0 goes to zero at lower dosages.

The plastic viscosity of cement suspensions at increasing the admixtures dosage is reduced, but not as significantly, as the yield point (Fig. 3). The largest influence on the plastic viscosity is exerted by SB-3, the lowest—by S-3, and the complex admixtures fall in between.

Comparison of the data, obtained from the rheological research and from the mini-cone method, shows that the dosage of admixtures, at which the highest value of mini-cone flow is achieved, correlates with the admixtures' dosages, at which the yield point equals to zero. This becomes understandable, if we take into account that the cone spreads under the action of gravity, and the flow of cement paste would proceed until the gravity force, acting on the cement paste lower layer, would exceed the shearing stress of the cement paste.

Comparing the yield point's (Fig. 2) and mini-cone flow's (Fig. 1) curves of dependence on admixtures dosage, we can see, that the greatest alteration of τ_0 is observed in the low dosage range, while the mini-cone flow in this concentrations range is changed only marginally. To explain this difference, the interrelation between the flow value and the yield point was considered. After the equilibrium is reached, the gravity force is balanced with the suspension's particles interaction force, which can be expressed through the yield point with the Eq. (1).

$$\frac{mg}{S} - \frac{A}{P} = \alpha \cdot \tau_0 \tag{1}$$

where m—suspension weight, kg; S—flow area, m^2; P—flow perimeter, m; A—constant, characterizing the superficial tension of suspension; α—change packing coefficient, equal to 4; τ_0—yield point, Pa.

As we can see from the equation, there is an inverse proportion between the cone flow and the yield point, characterized with the minor alteration of d at the considerable initial alteration of τ_0. This explains the observed S-shape of the curves, indicating the mini-cone flow's dependence on the admixtures' dosage.

The numerical value of constant A, found from boundary conditions, is equal to 15. After transformations, we obtain.

$$d = \frac{\sqrt{A^2 + 16m \cdot g \cdot \alpha \cdot \tau_0} - A}{2\alpha \cdot \tau_0} \tag{2}$$

In Fig. 4 a calculated curve, obtained after substitution of numerical values to the Eq. (2), and the experimental data of mini-cone flow are presented. As it is seen in

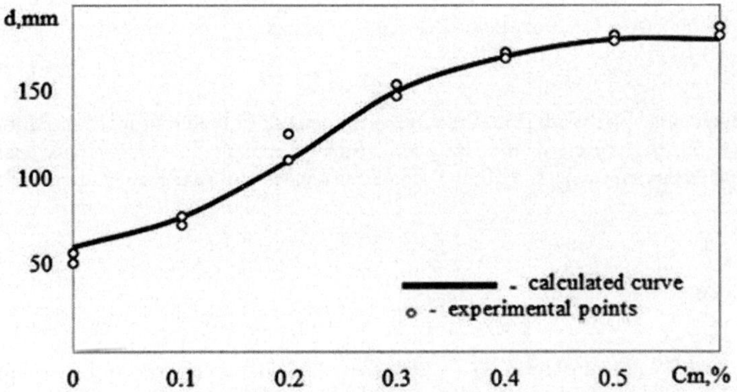

Fig. 4 The dependence of mini-cone flow on the SB-3 superplasticizer dosage

the diagram, there is a good agreement between the experimental and calculated data.

In the cement paste, as in any Bingham fluid, the deformation takes place, if τ is more than τ_0. The value of τ_0 is predetermined by combined particles bond forces at their contact points and allows calculating the individual contact energy by the Eq. (3), with the obtained experimental data and literature data:

$$U_K = P \cdot h = \frac{\tau_0 \cdot h}{1/(2r)^2} = \tau_0 \cdot h \cdot (2r)^2, \tag{3}$$

($\tau_0 = 40$ Pa, $r = 3$ μm, h—0.75 nm): $U_K = 40 \cdot 0.75 \cdot 10^{-9} \cdot (2 \cdot 3 \cdot 10^{-6})^2 = 1.08 \cdot 10^{-18}$ J $\approx$ 270 kT.

The obtained value is considerably larger than kT, which predetermines the thixotropic flow of cement suspension without admixtures. The decrease of U_K to values below kT, at the adsorption of admixtures on particles surface, results in the system's transition to liquideous state [15].

4 Conclusion

The carried-out rheological research allows us making a conclusion, that the cement paste fluidity is determined with its yield point value and achieves the maximum value at τ_0, equal to zero. The nature of a plasticizing admixture's hydrophilic groups influences such its properties, as plasticizing ability, plasticizing activity and selectivity; with that, hydroxyl groups predetermine the better properties of the admixture, in comparison with sulfo-groups. In the complex admixture, containing 70% of admixture with hydroxyl groups and 30% of admixture with sulfo-groups, the synergistic effect manifests itself to a large extent. This is observed in the decrease of the complex admixture dosage, necessary for achieving the maximum plasticizing effect.

Acknowledgements The work is realized in the framework of the Program of flagship university development on the base of the Belgorod State Technological University named after V.G. Shukhov, using equipment of High Technology Center at BSTU named after V.G. Shukhov.

References

1. Stepanova MP, Potamoshneva ND, Chernyshov EM, Bazhenov YM (2013) Nanostructured portlandite-aluminosilicate contact-condensational hardening systems and composites on their basis. Bull MSUCE 2:114–122
2. Usherov-Marshak AV (2014) Future outlook of concrete. Build Mater 3:4–5
3. Usherov-Marshak AV (2006) Admixtures for concrete: progress and challenges. Build Mater 10:8–12
4. Recommendations on evaluating the efficiency of concrete admixtures application. CRCRI, Moscow (2020)
5. Recommendations on physical-chemical control of the composition and quality of superplasticizer S-3. CRCRI, Moscow (2020)
6. Bazhenov YuM, Ratinov VB, Rosenberg TI et al (1986) Research of complex admixtures mechanism of action. Stroyizdat, Moscow
7. Kruglitsky NN, Boyko VN, Zagaychuk AS et al (1981) Dependence of mortar mixes consistency on SAA admixtures. Build Mater Struct 3:34–35
8. Dobrolyubov G, Ratinov VB, Rosenberg TI (1983) Forecasting the durability of concrete with admixtures. Stroyizdat, Moscow
9. Kalashnikov VI, Makridin NI, Ivanov IA (1986) The influence of «electrolyte-superplasticizer» complex on cement compositions' rheology. In: V symposium Rheology of concrete mixes and its technological tasks, pp 88–89. Riga
10. Kosukhin MM (1995) Regulating the properties of concretes and concrete mixes with complex admixtures, having various hydrophilic groups: diss. cand. of engineering. VGASU, Voronezh
11. Kosukhin MM, Shapovalov NA (2006) Theoretical aspects of superplasticizers' mechanism of action. Concr Reinforced Concr 3:25–27
12. Batrakov VG (1981) Methodology of optimizing the concrete composition with admixtures S-3 and SDB. In: All-Union conference improving the efficiency and quality of concrete and reinforced concrete structures production by means of chemical admixtures, pp 15–17. V.G. Vilnius
13. Ivanov FM (1985) Efficiency of using superplasticizers. In: Concretes with efficient modifying admixtures: collected papers. CRCRI, Moscow
14. Kosukhin MM, Kosukhin AM (2017) Theoretical and methodological basis for creating polyfunctional modifiers of cast-in-situ concretes for reconstruction works. In: Bulletin of BSTU named after V.G. Shukhov 1, pp 23–32
15. Kosukhin MM, Starostina IV, Kosukhin AM (2018) The research of clinker monominerals modified suspensions' rheological properties. Int J Eng Technol (UAE) 7(2):31–33

Methodological Principles and Algorithm for Forming Technogenic Fibrous Materials with Low Bulk Density in a Flat-Matrix Extruder

M. V. Sevostyanov and A. V. Osokin

Abstract The process of forming technogenic fibrous materials with a low bulk density is considered. The relevance of processing these materials in order to obtain products in a compacted form, for example, granular stabilizing additives of macadam-mastic asphalt concrete, is shown. The process of movement of a viscous-plastic material through a channel of a die with a variable cross-section is investigated. The basic equations that characterize this process are obtained, such as: the equation of the change in axial pressure along the length of the conical and cylindrical parts of the filament extrusion device, respectively; the pressing equation that takes into account the physical and mechanical characteristics of the forming material. The analysis of the equations allowed us to establish the general nature of the compaction process of the material in the entire range of molding pressures (until the final production of the formed bodies). Computational methods for determining the parameters of the dies for the design of flat-matrix granulators were developed. Based on theoretical and experimental studies, a patent-protected flat-matrix granulator for forming fibrous materials, as well as a design of a precast press matrix, including a set of removable forming elements, were developed. The flat-matrix granulator has advanced technological capabilities, provides an increase in the quality of products and productivity by equipping the device for pre-compacting the charge stock. An algorithm consisting of a sequence of computation of the parameters of the dies of flat matrices for forming technogenic fibrous materials is developed.

Keywords Technogenic materials · Extrusion · Granulation · Press matrix · Low bulk density

M. V. Sevostyanov (✉)
Department of Technological Complexes, Machines and Mechanisms, Belgorod State Technological University Named After V.G. Shukhov, Kostyukov St., 46, Belgorod 308012, Russia

A. V. Osokin
Joint-Stock Company, Research Institute of Parachute Engineering, Irkutsk Str., 2, building 1, Moscow 107241, Russia

S. V. Klyuev (ed.), *Digital Technologies in Construction Engineering*,
Lecture Notes in Civil Engineering 173, https://doi.org/10.1007/978-3-030-81289-8_5

1 Introduction

Recycling of waste, including the complex processing of technogenic materials, is particularly relevant now. Various industrial wastes, in one form or another, are involved in production processes and used as secondary raw materials [1, 2].

The granulated material obtained in this process, as a semi-finished product or finished product, is increasingly used in various industries: in construction, chemical, food, and others, as well as in drive engineering and heat power engineering [3].

It is particularly difficult to form technogenic materials with a low bulk density [4]. For example, the construction of roads with a modern road surface made of macadam-mastic asphalt concrete involves the use of stabilizing additives in its composition, usually in the form of granules, based on cellulose-containing or other fibrous materials [5, 6]. The bulk density of such stabilizing additives is approximately 600 kg/m^3, and the processed pulp is 150–200 kg/m^3, which is relatively low. As the technical cellulose used in the production of granular stabilizing additives from foreign manufacturers is quite complex in production and high in cost, there is a need to develop national technologies and aggregates for processing cellulose raw materials.

2 Materials and Methods

Cellulose-containing, basalt, fiberglass and other fibrous materials can be processed and formed into bodies of a certain geometric shape and size and have specified physical and mechanical characteristics, depending on the requirements of the consumer.

Therefore, the scientific and technical studies presented in the article, aimed at improving the efficiency of the process of extrusion of technogenic fibrous materials by improving the design and technological parameters of the developed design of the flat-matrix granulator, are very relevant and in demand.

The analysis of the current state of granulation methods for various materials (including those with low bulk density) allowed determining the most effective way to obtain products by pushing the charge stock through the dies of a flat-matrix granulator.

3 Results and Discussions

As a result of the conducted scientific and technical research, a design of a flat-matrix granulator [7] with a device for pre-compacting materials was developed, the general view of which is shown in Fig. 1. The device shown in Fig. 1, works as follows. The granular fibrous material is fed through the loading hopper *9* to the pre-sealing device. The high-frequency vibrations of the generator *10* contribute to the better bulk of the material. Through the holes in the fixed cone, a moistening liquid or steam is supplied to the material under the pressure developed by the pump *13*. For uniform distribution of liquid or steam over the entire volume of the material, the holes are drilled evenly around the circle. The material caught between the outer fixed *7* and the inner movable *8* cones, under the action of a compressing normal force from the working surfaces of the cones, is compacted when the material is rolled by the inner movable cone *8*. The degree of compaction depends on the value of the eccentricity "*e*" and the length of the working zone (parallelism zone). To capture the material better, the working surface of the inner movable cone *8* can have grooves of rectangular or segmental cross-section along the elements of the cone.

The pre-compacted material, under the pressure of the new material coming from above and gravity, passes between the cones at the moment of their maximum separation and is poured in a continuous stream onto the stationary matrix *5* of the granulator, falling under the pressing rollers *6*. To reduce the stagnant zone and

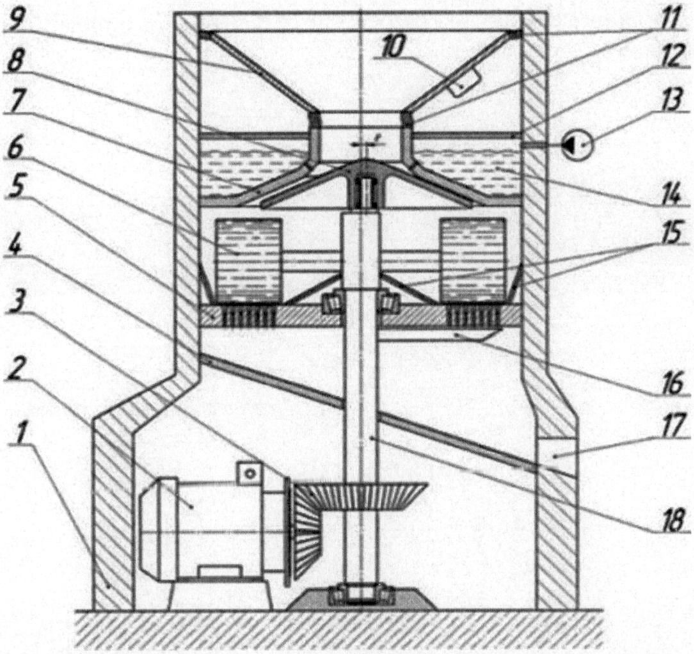

Fig. 1 Fiber granulator 1—body, 2—motor, 3—mechanical transmission, 4—tilting table, 5—flat fixed press-matrix, 6—pressing rollers, 7—external fixed cone, 8—internal movable cone, 9—loading hopper, 10—high—frequency generator, 11—elastic elements, 12 – pressure sealing plate, 13—pump, 14—tank, 15—inclined sides, 16—knife for cutting granules, 17—discharge hole, 18—shaft

direction of all the material under the pressing rollers, the upper bearing assembly of the vertical shaft *18* and the outer surface of the press matrix *5* are equipped with inclined sides *15*. Under the action of the pressing force from the side of the rollers *6*, the material is pushed through the moldable holes of the matrix *5*. The bundles coming out from the other side of the matrix are cut off with a knife *16* and fall on a tilting table *4*, along which they are directed to the discharge hole *17*. The tilt of the table is selected taking into account the self-pouring of granules under the influence of gravity. In this case, there is no need to use additional elements in the form of throwing knives.

During the operation of the granulator, the press matrix perceives large mechanical loads from the friction of the material and the pressing forces of the press rollers. As a result, the resource of flat matrices is very limited and depends on the physical and mechanical characteristics of the processed materials. Usually, the operating life does not exceed 2000 h.

Taking into account all mentioned above, in order to extend the service life of flat matrices, a design of a precast press matrix [8] was developed, including a set of removable forming elements—Fig. 2.

The precast matrix of the press granulator consists of a jig in the form of a disk *1*, in which cells are cut for removable forming elements *2* with dies 3. The fixation of removable forming elements *2* is carried out by means of a profiled flange *4* and contouring screws *5*.

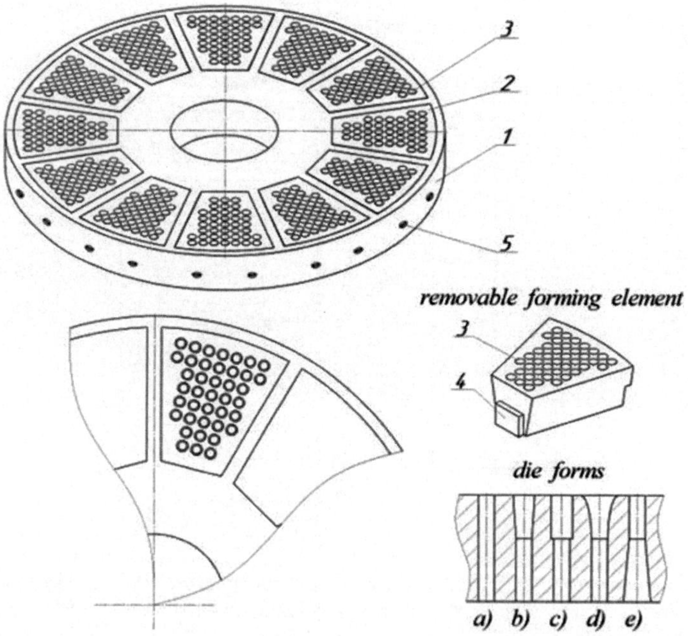

Fig. 2 Precast matrix of the press granulator 1—jig, 2—removable forming elements, 3—dies, 4—flange, 5—contouring screws

The number, size and shape of the dies *3*, as well as the material from which they are made, is determined by the technology of granule production and the overall dimensions of the matrix.

The sequence of calculation of the parameters of the dies of flat matrices for forming technogenic fibrous materials is presented in the form of an algorithm in Fig. 3.

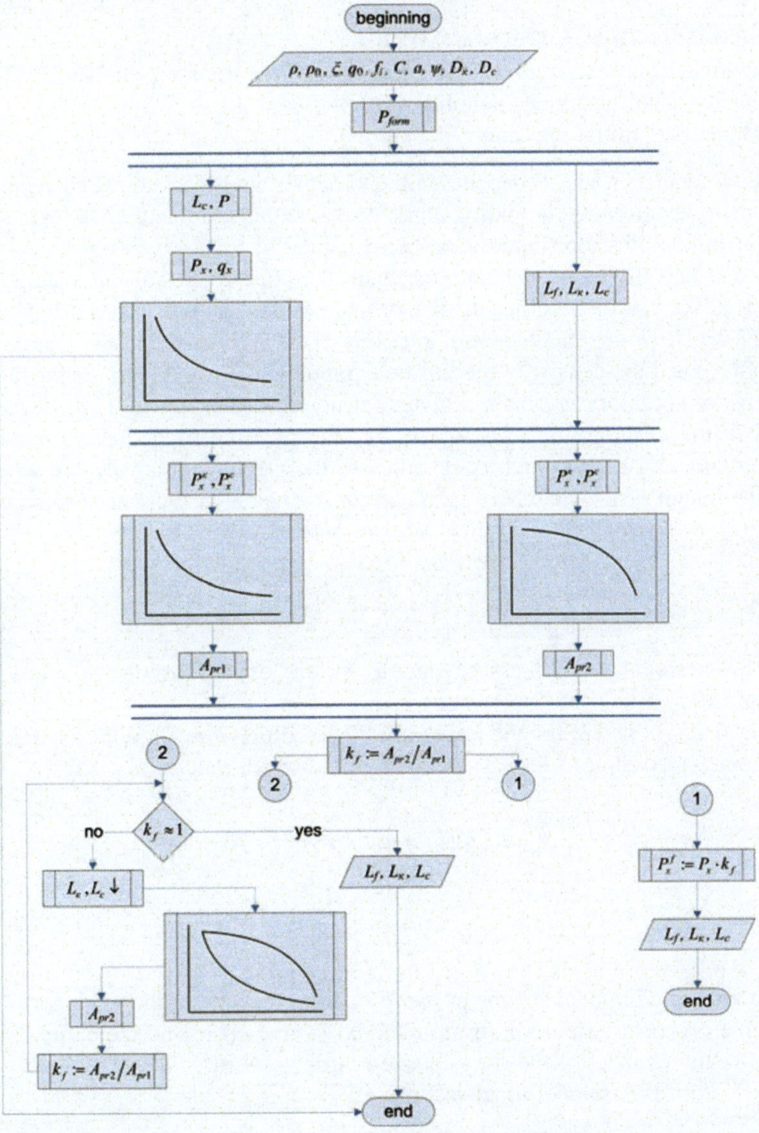

Fig. 3 Computation algorithm of die parameters

Professor S. V. Melnikov derived the basic equation (law) of pressing granular and fibrous capillary-porous materials [9], according to which:

$$P_{form} = C \cdot \left(e^{a(\rho-\rho_0)} - 1\right), \tag{1}$$

there

C— constant parameter, $C = b/a$, (MPa);
a— parameter that characterizes the properties of the material, (m^3/kg);
ρ— density of the resulting monolith, (kg/m^3);
ρ_0— Density of initial raw material, (kg/m^3).

The parameters a and C depend on the structural and mechanical characteristics of the material (strength, humidity, particle size) and determine the resistance of the material to compression (determined experimentally).

If we are given the physical and mechanical characteristics of the material to be formed (C, a, f_i—the coefficient of external friction of the material against the channel walls, ξ—the coefficient of side thrust), as well as the initial charge stock density (ρ_0) and the density of the finished granules (ρ_g), we determine the pressure P_{form}, MPa, necessary to obtain a given density according to Eq. (1).

According to the found value of the forming pressure P_{form}, having previously set the diameter of the cylindrical channel of the press matrix— D_c, m, we determine the length of the channel—L_c, m, of constant circular cross-section according to Eq. (2), which will provide this back pressure.

$$L_c = \frac{D}{4 \cdot \xi \cdot f_i} \ln\left[1 + \frac{C \cdot \xi}{q_0} \cdot \left(e^{a(\rho-\rho_0)} - 1\right)\right], \tag{2}$$

there q_0—residual back pressure due to elastic expansion of the compressed material, MPa.

After that, we determine the change in the axial pressure P_x along the length of the channel according to Eq. (3) and plot the obtained values "$P_x - L_c$".

$$P_x = \left(P + \frac{q_0}{\xi}\right) \cdot e^{-\xi \cdot f_i \cdot \frac{4}{D} \cdot x} - \frac{q_0}{\xi}, \tag{3}$$

there

P—the normal component of the pressure P_0 from the side of the working body, at which the compressed portion of the material is pressed in and the entire material located in the channel of the die is pushed through, N/m^2;
x—depth from the channel input section, m.

Then we compute the length of the die L_f with a conical inlet part, having previously set the diameter of the input section of the conical part of the channel of the press matrix—D_κ; D_c and the length of the conical part L_k, expressed as a fraction of the total length of the die L_f–n, according to the dependence (4), which will provide the necessary back pressure.

$$L_f = \frac{D_c}{(1-n)\cdot 4\cdot \xi \cdot f_i} \ln\left(\left[\frac{C\cdot\xi}{q_0}\cdot\left(e^{a(\rho-\rho_0)}-1\right)+1\right]\cdot\left(\frac{D_c}{D_\kappa}\right)^{\frac{2f_i\cdot\cos\psi}{tg\psi}\cdot\xi}\right) \tag{4}$$

Having determined the length of the conical L_κ and cylindrical L_c parts of the die, we construct the curves of the axial pressure drop P_x along the length of the channel during the viscous-plastic flow of the material according to Eqs. (5) and (6).

$$P_x^\kappa = \left(P+\frac{q_0}{\xi}\right)\cdot\left(\frac{D_\kappa - 2x_1\cdot tg\psi}{D_\kappa}\right)^{\frac{2f_i\cdot\cos\psi}{tg\psi}\cdot\xi} - \frac{q_0}{\xi}, \tag{5}$$

$$P_x^c = \left(P_1+\frac{q_0}{\xi}\right)\cdot e^{-\xi\cdot f_i\cdot\frac{4}{D_c}\cdot x_2} - \frac{q_0}{\xi}, \tag{6}$$

there

P—normal component of the pressure from the press roll side, MPa;
P_1—pressure at the inlet to the cylindrical part, MPa;
Ψ—the angle of inclination of the walls of the die to its axis, (deg).

The areas of A_{pr1} and A_{pr2} of the figures under the graphs (Fig. 3.) are proportional to the work of external forces (from the side of the press roller) when pushing the material through the channel of the die. Having determined the numerical value of the areas (for example, by plotting graphs or analytically), we calculate the proportionality coefficient k_f as a quotient of A_{pr2}/A_{pr1}, showing how many times the resistance of a channel with a conical inlet part in the molding of fibrous materials is greater than the resistance of the same channel in the viscous-plastic molding of materials with the same characteristics. This coefficient is necessary for further calculations of the consumed capacity.

Having determined the desired values (back pressure of the die P_f with a conical inlet part, the proportionality coefficient k_f and the length of the die sections L_f, L_κ, L_c) we complete the computation (① on Fig. 3).

To reduce the energy consumption of the forming process and the metal consumption of the press matrix, you can use a cycle with the selection of the parameters of the die (② on Fig. 3). At the same time, gradually reducing L_κ and L_c, we find their values such that $A_{pr1} \approx A_{pr2}$ ($k_f \approx 1$). After that, the computation ends. The obtained values of L_f, L_κ, L_c are used in the manufacture of the press matrix. The diameter of the cylindrical part of the die is selected based on the requirements for the geometric dimensions of the granules. As a rule, asphalt concrete plants use granules with a diameter of 4 … 7 mm [10].

4 Conclusion

Based on the conducted theoretical and experimental studies, computational methods for determining the parameters of the dies of flat matrices for the design of flat-matrix granulators were developed. Mathematical expressions describing this process are obtained. Mathematical expressions describing this process are obtained. The design of the flat-matrix granulator with the possibility of pre-compaction of the material and the design of the press matrix with an increased service life was developed. The obtained granular stabilizing additives have a bulk density of $\rho = 550 \pm 30$ kg/m^3, humidity $W = 8 \pm 1\%$. The runoff index of the binder was 0.07 – 0.14, which corresponds to the requirements of GOST 31015. The obtained values of productivity and power consumption were $Q_{pract} = 118 \pm 5$ kg/h and $N\Sigma \approx 24.7$ kW. The computational values of these parameters were $Q_{teor} = 107.15$ kg/h and $N\Sigma \approx 23$ kW, which corresponds to an error of $\Delta Q = 9.2\%$ and $\Delta N\Sigma = 7.02\%$, respectively.

Acknowledgements The work was carried out within the framework of project No. 10089447, the program of the REC "Innovative Solutions in the agro-industrial complex", the scientific and production platform "Rational Use of Natural Resources".

References

1. Shein N, Sevostyanov V, Obolonsky V et al (2019) Resource and energy-saving technologies of complex processing and utilization of technogenic materials. IOP Conf Ser: Mater Sci Eng 552(1):012042
2. Il'Ina T (2013) Classification of dispersed materials and recommendations for the processes of their agglomeration. Chem Oil Gas Eng 4:17–19
3. Sevostyanov M (2016) Resource-saving equipment for complex processing of man-made materials. Bull BSTU named after V.G. Shukhov 4:140–145
4. Sevostyanov M, Il'Ina T, Martakov I (2019) Process of charge prevention with low bulk density. In: III International scientific and technical conference on "energy system". IOP Conf Ser: Mater Sci Eng 552:012039
5. Yadykina V, Gridchin A, Tobolenko S (2012) Stabilizing additive for crushed-mastic asphalt concrete from industrial waste. Constr Mater 8:64–66
6. Kiryukhin G (2006) Experience in the construction of road surfaces made of crushed stone-mastic asphalt concrete in Russia. Herald HNADU 34:34–35
7. Sevostyanov V, Ilyina T, Osokin A et al (2013) Fiber material granulator. Pat. 135539 RU, B 01 J 2/20, (2013.07)
8. Glagolev S, Sevostyanov M, Gridchin A et al (2017) Precast matrix of the press granulator. Pat. 170904 RU, B 01 J, (2016.11)
9. Melnikov S (1978) Mechanization and automation of livestock farms. Training manual. L/Kolos, Leningrad Branch
10. GOST 31015-2002 (2003) Mixtures of asphalt concrete and asphalt concrete crushed stone-mastic. Technical specifications. Gosstroy of Russia, Moscow, GUP CPP 27

Multifunctional Building Material from Non-recycled Glass Waste

V. I. Onishchuk, A. S. Glivuk, E. V. Korobanova, and V. A. Doroganov

Abstract Modern innovations in the construction of buildings and structures for civil, special and agricultural purposes are determined by various vectors, the most significant of which is the widespread use of modern building materials with multifunctional properties. The article presents the results of research aimed at creating a technology for obtaining a structural and heat-insulating material based on glass scrap generated in the field of functional use of glass containers and industrial processing of flat window glass. Often, the quality of the glass fight does not allow it to be used both for repeated use in the production of various types of glass products, and in the production of a well-known thermal insulation material-foam glass. The developed method for obtaining a porous structural and thermal insulation material is based on the processing of a non-cyclical glass container fight to obtain, at the first stage, a glass binder suspension—a three-phase heterogeneous system that includes solid microparticles of glass with a hydrated surface, a transition—colloidal phase with micro-sized solid phase particles and a nanostructured vitreous binder. At the second stage, the resulting suspension is structured as a result of polycondensation processes, forming a monolith. The third stage is associated with the heat treatment of the obtained monoliths, resulting in the production of porous vitreous materials. The main significant results presented in the article are the established physico-chemical nature of the formation of a porous structure, which differs from today's generally accepted ideas of obtaining porous vitreous materials. Experimentally determined properties allow us to attribute the obtained porous vitreous materials to the group of structural and thermal insulation materials.

Keywords Glass waste · Mechanochemical activation · Glass binder suspension · Structure and properties · Curing · Formation of the porous structure of the material · Properties of porous materials

---

V. I. Onishchuk (✉) · E. V. Korobanova · V. A. Doroganov
Department of Glass and Ceramics Technology, V. G. Shukhov Belgorod State Technological University, Belgorod, Russia

A. S. Glivuk
LLC "Polymer Glass Materials", Belgorod, Russia

S. V. Klyuev (ed.), *Digital Technologies in Construction Engineering*,
Lecture Notes in Civil Engineering 173, https://doi.org/10.1007/978-3-030-81289-8_6

1 Introduction

Modern innovative building materials are mainly composite materials consisting of a certain number of components, each of which makes an additive contribution to the resulting group of properties that determine the functional direction of the application of such materials in practice. At the same time, the components that make up the composite material can not only have different physical and mechanical characteristics, but also represent substances of different nature—silicate crystalline and amorphous materials, polymers, etc. [1–7].

Among the innovative building materials and their technologies, a special place is occupied by such materials that have the prefix "nano" in their name. These are materials that are obtained using elements of nanotechnologies, or modify the compositions of mixtures in the process of obtaining multi-functional nanoadditives. Both in the first and in the second case, it is possible to obtain materials with a complex structure and, due to the processes of self-organization of substances at the nanoscale, provide them with improved properties and characteristics. Therefore, the use of nanotechnology elements in the production of multifunctional building materials and products in modern conditions is a new promising direction in science and high-tech production. According to many experts in the field of building materials science, every year the number of studies in this direction allows us to talk about the gradual introduction of nanotechnology in the production of building materials and products [8, 9].

2 Methods and Materials

The glass binder suspension (SHS) was obtained by mechanochemical activation of a mixed battle of flat glass and glass bottles of white, green and brown colors, the following chemical composition, wt.%: 71…72.5 SiO_2; 2.5…3.5 Al_2O_3; 11.0…12.5 CaO + MgO; 13.2…14.0 Na_2O; 0.3…0.5 SO_3. Mechanochemical activation was performed by liquid-phase grinding of glass waste in a porcelain drum mill with a capacity of 100 L.

The study of the rheological characteristics of the obtained GBS was carried out on a rotary viscometer with coaxial cylinders "Reotest-2" immediately after the suspension was produced from the mill.

Nanostructured vitreous binder (NVB) was isolated from GBS by centrifugation at a speed of 6000 revolutions per minute.

The dispersion composition of GBS was studied using the Analysette 22 NanoTec laser particle size analyzer.

The dispersion composition of NVB was studied using the Zetatrac particle size analyzer.

To identify the polymer state of silicic acid in GBS, a photocolorimetric determination was performed, based on measuring the intensity of the color of the resulting solution, during the formation of yellow-colored silicomolybdenic acid in the interaction of isopolymolybdate anions with silicic acid [10] at pH < 3. Changes in the optical density were recorded every minute for 1 h. The optical density values were measured on a FEC-photocolorimeter at a wavelength of 400 nm.

The microstructure of the GBS samples was studied using a high-resolution scanning electron microscope TESCAN MIRA 3 LMU with the functions of analyzing the elemental composition of the substance. The macrostructure of the samples of porous vitreous materials was carried out using an optical microscope Rolam p211.

Physical and mechanical properties were determined according to standard methods for determining the corresponding properties.

3 Results and Discussion

The production of a GBS provides for the creation of conditions for the effective liquid-phase mechanical activation of glass particles according to a certain regime, which provides the resulting product with predictable properties: density, dispersion composition, volume concentration of the solid phase, and as a result, high sedimentation stability, rheological characteristics, hardening time and strength of monoliths [10, 11].

The obtained GBS had a density in the range of 1.74…1.77 g/cm^3 and a volume concentration of the solid phase (C_v) in the range of 0.53…0.59, which subsequently ensures the production of porous vitreous materials with different values of physical and mechanical properties.

From the point of view of studying the macrostructure, GBS is a composite material in which the role of the matrix is performed by NSB, and as a filler—glass particles of various (<50 microns) sizes. The third, the transition phase, is actually a mixture of HBC with the inclusion of glass particles hydrated to different depths with a size of 200…2000 nm and is partly a matrix, and partly a filler phase. The complex, multi-level surface size and structure of solid and hydrated glass particles, after the end of the mechanochemical activation process, provides the SHS with a thixotropic flow pattern, close to the ideal model of a solid-plastic system, with the difference that the GBS exerts a high resistance to shear rates of 20…30 s^{-1}(Fig. 1).

The difference in the forward and reverse flow values is due to the fact that the suspension does not have time to recover and form aggregative structures that cause thixotropy.

At a gradient of 50 s^{-1} at the end point of the diagram, the destruction of the structure stops. With a subsequent decrease in the speed of rotation of the cylinder, the suspension is structured, its viscosity increases, and at a speed of rotation of 1…2 s^{-1}, the shear stress approaches the forward stroke indicators, which characterizes the system as relatively stable over time. In a static state, the suspension behaves like a solid with a certain shear modulus.

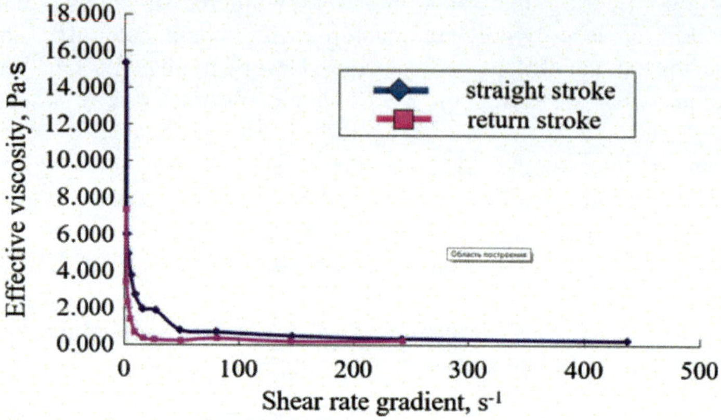

Fig. 1 Effective viscosity of glass binder suspension (GBS)

The process of GBS structuring can be represented as a series of structural transitions, accompanied by a sequential consolidation of solid phase particles with the active participation of NSV and a change in the types of communication in spatial structures from point and coagulation contacts between vitreous particles to wide accretion along the interphase boundaries by means of high-silica bridges formed from a gel of colloidal silica.

The study of the dispersion composition of SHS showed that the average particle size is 6.02 microns, the particles have an elongated spherical shape, and the specific surface area is 22,406.21 cm^2/cm^3.

Figure 2 shows the complex multi-level structure of the GBS.

Fig. 2 Micrographs of the GBS structure: 1—glass particles with a low degree of surface hydration, 2—transition phase, 3—NVB located on the surface of hydrated glass particles

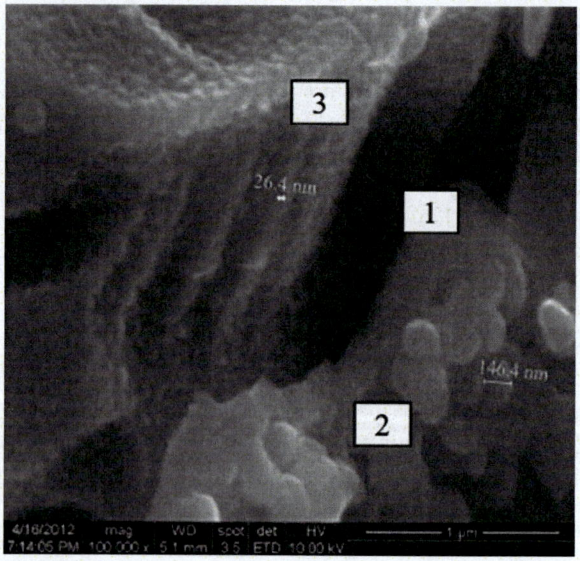

Table 1 Chemical composition of various GBS phases

Component phase	O	Na	Mg	Al	Si	S	K	Ca
Glass particles with a small degree of surface hydration	38.93	10.16	2.38	1.76	39.59	–	0.57	6.60
Transition phase	40.80	2.31	0.27	0.62	55.15	–	0.09	0.76
NVB	38.61	3.48	–	–	55.68	2.66	0.17	–

Microscopic studies of the solidified SHS monolith showed that it consists of particles with a size of 50…250 nm connected to each other by a solidified gel structured from smaller particles with a size of 20…35 nm.

Studies of the chemical composition of GBS showed its heterogeneity: large poorly hydrated particles have a composition close to that of the original glass, containing a larger amount of the alkaline component located in the thickness or adsorbed on the surface of the grains. Large vitreous particles are fused together with a high-silica gel, which almost does not contain sodium oxide (Na_2O) in its composition.

The chemical composition of the various components of the GBS in atomic form is shown in Table 1.

The production of porous materials was carried out by heat treatment of monoliths of suspensions in the temperature range of 780…830 °C. Figure 3 shows the macrostructure of the resulting porous materials. The size and pore gradient across the cross-section of the samples indicate that the thermal treatment of monoliths of suspensions of similar compositions at different temperatures ensures the production of materials with different properties (Table 2).

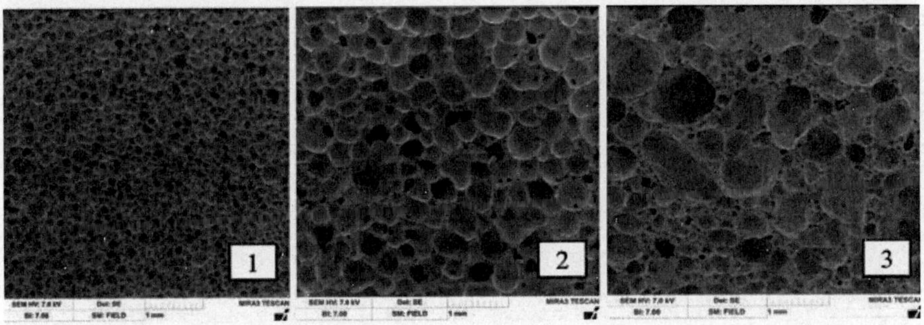

Fig. 3 Micrographs of porous materials obtained at different temperatures of formation of the porous structure, °C: 1—790; 2—810; 3—830

Table 2 Physical and mechanical properties of porous materials obtained from GBS

Property	Indicators for samples		
	1	2	3
Density, kg/m^3	650	380	140
Compressive strength, MPa	23.2	8.7	3.1
Water absorption, %	0.5	1.5	4.3
Thermal conductivity, W/m•°C	0.21	0.11	0.075
Water permeability	Absent	Absent	Absent

4 Conclusion

The vitreous binder suspension obtained by mechanochemical activation of non-recycled glass waste has a heterogeneous structure, which is prone to auto-curing to produce a monolith, the heat treatment of which without the use of traditional gas-forming agents provides the production of porous materials, the properties and functional purpose of which can be regulated by the parameters of suspension production and the temperature–time treatment regime. Experimentally determined properties prove that the materials obtained on the basis of GBS monolith can be attributed to the group of thermal insulation, structural-thermal insulation or structural materials.

Acknowledgements The work is realized in the framework of the Program of flagship university development on the base of the Belgorod State Technological University named after V. G. Shoukhov, using equipment of High Technology Center at BSTU named after V. G. Shoukhov.

References

1. Katunský D (2014) Progressive multifunctional building materials, constructions and technological methods—one step closer to green visegrad in V4 Countries. In: Zborník vedeckých prác, proceedings of scientific works. Stavebná fakulta—Technická univerzita v Košiciach, Czech Republic
2. Costa CM, Costa P, Lanceros-Mendez S (2021) Overview on lightweight, multifuntional materials. In: Woodhead Publishing in Materials, Advanced Lightweight Multifunctional Materials, pp 1–24. https://doi.org/10.1016/B978-0-12-818501-8.00002-0
3. Gilyazidinova N, Rudkovskaya N, Santalova T (2018) Nature conservation technology for producing slag-foam glass as a structural and thermal insulating material. E3S Web Conf 41:02017. https://doi.org/10.1051/e3sconf/20184102017
4. Sopegin G, Rustamova D, Fedoseev S (2019) Analysis of existing technological solutions of foam glass production. Vestnik MGSU 14:1584–1609
5. Haimei Z (2011) Heat-insulating materials and sound-absorbing materials. In: Woodhead publishing series in civil and structural engineering, building materials in civil engineering, pp 304–423. https://doi.org/10.1533/9781845699567.304

6. Mugoni C, Montorsi M, Siligardi C, Andreola F, Lancellotti I, Bernardo E, Barbieri L (2015) Design of glass foams with low environmental impact. Ceram Int 41(3):3400–3408. https://doi.org/10.1016/j.ceramint.2014.10.127
7. Francis AA, Abdel Rahman MK, Daoud A (2013) Processing, structures and compressive properties of porous glass-ceramic composites prepared from secondary by-product materials. Ceram Int 39(6):7089–7095. https://doi.org/10.1016/j.ceramint.2013.02.048
8. Bozsaky D (2016) Special thermal insulation methods of building constructions with nanomaterials. Acta Technica Jaurinensis 9:29–41
9. Bjørn PJ (2016) Nano-based thermal insulation for energy-efficient buildings. In: Start-up creation. Woodhead Publishing. https://doi.org/10.1016/B978-0-08-100546-0.00008-X
10. Onishchuk V, Zhernovaya N (2010) Physico-chemical aspects of the formation of the structure of ultrafine glass suspensions. In: Bulletin of BSTU named after V.G. Shukhov, vol 3, pp 104–108
11. Doroganov V, Doroganov E, Peretokina N, Onishchuk V, Trepalina Yu, Gavshina O (2016) Corundum and zirconia composites based on artificial ceramic binders. Refract Ind Ceram 57:92–97. https://doi.org/10.1007/s11148-016-9934-5

Prediction of Sound-Insulating Properties of Cellular Concrete Composites

V. N. Tarasenko, J. V. Denisova, and N. D. Chernysh

Abstract Currently, the problem of sound insulation in residential buildings is one of the main ones when the project passes the examination. Increased comfort in terms of sound insulation in category A houses is the main one, along with insolation and energy efficiency. Cellular concrete composite materials are widely used, often used in the construction of residential prefabricated buildings as a construction of wall filling and partitions, as well as walls between apartments and stairwell-elevator halls, but their sound insulation is little studied and is not a value fixed when releasing small wall stones. The paper proposes the principles of calculating the sound insulation of small wall stones made of cellular concrete with a commercial density of 500 and 600 kg/m^3 with options for interior finishing in the form of grout or plastering. At the same time, the thicknesses of these layers varied from 2 to 10 mm. The results of the calculations gave a good convergence, confirmed by laboratory studies. The variants of performing calculations of sound insulation of partitions made of cellular concrete small wall stones with the use of commercial plaster compositions made of cement-sand and lime-sand mortar are presented.The above methods for calculating the air noise insulation index provide good convergence with the results of laboratory studies, which can be used in the future when predicting the air noise insulation index for new modern wall materials that are at the testing stage.

Keywords Sound insulation · Cellular concrete · Efficiency · Noise protection

1 Introduction

Currently, the problem of sound insulation of enclosing structures is quite acute in the exploited housing stock and newly erected buildings. Modern wall materials are often used in standard solutions of fences without the necessary additional sound insulation. In the construction of high-comfort residential buildings, standard

V. N. Tarasenko (✉) · J. V. Denisova · N. D. Chernysh
Belgorod State Technological University Named After V.G. Shukhov, Belgorod, Russia

S. V. Klyuev (ed.), *Digital Technologies in Construction Engineering*,
Lecture Notes in Civil Engineering 173, https://doi.org/10.1007/978-3-030-81289-8_7

Table 1 Estimated indices of air noise insulation for walls and partitions made of aerated concrete blocks

Grade of aerated concrete by density	Average masonry density, used to calculate loads from its own weight ρ, kg/m^3	Thickness of walls or partitions h, m	Estimated air noise isolation index R_w, dB
D500	570	0.080	31
		0.100	35
		0.120	38
		0.160	43
		0.200	46
		0.250	49
		0.300	52

solutions of wall enclosing structures should be reviewed taking into account additional sound insulation in order to ensure the conditions of comfortable stay.

To date, in the construction of residential buildings, a number of factors should be taken into account to ensure the conditions of comfortable stay. In accordance with modern requirements [1–5], the conditions for providing insolation, the sufficiency of natural light, air and vapor permeability, and the thermal efficiency of enclosing structures are the main factors and are considered first. However, we should not forget about sound insulation, which is especially important for high-comfort buildings.

Noise protection in the building is of great importance; especially the role of sound insulation of living accommodation should be noted. The rated parameter of internal enclosing structures (walls, interior partitions) of residential and public buildings is the air noise insulation index R_w, dB. The calculated values of the air noise insulation indices for internal enclosing structures R_w are given in Table 1.

A large share of the market for cellular concrete wall materials is occupied by small aerated concrete wall stones, foam and expanded clay concrete small blocks.

2 Methods and Materials

As the analysis of the sound insulation values of aerated concrete blocks [3, 6, 7] showed, the scope of their use is limited for walls with a thickness of 250…300 mm; for partitions of 160…200 mm. When using blocks of other thicknesses in the construction of partitions, it is recommended to provide a number of technical measures to improve sound insulation, namely, the device of additional sound insulation on the frame. Such measures are quite time-consuming and are rarely envisaged at the time of design.

The development of the production of lightweight concrete based on porous aggregates, which were originally intended for use mainly in external fences, led to the use of these concretes in the internal construction of residential buildings. With certain raw materials and industrial bases of prefabricated housing construction, the complex use of light concrete for the manufacture of residential building structures is economically feasible. However, the use of these concretes in internal fences is constrained by sound insulation requirements [4, 8, 9].

According to the calculation methods [1, 2, 5], to provide the required sound insulation, an acoustically homogeneous light concrete structure must have the same surface density as a fence made of heavy concrete. This is due to a significant increase in the thickness of light concrete elements compared to the thickness of heavy concrete structures, and a sharp decrease in their efficiency. At the same time, there is evidence of increased sound insulation qualities of light concrete enclosing structures.

3 Results and Discussion

Using the calculation methods established to date [1, 2, 7], the following sound insulation indicators were obtained for foam concrete partitions with a density of 800, 900 and 1000 kg/m^3. A partition without doors between the kitchen and the living room with a thickness of 100 mm was taken into consideration (Table 2).

$$m_{\ni} = \delta \cdot \rho \cdot K, \tag{1}$$

where $m_{\ni}$—surface density of the enclosing structure material, kg/m^2; δ—enclosing structure thickness, m; ρ—density, kg/m^3; K—a coefficient that takes into account the relative increase in the flexural stiffness of a enclosing structure made of concrete on light aggregates, porous concrete, etc. in relation to structures made of heavy concrete with the same surface density [2, Table 10].

The necessary air noise isolation index of 65 dB is assigned to the boundary conditions, which ensures isolation from household noise of electrical appliances and equipment. According to the diagram shown in Fig. 1, using curve 4, the

Table 2 Estimated air noise insulation indices for cellular concrete partitions

№	Type of single-layer material used in fencing	Density of the used material, kg/m^3	Thickness of walls or partitions h, m	Estimated air noise isolation index R_w, dB
1	Foam concrete, aerated concrete, expanded clay concrete	500	0.10	31
2		500	0.19	35
3		600	0.10	38
4		600	0.19	43
5		800	0.10	46
6		800	0.19	52

estimated surface density of the enclosing structure is specified, it is 800–820 kg/m^2 for this calculation. Then, for structures made of foam concrete, aerated concrete, expanded clay concrete and other cellular concrete composites with the same density of 500 kg/m^3, the index of isolation from air noise can be calculated as follows:

$$800 = \delta \cdot 500 \cdot 1.7, \text{ then } \delta = 0.94\ (\text{M}).$$

Calculations confirm that for these types of materials, the minimum thickness of the partitions should be 0.94 m, which is impractical.

However, in accordance with the regulatory data, 41 dB of air noise isolation is sufficient [1, 2]; then, in accordance with Fig. 1, it is sufficient to provide a surface density of 162–168 kg/m^2.

$$170 = \delta \cdot 500 \cdot 1.7, \quad \text{then } \delta = 0.2\ (\text{m}).$$

Cellular concrete materials with a density of 600 kg/m^3 can be used as partitions, and then their thickness should be 0.166 m or more. In this case, the sound insulation index will be 41 dB and will increase slightly with an increase in the thickness of the partition to 190 mm (Table 3) [7, 9, 10].

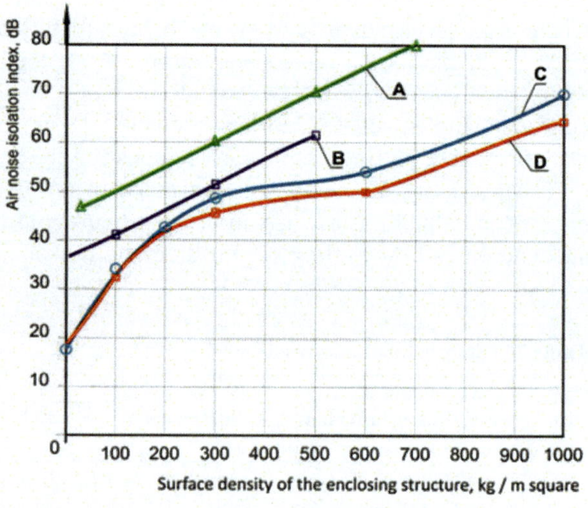

Fig. 1 The value of sound insulation of the enclosingstructure depending on the surface density: A—maximum values for structures with two dense layers with damping due to the air layer; B—wood and wooden materials in a single-layer structure; C—masonry, concrete, gypsum in a single-layer structure [7]; D—estimated values of sound insulation for small wall stones made of cellular concrete

Table 3 Air noise insulation indices for foam concrete partitions with a thickness of 100 mm, obtained under laboratory conditions

№	Type of single-layer material used in enclosing structure	Density of the used material, kg/m^3	Estimated air noise isolation index R_w, dB
1	Plastered partitions (5 mm on each side)	800	41.0
2		900	41.5
3		1000	42.0
4	Stuccoed partitions (10 mm on each side)	800	42.0
5		900	42.5
6		1000	43.0

4 Conclusion

Accordingly, it can be concluded that all partitions made of foam, aerated concrete with a density of 800 kg/m^3 or more meet modern requirements for sound insulation and stuccoed partitions allows increasings lightly their sound insulation index by increasing the surface density of the structure.

It should be noted that the above methods of approximate calculation of the air noise insulation index give good convergence with the results of laboratory studies, which can be used in the future when predicting the air noise insulation index for new modern wall materials that are at the testing stage.

Acknowledgements The work is realized in the framework of the Program of flagship university development on the base of the Belgorod State Technological University named after V.G. Shukhov, using equipment of High Technology Center at BSTU named after V.G. Shukhov.

References

1. Tarasenko VN (2015) Cellular concrete in low-rise residential construction. In: 10th international scientific and practical conference, pp 142–144
2. Suleymanova LA, Kara KA (2011) Energy-saving technologies of highly porous concrete. In: Belgorod region: past, present, future. Mat. reg. research and practical, pp 98–102
3. Rakhimbayev ShM, Degtev IA, Tarasenko VN, Anikanova TV (2007) On the issue of reducing shrinkage deformations of foam concrete products. News of higher educational institutions. Construction 12:41–44
4. Tarasenko VN (2016) Non-destructive methods of control of cellular concrete construction materials. In: Collection of reports of the international scientific and practical conference, pp 194–198
5. Strokova VV, Netsvet DD, Nelubova VV, Serenkov IV (2017) Properties of composite binder based on nanostructured suspension. Constr Mater 1–2:50–54
6. Shahova LD, Chernositova ES, Denisova JV (2017) Flowability and durability of cement containing technological additives during grinding process. AER-Adv Eng Res 133:162–167
7. Tarasenko VN (2018) Impact of foamed matrix components on foamed concrete properties. MEACS 2017. IOP Conf Ser: Mater Sci Eng 327:032054

8. Tikhomirov VK (1983) Foam. In: Theory and practice of their production and destruction, 2nd ed. Chemistry, Moscow
9. Svatovskaya LB, Ovchinnikov VP, Solov'eva SV et al (1996) Management of the activity of cement mixtures using additives of the "Elbi" type. Cement 2:28–32
10. Tarasenko VN (2016) Predicting the sound-proofing properties of cellular concrete composites. In: Intelligent building composites for green construction. Collection of reports of the international scientific and practical conference dedicated to the 70th anniversary of the honored scientist of the Russian federation, corresponding member of the Russian Academy of Sciences, Doctor of Engineering Sciences, Professor Valery Stanislavovich Lesovik, pp 135–140. Belgorod State Technological University named after V.G. Shukhov, Belgorod

Features of Titanium Hydride Crystal Oxidation at Heat Treatment

V. I. Pavlenko, R. N. Yastrebinsky, Z. V. Pavlenko, and N. I. Bondarenko

Abstract The use of hydrogen-containing materials in the nuclear power industry as a protection against neutrons is due to the large number of hydrogen atoms in 1 cm^3 of the material. According to the data of differential thermal analysis (DTA), it was found that the exothermic process caused by the oxidation of the initial and modified titanium hydride crystals occurs in the temperature range of 400–950 °C, and the intensive dissociation of titanium hydride for the initial crystals—at 757 °C and 807 °C—for the modified crystals. The microstructure of the titanium hydride crystals was studied by scanning electron microscopy (SEM) in the modes of reflected (backscattered) and secondary electrons. The high-resolution scanning electron microscope TESCAN MIRA 3LMU is used in the secondary electron shooting mode. It was found that the depth of oxygen diffusion into the internal volume of the titanium hydride crystals at a temperature of 950 °C reaches 120–130 microns for both the initial titanium hydride crystals and the modified crystals. It was found that the average content of atomic oxygen on the surface of the initial titanium hydride crystals is 17.23%, and at a depth of 100 microns—2.45%, while on the surface of the modified titanium hydride crystals the oxygen content is 11.80%, and at a depth of 100 microns—1.81%.

Keywords Crystals · Titanium hydride · Oxidation · Oxide film · Diffusion

1 Introduction

The use of hydrogen-containing materials in the nuclear power industry as a protection against neutrons is due to the large number of hydrogen atoms in 1 cm^3 of the material [1–3]. At the same time, if in the free state the cross-section of the scattering of thermal hydrogen neutrons is 38 barns, in paraffin—about 80 barns, then in boron—3840 barns (unit of effective cross-section, $1b = 10^{-24}$ cm^2).

V. I. Pavlenko · R. N. Yastrebinsky · Z. V. Pavlenko · N. I. Bondarenko (✉)
Belgorod State Technological University Named After V.G. Shukhov, Belgorod, Russia
e-mail: bondarenko-71@mail.ru

S. V. Klyuev (ed.), *Digital Technologies in Construction Engineering*,
Lecture Notes in Civil Engineering 173, https://doi.org/10.1007/978-3-030-81289-8_8

A significant disadvantage of these materials is the low operating temperature range. In this regard, metal hydrides are preferred, which have a significant range of temperature stability (up to 600–800 °C). Titanium hydride is the most attractive in this respect, as the content of hydrogen atoms in 1 cm^3 of the metal is 9.2×10^{22} in it and it is second only to vanadium hydride—11.4×10^{22} [4].

The formation of a protective oxide film with a thickness of up to 2 nm on the surface of titanium occurs at room temperature in air for several hours and its composition is close to that of titanium oxide. Further oxidation of titanium at room temperature occurs very slowly (for 4 years up to 5 nm) [5].

High temperatures contribute to the good chemical affinity of titanium for oxygen. Intensive absorption of oxygen begins at a temperature of 400 °C, nitrogen—at 800 °C. With a further increase in temperature, the activity of titanium increases sharply [6].

The oxidation of titanium at temperatures below 800 °C is prevented by a surface film consisting of titanium oxides, nitrides, and hydrides [7, 8]. With an increase in temperature, the film begins to dissolve in titanium, dissociation of titanium hydride and intensive diffusion of air oxygen into the metal occurs. The kinetics of the interaction of titanium with oxygen is complex, as several layers of oxides are formed on the surface of titanium, between which exchange processes occur, causing continuous oxidation of the metal [3, 9, 10].

2 Methods and Materials

For the experiment, titanium hydride of the GTK-D series was used in the form of spherical granules (crystals) with a diameter of 1–4 mm, the ultimate composition of which is presented in Table 1.

Chemically pure boric acid H_3BO_3 was also used. Sodium methylsilicate with a density of 1190 kg/m^3, with a silicon (Si) content of 4% by weight, pH = 12.

The microstructure of the titanium hydride crystals was studied by scanning electron microscopy (SEM) in the modes of reflected (backscattered) and secondary electrons.

The contrast in the reflected electron mode depends largely on the crystallographic orientation of the grains relative to the sample surface and the chemical composition of the phases. The image obtained in this mode allows observing the microstructure of the material without etching the polished section first.

Studies by the method of diffraction of reflected (back-scattered) electrons (DBSE) were carried out on a scanning electron microscope Quanta 600 FEG. This

Table 1 Ultimate composition of titanium hydride crystals

Element	Ti	H	Al	Fe	O	Si	C	N
wt. %	95.63	3.50	0.40	0.20	0.10	0.08	0.05	0.04

microscope is equipped with an integrated Pegasus 2000 system, which allows determining the characteristics of grains at working magnifications of 200–100,000 times. The shooting was conducted at an accelerating voltage of 20 kV and an electron beam current of 36 nA.

In the case of secondary electrons, the best resolution is provided. As the output of secondary electrons depends on the angle of incidence of the primary electrons on the surface of the object, to obtain a high-quality image, it is necessary to obtain a developed surface. A high-resolution scanning electron microscope TESCAN MIRA 3LMU, including an energy-dispersion spectrometer (EMF) X-MAX 50 Oxford Instruments Nano Analysis for electron-probe micro-analysis, was used in the secondary electron shooting mode.

3 Results and Discussion

According to the data of differential thermal analysis (DTA), it was found that the exothermic process caused by the oxidation of the initial and modified titanium hydride crystals occurs at the temperature range of 400–950 °C, and the intensive dissociation of titanium hydride occurs at 757 °C (for the initial crystals) and 807 °C (for the modified crystals). The following scheme of the process of oxidation of the outer surface of the titanium hydride crystals by air oxygen during heat treatment can be presented:

$$\mathrm{Ti, TiH_x} \xrightarrow{400\,^\circ\mathrm{C}} \mathrm{Ti, TiH_x, TiO} \xrightarrow{750\,^\circ\mathrm{C}} \mathrm{Ti, TiO} \xrightarrow{900\,^\circ\mathrm{C}} \mathrm{TiO_2}.$$

It is known that almost all of the hydrogen in α-Ti is present in the hydride form. Its concentration goes mainly along the grain boundaries, and this in turn leads to the embrittlement of titanium. In β-Ti, the solubility of hydrogen is several times greater than in α-Ti, but its effect on embrittlement is weaker, as in this case it is dissolved evenly throughout the entire volume of the material.

It is of practical and theoretical interest—to what depth of the titanium hydride crystals air oxygen diffuses from the external environment at the maximum temperature of the exothermic effect (the process of titanium oxidation according to DTA data), i.e. at a temperature of 900 °C. It should be taken into account that the rate of interaction of titanium with air oxygen and with pure oxygen is approximately the same up to a temperature of 1000 °C. To solve this problem, the electron probe microanalysis of high-resolution scanning electron microscopy Tescan Mira 3 LMU was used. It was found that the depth of oxygen diffusion into the internal volume of the titanium hydride crystals at a temperature of 950 °C reaches 120–130 microns in both types of crystals (initial and modified) (Figs. 1 and 3). With increasing depth from the surface of the crystals, the atomic oxygen content continuously decreases (Figs. 2 and 4), (Tables 1, 2 and 3).

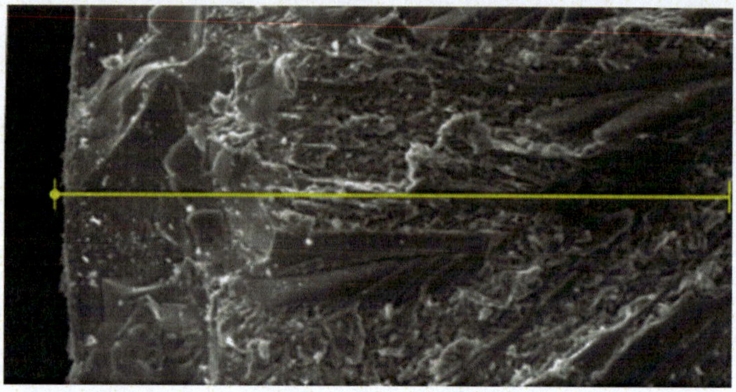

Fig. 1 The surface of the initial crystals of titanium hydride after heat treatment at 900 °C

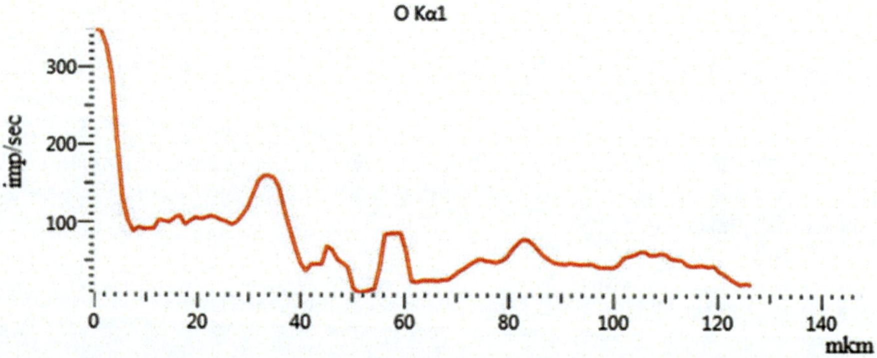

Fig. 2 Electron probe microanalysis of the surface of the initial hydride crystals after heat treatment at 900 °C

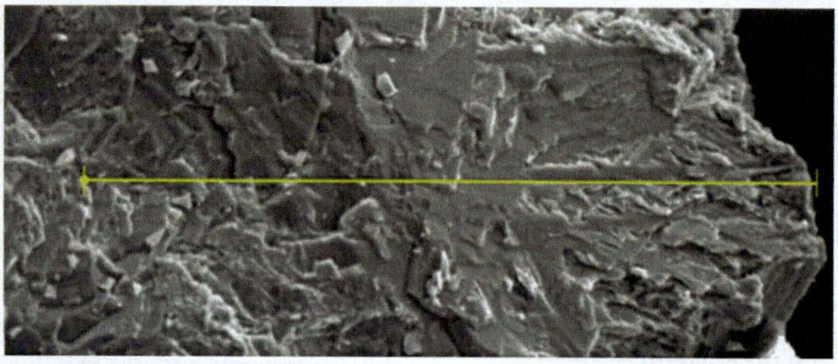

Fig. 3 The surface of the modified titanium hydride crystals after heat treatment at 900 °C

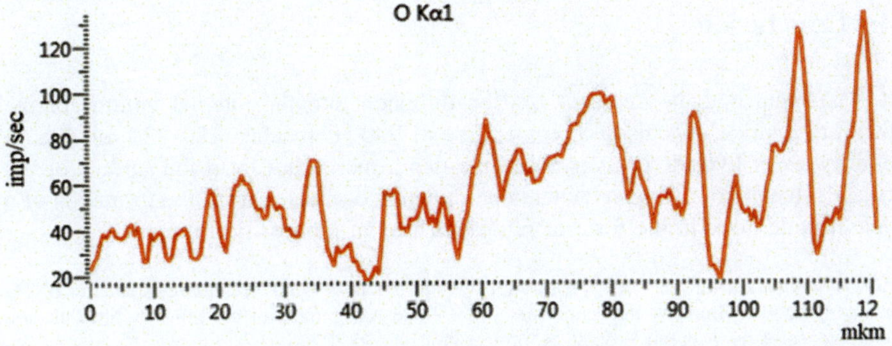

Fig. 4 Electron probe microanalysis of the surface of the modified hydride crystals after heat treatment at 900 °C

Table 2 Energy-dispersion analysis of the surface of the initial hydride crystals after heat treatment at 900 °C

Total spectrum on the line	Wt. %	Sigma wt. %	Atom. %
Ti	93.50	0.19	82.77
O	6.50	0.19	17.23

Table 3 Energy-dispersion analysis of the surface of the modified hydride crystals after heat treatment at 900 °C

Total spectrum on the line	Wt. %	Sigma wt. %	Atom. %
Ti	95.60	0.26	88.09
O	4.28	0.26	11.80
Si	0.07	0.02	0.11

The average atomic oxygen content on the surface of the initial titanium hydride crystals is 17.23%, and at a depth of 100 microns—2.45%, whereas on the surface of the modified titanium hydride crystals the oxygen content is 11.80%, and at a depth of 100 microns—1.81%.

Hydrogen stabilizes the β-phase of titanium, and oxygen, on the contrary, stabilizes the α phase of titanium, which has a hexagonal lattice with a dense packing of atoms. As a result, the diffusion of oxygen into the internal volume of the crystals should be difficult.

The reduced oxygen content in the deep layers of the modified titanium hydride crystals is also due to the insulating role of the borosilicate shell. As it was previously noted, the borosilicate shell shields the porous layer on the outer surface of the titanium hydride.

4 Conclusion

It was found that the depth of oxygen diffusion into the internal volume of the titanium hydride crystals at a temperature of 900 °C reached 120–130 microns in both types of hydride crystals. With the depth increasing from the surface of the crystals, the atomic oxygen content continuously decreased, with the formation of a dense oxide layer in the form of rutile, vitrified in a borosilicate matrix.

Acknowledgements The work is realized using equipment of High Technology Center at BSTU named after V.G. Shukhov the framework of the State Assignment of the Ministry of Education and Science of the Russian Federation, project No. FZWN-2020-0011.

References

1. Sickafus KE (2007) Radiation-induced amorphization resistance and radiation tolerance in structurally related oxides. Nat Mater 6(3):217–223
2. Cherkashina NI, Pavlenko AV (2018) Synthesis of polymer composite based on Polyimide and $Bi_{12}SiO_{20}$ Sillenite. Polym-Plast Technol Eng 57:1923–1931
3. Pavlenko VI, Kuprieva OV, Cherkashina NI, Yastrebinskii RN (2015) Defects in modified titanium hydride crystals subjected to heat treatment. Russ Phys J 58(5):724–729
4. Pavlenko VI, Yastrebinsky RN, Kuprieva OV, Pavlenko ZV, Bondarenko NI, Shapovalov NA (2021) Adsorption modification of titanium hydride (II) with sodium alkylsiliconate. Bull Technol Univ 24(1):10–13
5. Pavlenko AV, Cherkashina NI, Noskov AV (2017) Calculation of the frequency electronic transmission factors at the passage through the polymeric polyimide composite material filled by bismuth silicate. Probl Atomic Sci Technol 5:21–26
6. Cherkashina NI, Pavlenko AV (2018) Modification of optical characteristics of a polymer composite material under irradiation. Tech Phys 63(4):571–575
7. Cherkashina NI, Pavlenko AI (2016) Investigation of the mechanisms of modification of the surface of titanium hydride by organosiloxane structures. In: Bulletin of BSTU named after V. G. Shukhov, vol 9, pp 164–168
8. Pavlenko ZV, Cherkashina NI, Demchenko OV (2016) Creating a protective coating on the surface of titanium hydride shot. Bull BSTU named after V.G. Shukhov 10:166–171
9. Cherkashina NI, Pavlenko ZV, Kashibadze NV (2020) Increasing the stability of the polyimide radiation-protective composite to the effects of atomic oxygen. In: Lecture notes in civil engineering, vol 95, pp 79–85
10. Jang ES, Khan SB, Seo J, Akhtar K, Choi J, Kim KI, Han H (2011) Synthesis and characterization of novel UV-Curable PU-Si hybrids: influence of silica on thermal, mechanical, and water sorption properties of polyurethane acrylates. Macromol Res 19:1006

Assessment of Best Available Techniques for Production of Building Materials

A. S. Ptuskin

Abstract A sustainable society cannot be realised without efficient technologies, green technologies have spawned positive externalities in the form of improvements of the environment quality. Modern principles of ecological regulation of industrial enterprises activity are based on the concept of the best available techniques, i.e. the most effective new techniques that provide the highest level of environmental protection and have reached the level that makes their implementation in the relevant sector of the industry being possible. Development and improvement of formal models and methods of the best available technique identification remains an actual task. The present work suggests an entropy approach for estimation and selection of the best available technology in production of building materials. New heuristic entropy-based methods are developed to solve the problem of comparison and selection of the best alternative. The method is effective, user-friendly, and provides a stable solution. Numerical example illustrates the application of the new entropic methods for solving practical environment-protection problems.

Keywords Production of building materials · Best available technique · Selection of the best available techniques · Information entropy · Multi-criteria decision making

1 Introduction

Sustainable development in production has become a crucial issue in recent years and it is imperative to safeguard the environment, safety and welfare of those alive today and prepare for generations yet to come. A sustainable society cannot be realised without efficient algorithms and technologies; green technologies have spawned positive externalities in the form of improvements of the environment quality [1].

A. S. Ptuskin (✉)
Bauman Moscow State Technical University, Kaluga Branch, Kaluga, Russian Federation

S. V. Klyuev (ed.), *Digital Technologies in Construction Engineering*,
Lecture Notes in Civil Engineering 173, https://doi.org/10.1007/978-3-030-81289-8_9

In this respect, it is extremely important to design methods for regulating environmental protection that stimulates the rapid development of innovative environmental technologies. Environmental initiatives should include pollution prevention, recycling of materials and waste reduction. Modern principles of regulation of industrial enterprises that have significant negative impact on the environment are based on the application of the best available techniques (BAT) [2].

BAT corresponds to the techniques with the best environmental characteristics that can be introduced at a reasonable cost. The permit conditions defined in accordance with the BAT are an effective environmental tool and allow to minimize the main types of the negative impact.

The evaluation of the BAT and the comparison of alternative technological variants is a complex task and the application of optimization models provides an objective basis for reasonable decisions making. The present paper is devoted to the use of the entropy-based methods for main stage of the BAT selection process for production of building materials.

2 Materials and Methods of Research

Nowadays, the number of publications devoted to the problem of the BAT estimation and selection is growing rapidly [3]. The most popular approach to the determination of the BAT is the multi-criteria decision-making (MCDM) methodology [4]. However, the formal methods most often used to select BAT not always efficient in practice. Hence, variants to improve these methods are proposed for various tasks (see, e.g., [5–7] among many others). Accordingly, the design of new effective methods for determination of the BAT remains a challenging research direction [8]. The entropic approach permits to overcome some of the above disadvantages of existing BAT selection methods.

The entropy has found applications in a wide range of environmental problems (see [9]). Information, or Shannon's, entropy measures the level of uncertainty in the state of a system [10]. Given a set of events $E = \{e_1, \ldots, e_n\}$ with a priori probabilities of event occurrence $P = \{p_1, \ldots, p_n\}$, $p_1 \geq 0$, such that $p_c + \cdots + p_n = 1$, the entropy function H is defined as follows:

$$H = -\Sigma_c p_c \log p_c.$$

A decrease of entropy indicates a decrease in uncertainty, or in other words, an increase in decision maker's knowledge of the system. Accordingly, the growth of entropy indicates a decrease in the knowledge [11]. The authors of the articles [12, 13] among many others consider the entropy models of complex systems and various applications in the analysis and management of production, social and business processes.

We propose a new heuristic entropy method for solving the MCDM problem of comparison of options and selection of the best alternative for production of building materials. It provides stability of the solution in the large range of varying parameters.

Each alternative is evaluated according to several criteria; these estimates are assumed to be independent for various criteria. Based on the assessments for all the criteria we will select the alternative that provides the minimal negative impact on the environment, that is, our goal of BAT selection is to minimize the impact of the harmful factors (i.e., negative criteria). The idea is that we estimate the level of our confidence for each technology using the entropic measure of our knowledge for a particular criterion and it will be recognized as the best one. Finally, we shall combine these individual estimates for all available criteria.

Consider the above arguments in the formal terms. There is a set of m alternative technologies. Each technology is evaluated according to n criteria; each criterion is determined by the estimate a_{ct}; the weight of each criterion W_c is assumed to be given by the experts and known to the decision maker; $c = 1, \ldots, n$; $t = 1, \ldots, m$.

Comparing each of available technologies by a separate criterion permits to either decrease or increase our confidence that a particular technology can be accepted as the best one.

Let us fix the criterion c and the technology t. Consider the following event EV_{ct}: "In the best technology, the value of the criterion c will be equal to the value of this criterion for the technology t". Further, the probability of this event is $p_{ct} = (1/m)$ if the value a_{ct} is not known. It means that each technology can equally claim that it will be recognized as the best, i.e. in the best technology the value of the criterion is equal to the value of this criterion for this technology. The level of our knowledge about the event EV_{ct} may be estimated by the information entropy is as follows:

$$H_{ct}^{0} = -(1/m)\log(1/m) - (1-(1/m))\log(1-(1/m)).$$

The probabilities p_{ct} for all t vary and, accordingly, the level of knowledge about the event EV_{ct} is change when we know values a_{ct}; $c = 1, \ldots, n$; $t = 1, \ldots, m$. In our case, the goal of the choice is to minimize the harmful impacts of the environmental criteria (the smaller the a_{ct} is, the more this technology corresponds to the BAT). Then the p_{ct} values are estimated as follows:

$$p_{ct} = (\min\{a_{cu}\}/a_{ct})/\Sigma_u(\min\{a_{cu}\}/a_{ct}); \quad u = 1, \ldots, m.$$

Given a_{ct}, the entropy measure of the uncertainty level of the event EV_{ct} is calculated as follows:

$$H_{ct} = -p_{ct}\log p_{ct} - (1-p_{ct})\log(1-p_{ct}).$$

The change in the level of knowledge (that is, our confidence that technology t is the best) can be determined by correcting the entropy measurement ΔH_{ct} after obtaining the new information on the value a_{ct}. We believe that in the case when $p_{ct} < (1/m)$ our confidence that the technology t will be the best decreases:

$$\Delta H_{ct} = -\left|H_{ct}^{0} - H_{ct}\right|,$$

and in the case when $p_{ct} > (1/m)$ the confidence that the technology t will be the best increases:

$$\Delta H_{ct} = \left|H_{ct}^{0} - H_{ct}\right|.$$

In this manner, we define ΔH_{ct} for all $c = 1, \ldots, n$; and $t = 1, \ldots, m$. Then the degree of confidence that the technology t is the best, with all the criteria being taken into account can be found as follows:

$$S_t = \Sigma_c W_c \Delta H_{ct}.$$

The best technology selected is the technology b with the highest value of S_t:

$$S_b = \max\{S_t | t = 1, \ldots, m\}.$$

The method can be used both in the case when the purpose of the choice is to minimize criteria or to minimize criteria. In the latter case the estimates of the values of p_{ct} and ΔH_{ct} are taken with an opposite sign.

3 Results and Discussion

Let us explain the proposed method for objectively evaluating the alternative options for the example of the lime production, limiting our procedure for selecting the BAT technology with seven criteria (related to the emissions to the air) and considering four alternative technologies. The criteria a_{ct} and the criteria weights of the criteria W_c; $c = 1, \ldots n$ are given in Table 1.

Table 1 The criteria and the weights of the criteria for the numerical example

Criteria	Technology, t				W_i
	1	2	3	4	
	Flows (mg/m^3)				
Dust	105.8	149.3	83	58	0.15
Nitrogen dioxide (NO_x)	331.6	122	861.6	1123	0.14
Carbon monoxide (CO)	1294	504.6	1728.3	1328.6	0.07
Carbon black (C)	0.63	0	51.3	53	0.13
Sulfur dioxide (So_x)	153.3	0	547	809	0.15
Hydrogen chloride (HCl)	0	0	20.5	17.1	0.19
Hydrogen fluoride (HF)	0	0	27	0	0.17

Table 2 shows the results, illustrates the procedure for the BAT determination and includes all steps of the method. The best technology obtained is 2, for which the value of S_t is maximal.

The proposed entropic method for multi-criteria selection of the BAT is applied to the case study considered in the paper [4]. The evaluation of the best available techniques to control the emission of particles into atmosphere during powder processing in the tile industry have presented. Three technologies are compared. The work proposes a decision-making process for assessing the BAT based on two multi-criteria decision models—AHP and ANP. The resulting matrix of the weights and estimates is given in Table 3. The result of the solution by the entropic method is shown in Fig. 1.

Table 2 The values S_t for numerical example

Technology, t	S_t
1	−0.19
2	**0.28**
3	−0.58
4	−0.41

Table 3 The matrix of weights and estimates for the numerical example of [4]

Criteria	c	Weights, Wc	Technology, t		
			1	2	3
Implementation costs	1	0.142	0.73	0.18	0.09
Resource consumption	2	0.199	0.69	0.19	0.11
Energy efficiency	3	0.4	0.28	0.05	0.66
Wastewater management	4	0.059	0.44	0.2	0.36
Air emissions management	5	0.072	0.16	0.07	0.77
Waste management	6	0.054	0.29	0.07	0.63
Workers health	7	0.074	0.29	0.07	0.63

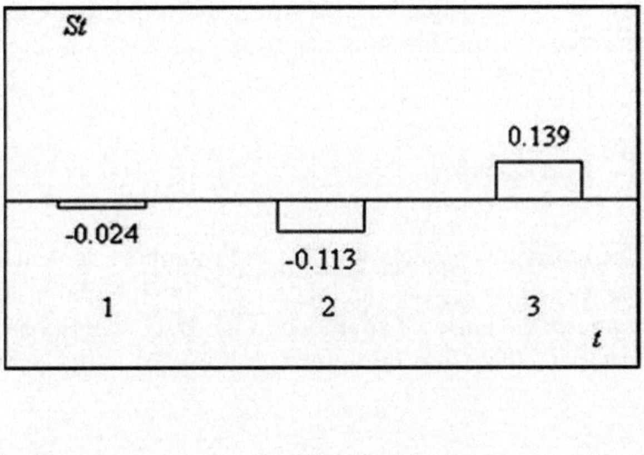

Fig. 1 The value of S_t for the numerical example of [4]

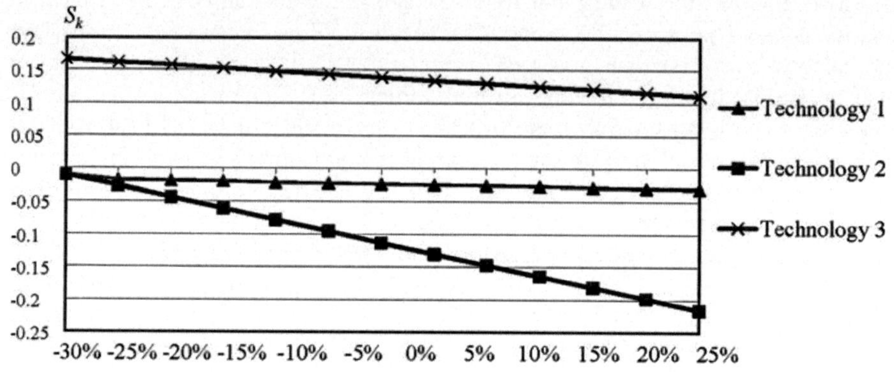

Fig. 2 Results of the sensitivity analysis for the criterion *Energy Efficiency* for the numerical example of [4]

Our results coincide with the results of the cited paper: technology 3 is the best, for which the value of S_t is maximal. However, the entropic solution method requires a smaller number of operations. In addition, it provides a more stable solution in a larger range of varying input data. Specifically, in the case of [4] where the ANP method was used, it was shown that, when the weight of the criterion Energy efficiency is increased in 25% (from 0.4 to 0.5), the alternative 3 reaches the same priority as the alternative 1. At the same time, the entropic method provides the range 30%, see Fig. 2. Also, an important additional advantage is that both the minimization and maximization criteria can be handled by the entropy-based method.

The entropic method leads to effective computational procedures and provides stable results. Numerical examples show that the proposed procedure is reliable and effective. The selection process considered in this paper is greatly facilitated by the use of mathematical and instrumental tools that create the basis and information structure for making final decisions, especially in the case of large-scale problems.

The entropic method leads to effective computational procedures and provides stable results. Numerical examples show that the proposed procedure is reliable and effective. The selection process considered in this paper is greatly facilitated by the use of mathematical and instrumental tools that create the basis and information structure for making final decisions, especially in the case of large-scale problems.

4 Conclusion

The reasonable selection of the BAT supports a cleaner production, minimizing the waste and emissions and ensuring the rational usage of resources. This paper suggests the entropic approach to the BAT identification for production of building materials. The illustrative example confirms the productivity of the method. We believe that the suggested methods can be considered as an extension of the tools for solving practical problems and are applicable for other types of the BAT estimation problems.

Acknowledgements This work was supported by the Russian Foundation for Basic Research (grant No. 18-410-400001).

References

1. Sbardella A, Perruchas F, Napolitano L, Barbieri N, Consoli D (2018) Green technology fitness. Entropy 20:776
2. Ptuskin A, Zhukova J (2020) A fuzzy model of the selection of alternative operations of the best available techniques at the installation level. Econ Math Methods 56(4):78–87
3. Ibáñez-Forés V, Bovea MD, Pérez-Belis V (2014) A holistic review of applied methodologies for assessing and selecting the optimal technological alternative from a sustainability perspective. J Cleaner Prod 70:259–281
4. Giner-Santonja G, Aragonés-Beltrán P, Niclós-Ferragut JJ (2012) The application of the analytic network process to the assessment of best available techniques. J Cleaner Prod 25:86–95
5. Abbas V, Chergu Z (2019) The impact of using new significant reference point with TOPSIS methods: study and application. Int J Inf Decis Sci 11(2):95–115
6. Liu Y-N, Wu S-Y (2020) A rule-based approach for dynamic analytic hierarchy process decision-making. Int J Inf Decis Sci 12(1):36–74
7. Pereira JS, Fayer GC (2020) Strategic decision making to maximise the efficiency of water usage in steel manufacturing process via AHP and BBN: a case study. Int J Inf Decis Sci 12 (4):328–347
8. Bréchet T, Tulkens H (2009) Beyond BAT: selecting optimal combinations of available techniques, with an example from the limestone industry. J Environ Manage 90:1790–1801
9. Cui H, Sivakumar B, Singh VP (2018) Entropy applications in environmental and water engineering. Entropy 20(8):598
10. Shannon CE (1948) A mathematical theory of communication. Bell Syst Tech J 27:379–423
11. Levner E, Ptuskin A (2017) Entropy-based model for the ripple effect: managing environmental risks in supply chains. Int J Prod Res 56(7):2539–2551
12. Levner E, Ptuskin A (2015) An entropy-based approach to identifying vulnerable components in a supply chain. Int J Prod Res 53(22):6888–6902
13. Ruiz-Hernández D, Menezes MBC, Amrani A (2019) An information-content based measure of proliferation as a proxi for structural complexity. Int J Prod Econ 212:78–91

Fine-Grained Concrete as a Structural Construction Material of EDT Drill-Injection Piles

N. S. Sokolov

Abstract The issue of ensuring the quality of performance of buried reinforced concrete structures is the main task of geotechnical construction. At the same time, the strength of fine-grained concrete of the cross-section of EDT drill-injection piles is a fundamental indicator for determining its bearing capacity on the ground and on the body. Electric discharge technology allows increasing the strength of fine-grained concrete. At the same time, it can exceed the strength of untreated concrete by the electrohydraulic method by 40–50%. An important role in the process of gaining the strength of concrete is played by compliance with the technological regulations for the manufacture of EDT piles. In geotechnical construction, cases of non-compliance of the concrete strength of finished piles with the design values are very common. The paper presents a case from geotechnical practice.

Keywords Fine-grained concrete · Electric discharge technology EDT · EDT drill-injection piles · Structural construction material · Buried reinforced concrete structures

1 Introduction

The construction of any structure involves operational technical control of the quality of construction, which allows ensuring the safe operation of the object under construction [1–17]. Of particular concern is the quality of manufacturing of drilling piles. It is known that the technology of manufacturing drilling piles is a complex technological chain consisting of drilling, concreting and reinforcement of the trunk. If we consider a drill-injection pile made according to the electrohy-

N. S. Sokolov (✉)
Chuvash State University Named After I. N. Ulyanov, Moskovskiy prosp., 15, Cheboksary 428015, Russian Federation

N. S. Sokolov
NPF (LLC SPC) "FORST", ul. Kalinina, 109a, Cheboksary 428000, Russian Federation

S. V. Klyuev (ed.), *Digital Technologies in Construction Engineering*, Lecture Notes in Civil Engineering 173, https://doi.org/10.1007/978-3-030-81289-8_10

draulic method (EDT piles), then the electrohydraulic treatment of the well walls in the body of freshly laid concrete is added to the technology of its manufacture. For EDT piles, the technological chain looks like: "well drilling—concreting—electrohydraulic treatment—trunk reinforcement".

In this paper, the influence of the quality of the trunk concreting on the load-bearing capacity of the pile is considered.

Below there is an example of the algorithm for selecting the composition of fine-grained concrete mix (CBM) for the manufacture of EDT piles.

This article examines the influence of the quality of the concreting of the trunk on the bearing capacity of the pile.

Below is an example of an algorithm for selecting the composition of a fine-grained concrete mixture (BMC) for the manufacture of ERT piles.

2 Methods and Materials

Drill-injection piles of EDT related to "Micro-piles" without compacting the walls of wells are ineffective. One of the ways to increase F_d is to compact the walls of drilling wells using electric discharge technology (EDT technology). By arranging widenings in specific areas along the length of the EDT pile, it becomes possible to replace the load-bearing capacity of the piles on the side surface with the load-bearing capacity of the pile under the widening. The latter is ten times higher than the former. The experience of EDT drilling-injection piles with several widenings shows their increased F_d values in comparison with EDT piles without widenings. The results of static tests with three widenings at one of the facilities in the city of Nizhny Novgorod showed twice the load-bearing capacity compared to the F_d of the EDT pile without compacting the walls.

Geotechnical construction in cramped conditions has recently become particularly relevant. This type of activity involves the implementation of the following activities:

(1) calculation, design, development and construction of the types of foundations providing the minimum draft of the built-up building or structure;
(2) calculation, design, development and arrangement of enclosing structures that minimize the impact of the built-up object on existing buildings and structures within the zone of influence;
(3) calculation, design, development and implementation of constructive measures to increase the operational reliability of buildings and structures also within the zone of influence.

In geotechnical construction, the most effective types of foundations are drilling-injection piles on fine-grained concrete.

3 Results and Discussion

One of the stages of structural design of concrete structures trunk of drilling-injection EDT piles is a selection of fine-grained concrete mix (CCM) according to GOST 7423-2010 "Concrete Mixtures. Technical conditions".

The algorithm for selecting the composition of the CCM is presented in the following sequence:

1. According to the values of the design load-bearing capacity of the EDT pile F_d on the ground, the class (grade) of concrete for compressive strength is assigned. At the same time, according to GOST 26633-91 "Heavy and fine-grained concretes", the average strength of concrete is laid at a coefficient of variation V = 13.5%, the security of at least 95% of the assigned value. For example, for the design grade of fine-grained concrete M400, the cubic strength value should be R = 38.5 MPa (392.5 kg/cm^2).
2. According to GOST 7473-2010 "Concrete mixes. Technical conditions" the brand is selected according to the workability of the concrete mixture P and the mobility index (cone slump).

   For example, the symbol P4 means the cone slump of 20 cm.

3. The hardening conditions are assigned. At the base below the freezing depth, the hardening conditions are natural. When carrying out geotechnical work in conditions of negative temperatures, either chemical hardening with the use of sodium formate or an electric method of heating with heating wires are used. It should be noted that electric heating from the experience of work is undesirable. It is possible that the occurrence of shrinkage cracks in the concrete body as a result of rapid strength gain and as a result of the separation of the part of the heated EDT pile hardening from the part naturally.
4. The components for fine-grained concrete are selected—cement, fine aggregate, concrete additives and water.

   4.1 Portland cement is usually delivered to the facility from the nearest cement plant. In the Middle Volga region, cement produced by JSC "Mordovcement" is used. According to GOST 31108-2003 "General construction cements", the controlled parameters are:

   (1) compressive strength at the age of 28 days R = 50 MPa;
   (2) normal density of cement paste 27%;
   (3) setting time: start 2 h 35 min, end 4 h 25 min;
   (4) true density ρ = 2.63 g/cm^3.

   4.2 Natural river sand according to GOST 8736-2014 "Sand for construction works" is accepted as a fine aggregate. Technical specifications (as amended)" with a model size of no more than Mk = 2.0. The percentage of fractions larger than Mk $\geq$ 2.0 mm and the density of mineral particles ρ_s are determined.

4.3 Additives are used to increase the strength of concrete and increase mobility. For example, the additive EMBELIT 8-100—a concrete modifier according to TU 5870-176-46854090-04, manufactured by LLC "Enterprise Master Concrete", Moscow, is both a plasticizer and a modifier.
4.4 Water is also subject to special requirements in accordance with GOST 23732-79 "Water for concrete and mortar".

5. In the construction laboratory under the assigned strength, mobility, workability, hardening conditions according to GOST 27006-86 "Concrete. Rules for selecting the composition of concrete" are designed.

5.1 Water-cement ratio, for example, W/C = 0.51, where W is the mass of water;
5.2 The ratio of materials by weight, for example, C:S = 1:2.1, where C is the mass of cement; S is the mass of sand;
5.3 The content of additives in % by weight of cement, for example, the content of EMBELIT 8-100 = 10;
5.4 Material consumption per 1 m^3 of concrete mix;

For example, one of the objects used: cement—850 kg; sand—810 kg; additive EMBELIT 8-100—85 kg; water—465 kg.

In addition to the characteristics of the nominal composition of fine-grained concrete, the composition selection algorithm provides a section of the actual possible material consumption per 1 m^3 of concrete mix.

6. The physical and mechanical properties of concrete that are required to confirm the correct selection of the composition at the object are the average density of concrete in a series of samples with a size of 10 × 10 × 10 cm, ρ (g/cm^3) and the compressive strength at the age of 7 and 28 days.

Next, we consider a case from construction practice. The project for the ten-story building of the hotel provided for the installation of drill-injection piles made using electric discharge technology (EDT). This technology includes drilling, concreting, electrohydraulic treatment, reinforcement and topping up of concrete. Due to the fact that these stages of work were performed by four contractors and there was no proper step-by-step control of the concrete strength set, the technological cycle was disrupted, and more than 50% of the piles did not reach the design load-bearing capacity. Therefore, there was a need to redesign the pile field.

The construction of the facility was carried out in difficult engineering and geological conditions in the old channel of the Volga River. The geological section on this site is represented by the following engineering-geological elements (EGE) (top to bottom) (Fig. 1): EGE-1—bulk soil (non-packed loam with sandy loam and debris); EGE-2—unsettled tight and smooth loess loam; EGE-3—unsettled fluid plastic loess loam; EGE-4—tight and smooth loam; EGE-5—solid and semi-solid variegated clay; EGE-6—clay polimict sand.

The construction site is characterized by a high level of ground (non-pressure) water.

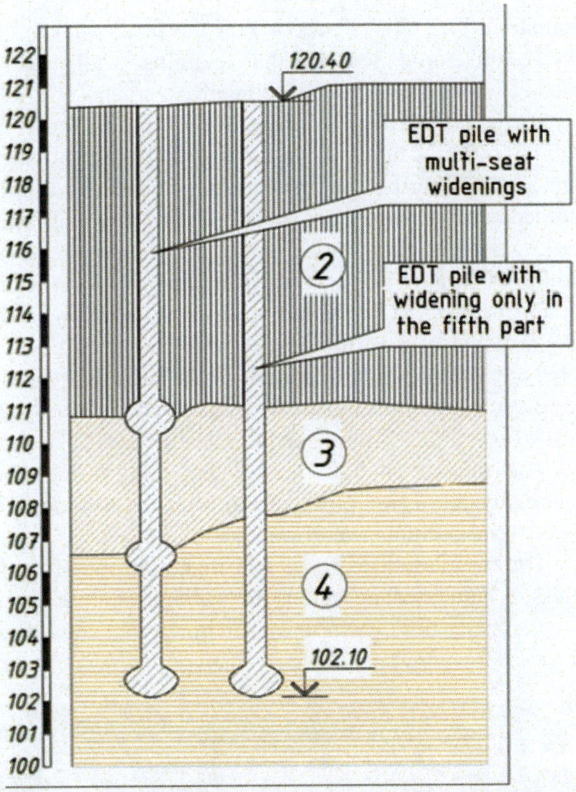

Fig. 1 Engineering and geological section of the test site and piles made using electric discharge technology with multi-seat widenings (left), and only with a widened fifth (right). The vertical axis on the left—the height mark, m

The construction of the facility was started 5 years before the start of the main construction with the construction of a pit fence (9.0 m deep) from two rows of drill-injection piles with a diameter of 450 mm in increments of 1.0 m. The retaining wall of the pit was built along the streets. The 10-storey large-panel residential building built on driven piles is directly adjacent to the pit.

The disadvantage of the constructed fence was the lack of a monolithic binding reinforced concrete belt on the top of the drill-injection piles. This was revealed only during the excavation of the pit. A row of fence piles on the side of the adjacent building tilted towards the pit (the maximum horizontal movement reached 55 mm). As a result of this situation, deformation cracks appeared on the exterior walls of the residential building. At the same time, the installed plaster beacons broke and continued to tear.

The emergency commission created in this regard instructed the head design organization to develop urgently emergency measures to stabilize the deformations of both the erected retaining wall and the adjacent building. As such measures, a scheme for strengthening the retaining wall was developed in the form of spacer structures made of pipes with a diameter of 1,000 mm, located on two levels in mutually perpendicular directions (Fig. 2). These measures helped to stabilize the situation. The plaster beacons on the residential building stopped tearing, and the horizontal movement of the retaining wall was paused. At the same time, geotechnical monitoring continued.

At the design depth of the pit of 9.0 m, the thrust mounts were placed at a depth of 4.5–6.5 m. Therefore, in order to avoid negative consequences for the adjacent houses with further pit excavation project of the arrangement of monolithic reinforced concrete buttresses on additional EDT drill-injection piles with a diameter of 0.35 m and a length from 12 to 19 m, depending on the geological conditions in different parts of the construction site was developed. Work on the installation of EDT piles had to be carried out in very difficult conditions between the pipes, and the removal of soil from the pit was carried out only manually.

To ensure the safe operation of the retaining wall during the construction of the zero cycle, as well as to create conditions for the dismantling of steel pipes of thrust structures, an algorithm for the construction of buttresses was developed.

The implementation of the above algorithm made it possible to dismantle gradually the thrust pipes. No further deformations of the retaining wall and the residential building were found.

The installation of EDT drill-injection piles [1–14] for the foundations of buttresses, buttresses, as well as the grillages themselves were made by one contractor. At the same time, monitoring of horizontal movements of the retaining wall and

Fig. 2 Fragments of the made buttresses

deformations of the sedimentary marks of the residential building was carried out daily, so that there were no violations in the technological chain "drilling—concreting—electrohydraulic treatment of the walls and the heel of the well—installation of armored frames" at this site.

The mandatory stages that confirmed compliance with the project of the projected EDT piles for the foundations of the buttresses were.

Strength tests of pre-made cubes of fine-grained concrete intended for the manufacture of piles, according to the algorithm given above.

Static load tests of experimental EDT piles.

At the test site (within the buttress manufacturing area), two pile bushes were made using electric discharge technology. In one of these piles were made without broadening, and the other with multiple caps (vertical binding piles, see Fig. 1). The test results of EDT pile bearing capacity with a static load are shown in Fig. 3.

As the thrust structures were removed, a significant part of the pit area was freed up for the construction of the pile field.

In connection with the reduction of the investor's terms of construction of the building, the customer decided to increase the speed of construction of the zero part, dividing the device of EDT drill-injection piles into stages. At the same time, one construction company contracted to perform drilling operations, another—concreting with fine-grained concrete, the third—electrohydraulic treatment of the walls and the heel of the well, the fourth—the manufacture and immersion of armored frames in finished wells filled with fine-grained concrete and processed by electric discharge technology.

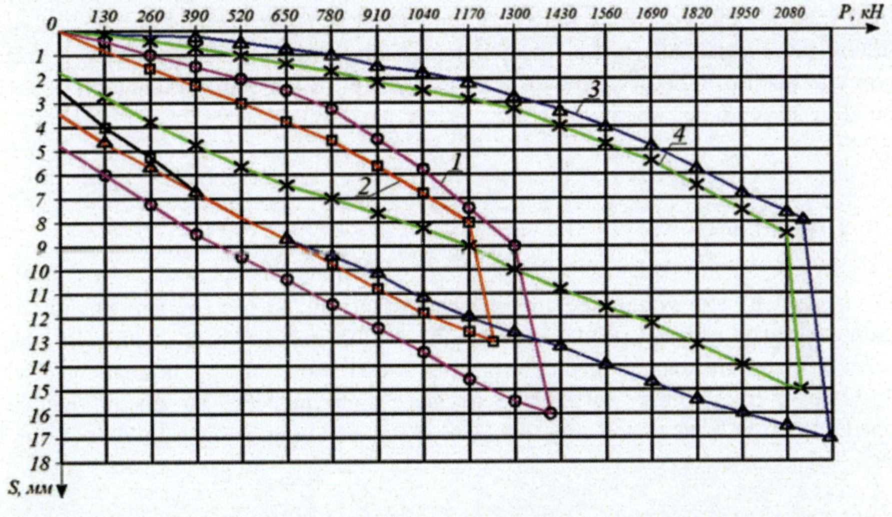

Fig. 3 Graphs of the results of static load tests of EDT drill-injection piles: 1, 2—only with a widened fifth; 3, 4—with multiple widenings. Letter designations: P—load on the pile; S—vertical movement of the pile

Table 1 The results of the tests of EDT piles on the static load

№ of pile	Bearing capacity F_d (kN)	Design load, N (kN)	Concrete pile body strength (kPa)	The reason for the insufficient load-bearing capacity of the pile
789	62.8	52.3	60	Low concrete strength
710	75.9	63.3	65	
579	95.0	79.1	78	
822	251.2	209.1	1200	Pile failure
728	565.2	471.0	2500	
767	376.8	314.0	2600	
803	251.2	209.3	1300	

Accordingly, the quality of work performed at different stages was controlled by different contractors, and as a result, the probability of its reduction sharply increased. The results of static load testing of the piles confirmed these concerns (Table 1). For more than 50% of the tested piles, the design load-bearing capacity was not achieved. The main reason for this was the low strength of fine-grained concrete due to the lack of proper supervision of its set. It should be noted that one of the contractors responsible for concreting piles used the RM-750 concrete mixing plant, which at high speeds saturates the concrete with air, which led to a shortfall in the design strength values. At the same time, the selection of the CCM composition was made correctly.

To ensure the design load-bearing capacity of grillages with defective piles, it was decided to supplement them with piles with multiple widenings. The entire pile field (and all the grillages) were redesigned, taking into account the results of testing the experimental piles for their load-bearing capacity. Due to this, it was possible to ensure the design load-bearing capacity of the entire foundation of the building as a whole.

4 Conclusion

In conclusion, we would like to emphasize that due to the implementation of different stages of work on the installation of EDT piles by four contractors, the necessary step-by-step quality control was not provided and additional material resources were spent to correct the negative consequences. To ensure reliable operation of the constructed structures, customers and contractors should not allow such situations to occur.

References

1. Sokolov NS, Tavrin VYu, Abramushkin VA (2005) Patent RF 2250958. The device for production of a stuffed pile. Declared 14.07.2003. Published 27.04.2005. Bulletin No. 12
2. Sokolov NS, Tavrin VYu, Abramushkin VA (2005) Patent RF 2250957. The method of production of a stuffed pile. Declared 14.07.2003. Published 27.04.2005. Bulletin No. 12
3. Sokolov NS, Pichugin YuP (2006) Patent RF 2282936. Generator of pulse currents. Declared 4.02.2005. Published 27.08.2006. Bulletin No. 24
4. Sokolov NS (2008) Patent RF 2318960. The method of production of a stuffed pile. Declared 26.12.2005. Published 10.03.2008. Bulletin No. 7
5. Sokolov NS (2008) Patent RF 2318961. Discharge device for production of a stuffed pile. Declared 10.07.2007. Published 10.03.2008. Bulletin No. 7
6. Deckner F, Viking K, Hintze S (2017) Wave patterns in the ground: case studies related to vibratory sheet pile driving. Geotech Geol Eng 35(6):2863–2878
7. Korff M, Meijers P, Wiersma A, Kloosterman F (2019) Mapping liquefaction based on CPT data for induced seismicity in Groningen. Earthquake geotechnical engineering for protection and development of environment and constructions-proceedings of the 7th international conference on earthquake geotechnical engineering, Rome, pp 3418–3425
8. Deckner F, Viking K, Guillemet C, Hintze S (2015) Instrumentation system for ground vibration analysis during sheet pile driving. Geotech Test J 38(6):893–905
9. Brinkgerve RBJ (2006) Plaxis: finite element code for soil and rock analyses. Balkema
10. Denies N, Holeyman A (2017) Shear strength degradation of vibrated dry sand. Soil Dyn Earthq Eng 95:106–117
11. Karol Reuben H (2003) Chemical grouting and soil stabilization. American Society of CivilEngineers
12. Moseley MP (2004) Ground improvement. Spon Press, London
13. Sokolov NS, Ryabinov VM (2015) About one method of calculation of bearing capacity of bored-injection EDT-piles. Osnovaniya, fundamenty i mekhanika gruntov 1:10–13
14. Sokolov NS (2014) Method of calculation bearing capacity of the bored-injection EDT-piles taking into account «thrust bearings». In: Materials of the 8th all-Russian (the 2nd international) the «new in architecture, designing of construction designs and reconstructions» conference (NASKR-2014), Cheboksary, pp 407–411
15. Sokolov NS, Viktorova SS, Fedorova TG (2014) Piles of increased bearing capacity. In: Materials of the 8th all-Russian (the 2nd international) the «new in architecture, designing of construction designs and reconstructions» conference (NASKR-2014), Cheboksary, pp 411–415
16. Sokolov NS, Petrov MV, Ivanov VA (2014) Calculation problems of bored-injection piles manufactured with the use of electric discharge technology. In: Materials of the 8th all-Russian (the 2nd international) the «new in architecture, designing of construction designs and reconstructions» conference (NASKR-2014), Cheboksary, pp 415–420
17. Sokolov NS, Sokolov SN, Sokolov AN (2016) Experience of restoration of an emergency building of Vvedensky cathedral in Cheboksary. Geotechnics 1:60–65

Methods for Reducing the Uncertainty of Risks of Failures of Elements in Technological Machines

N. S. Sevryugina, A. S. Apatenko, and N. I. Kozhukhova

Abstract The article proposes to consider the problem of assessing the technical condition of technological machines and equipment not at the practical and scientific—research levels, but to use interdisciplinary scientific and technical, practice-oriented research. It has been substantiated that, in accordance with the theory of probability, there can be only one failure in the unit at a time. The concept of an integrated parameter is introduced, which to a greater extent characterizes the state of the unit. An algorithm for recognizing the area of displacement of uncertainty zones has been developed, the translation of which into a digital format with the introduction of a recognition criterion will provide a defect identification (risk-failure). The practical implementation of the risk-failure identification mechanism is carried out with a simulation model of a change in the state of a technical system using the example of a cardan joint of a technological machine, which is in severe operating conditions during reclamation work. The inclusion of a digital module for monitoring the technical condition of the unit in the design of technological machines is substantiated.

Keywords Risk-failure identification · Technological machine · Algorithm

1 Introduction

The representation of a technical system as interconnected parameters is one of the methods to establish not only the degree of influence, but also to determine the technical condition by of diagnostic recognition of the intensity of parameter changes.

N. S. Sevryugina · A. S. Apatenko
Russian State Agrarian University—Moscow Timiryazev Agricultural Academy, Moscow, Russia

N. I. Kozhukhova (✉)
Belgorod State Technological University Named After V.G. Shukhov, Belgorod, Russia

S. V. Klyuev (ed.), *Digital Technologies in Construction Engineering*,
Lecture Notes in Civil Engineering 173, https://doi.org/10.1007/978-3-030-81289-8_11

As it is known, the unit is a complex technical system characterized by geometric, structural and functional parameters, the values of which were determined in the pre-operational period, after commissioning, go into the category of partially determined in the presence of a built-in control and measuring apparatus for assessing the state or uncertain indicators, in case of their absence [1–7].

Uncertainty of the state of a technical system occurs in the following cases: lack of information on the reliability of equipment (high error values); human factor (operator errors), etc.

Scientific developments of the solution to the indicated problem can be represented in the following areas: functional efficiency of technical systems [8–11], theoretical solution by developing mathematical models [12, 13], innovations [14–16].

Considering the system as a combination of elements structured to perform a separate function, when solving problems of performance assessment, the variability of the state is taken, which in the model is characterized both quantitatively and qualitatively by comparing with the normalized values of the parameters [17–20].

The purpose of the analysis of the uncertainty of the state of a technical system is to translate the uncertainty of the initial parameters and assumptions used when the risk assessment into the uncertainty of the results.

2 Materials and Methods

Modal analysis of the uncertainty of the state of a technical system. Each of the specified parameters over time transfers the system from normal operation, to a state of non-critical parameter change, and then to a pre-failure state and system failure: $D_0 \underset{i}{\overset{x}{\Rightarrow}} D_1$. Diagnosis of the unit state by a single parameter X obeys the following rule.

$$X_{\min} \leq X_{\text{nom}} \leq X_{\max} \Rightarrow x \in D_0$$

The good condition of the unit is represented as D_0, which is parametrically set by the normalized value when designing the unit as functionally substantiated and confirmed by certification tests. The presence of a defect leading to a failure—D_1, parametrically expressing the maximum deviation of the normalized value. The complexity of monitoring the change in values is solved by compiling a table of the parameter significance in terms of the risk-failure effect on the functionality as a whole. In accordance with the theory of probability, it is generally accepted that there can be only one failure in the unit at a time, even if several damages are diagnosed, then there is always a single cause, i.e. failure that caused a chain reaction of damage. As noted above, functional and structural relationships are characterized by a different number of parameters, therefore, the concept of an integrated parameter is introduced, which will largely characterize the state of the unit. In logarithmic form, it can be presented by following:

$$\ln\frac{f\left(\frac{x_0}{D_0}\right)}{f\left(\frac{x_0}{D_1}\right)}=\ln\frac{P_{D_1}}{P_{D_0}} \tag{1}$$

$$f\left(\frac{x}{D_0}\right)=\frac{1}{\sigma\sqrt{2\pi}}e^{\frac{(x-x_1)^2}{2\sigma^2}}\; f\left(\frac{x}{D_1}\right)=\frac{1}{\sigma\sqrt{2\pi}}e^{\frac{(x-x_2)^2}{2\sigma^2}}, \tag{2}$$

where x_1, x_2 is the mathematical expectation of the parameter x to the value of D_0, and D_1, which corresponds to the normalized state or pre-failure (failure).

σ is the standard deviation of the parameter x. The choice of this value is also influenced by the availability of measuring instruments; at present, this issue is being solved by installing digital monitoring and control modules. Any digital block includes a logical scheme for receiving information, comparing it with normalized values and making a decision on the criticality of the deviation of the result. I.e., when constructing a diagnostic model, the basic provisions of information theory are used, which characterizes an aggregate as a system containing uncertainty, which is distributed over all structural elements of the aggregate. The task of constructing a diagnostic model is theoretically aimed at revealing the uncertainty of the state of the model. As information is obtained during diagnostics, the level of uncertainty decreases by the amount of exclusion of the number of certain states of the model elements. Knowledge of the mechanism of distribution of uncertainty over structural elements can be used to evaluate various ways of searching for a defect. Mathematically, this mechanism is described by the following equations.

Digital description of the criteria for recognizing the state of a technical system Digital systems allow creation an algorithm for a defect identification (risk-failure) by providing a recognition criterion. Graphically, the options for obtaining a statistical solution in the presence of a zone of uncertainty are presented in Fig. 1.

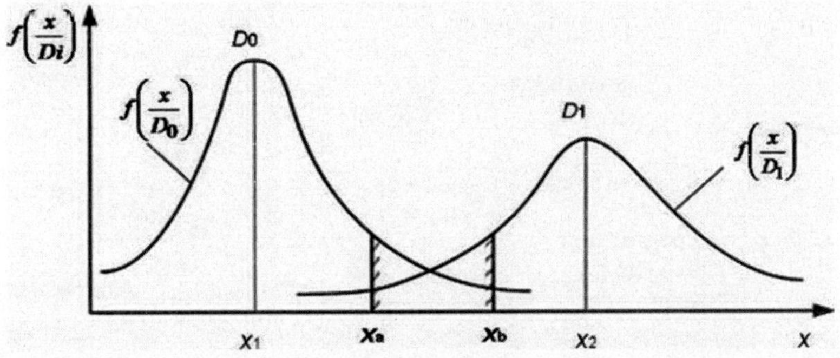

Fig. 1 Areas of displacement of the zone of uncertainty

The algorithm of the recognition rule for the area of displacement of uncertainty zones (Fig. 2) takes the following form:

$$\text{If}\begin{cases} if \cong x \leq x_a \Rightarrow x \in D_0 \\ if \cong x \leq x_a \Rightarrow x \in D_1 \\ if \cong x \leq x_a \leq x_b \Rightarrow H \in D_0 \Rightarrow H \in D_1 \end{cases} \tag{3}$$

where H is the uncertainty function, which for the digital control module is a signal for failure to recognize the system location: in good working order—H_{D0}; in failure state—H_{D1}

$$P(H_{D_0}) = P_{D_0} \int_{x_a}^{x_b} f\left(\frac{x}{D_0}\right)dx \quad P(H_{D_1}) = P_{D_0} \int_{x_a}^{x_b} f\left(\frac{x}{D_1}\right)dx \tag{4}$$

The condition for the minimum average risk is Eq. (5):

$$\frac{f^j\left(\frac{x_a}{D_0}\right)}{f^j\left(\frac{x_a}{D_1}\right)} \leq \frac{P_{D_1}}{P_{D_0}} \frac{f^j\left(\frac{x_b}{D_0}\right)}{f^j\left(\frac{x_b}{D_1}\right)} \leq \frac{P_{D_1}}{P_{D_0}} \tag{5}$$

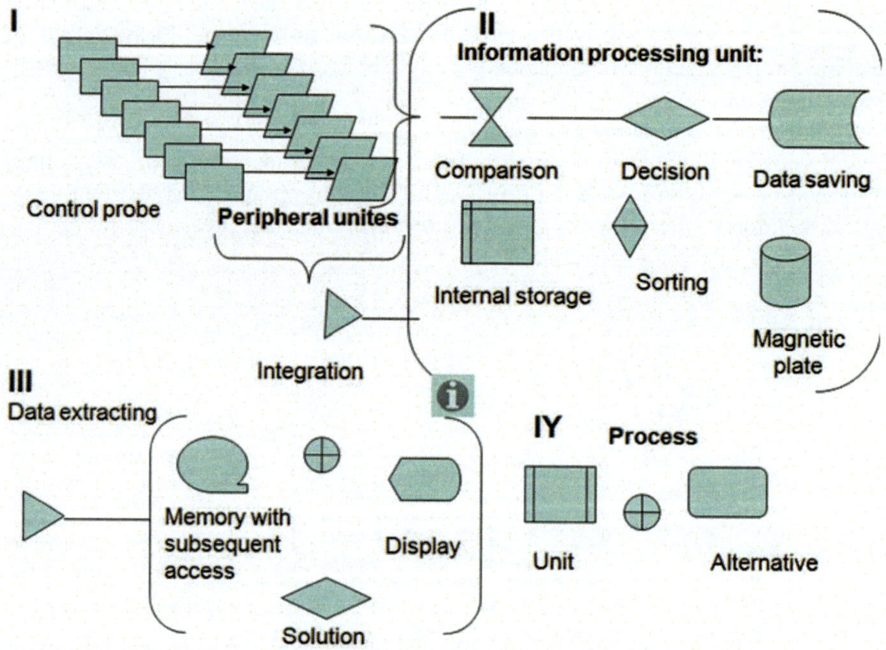

Fig. 2 Algorithm for collecting information and controlling a digital system according to [19]

3 Results and Discussions

Simulation modeling of a change in the state of a technical system using the example of a universal joint. The system for maintaining the operability of the structural element of the machine by the manufacturer is standardized in the format of proactive measures of service impact and the technology of carrying out repair work as needed. Structurally, the technical system—cardan joint, is represented by a typical set of interconnected elements with a series connection. The results of calculations of variational modeling of changes in the reliability of individual elements, taking into account a uniform, element-by-element decrease in the probability of no-failure operation (FBR) (by 1% for each element from 0.99 to 0.95; and for 1, 2, 3, 4 elements, respectively) at risk-failure of the entire system is shown in Fig. 3.

As shown in Fig. 3, taking into account the uncertainty in the intensity of changes in the parameters that reduce the level of uptime in the model, it is possible, already at the level of simulation modeling, to reduce the variability of recognition of the probability of a risk-failure of an element, setting a theoretically justified range of the parameter of the functional structure as a whole. For clarity, simulation of the cardan joint elements was performed with the imposition of deformation loads and the display of the fields most susceptible to wear, allowing to establish the zones of surfaces prone to risk-failures (Fig. 4).

Having a priori information about the rate of change in the state of each part of the universal joint, a logical analysis of possible failures is carried out. In particular,

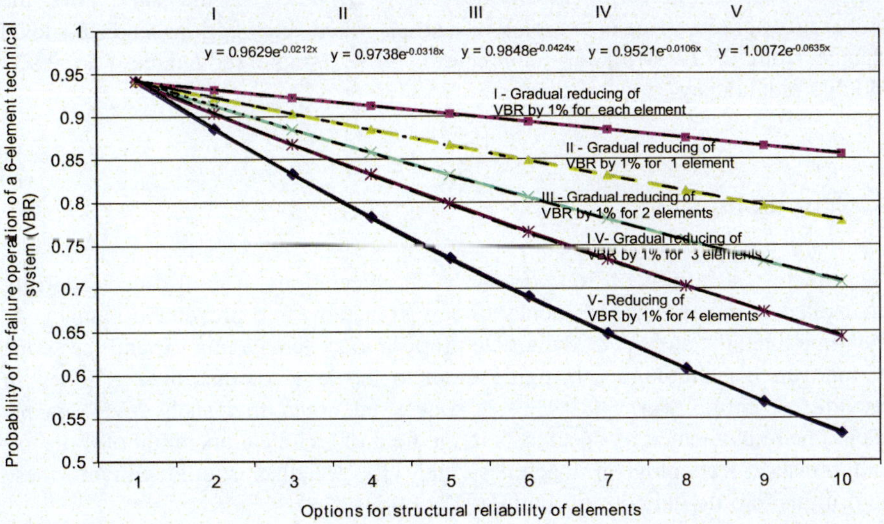

Fig. 3 Effect of changes in the reliability of individual structural elements on the risk-failure of the entire system

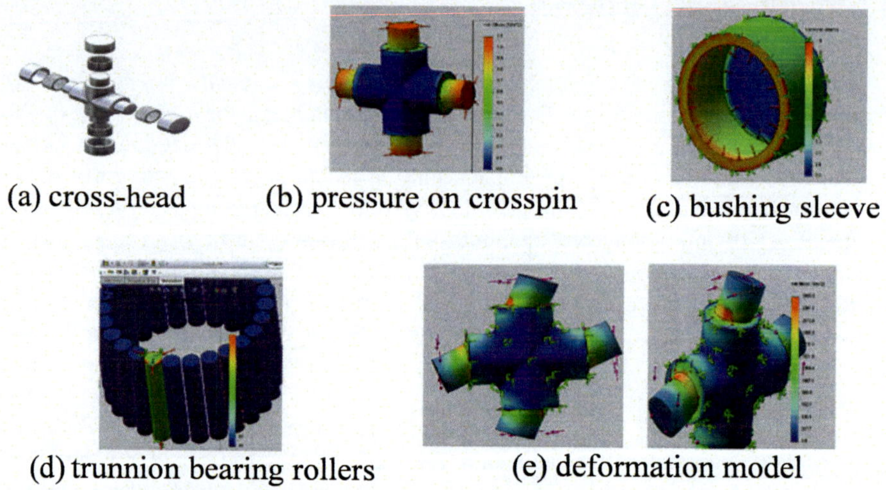

Fig. 4 Computer model of the cardan joint

it was found that the working surface of the bearing assembly is exposed to the greatest intensity of wear; to reduce the risks of its failure, preventive cleaning of the joint and replacement of the lubricant are carried out as a preventive operation; these actions can be qualified as a one-time light redundancy. In the transmission of technological machines performing reclamation work, a large number of cardan joints are used, so if there are 3 cardan shafts, the number of joints increases to 6, and the number of bearing assemblies up to 24 units. At the same time, the probability of failure-free operation with various redundancy options leads to a level change from 0.729 (without redundancy), 0.898 (single redundancy) to 0.937 (double redundancy).

4 Discussions

In a practice, it is difficult to carry out such calculations, it is also economically unjustified to increase the frequency of service actions in a preventive manner. At the same time, the authors of the article propose to systematically expand the scope of innovative technologies, bringing them to the level of dual-use, as noted in previous scientific works [19–23]. It should be noted that such developments should be implemented by specialists in the field of technical operation of transport and transport-technological machines, and IT specialists should convert these algorithms into digital control modules [24–28].

5 Conclusion

It has been substantiated that, in accordance with the theory of probability, there can be only one failure in the unit at a time, even if several damages are diagnosed, then there is always a single cause, i.e. failure that caused a chain reaction of damage.

It has been established that the mechanism for recognizing the uncertainty of the state of the system is based on functional and structural multi-parameter relationships, in connection with which the concept of an integrated parameter is introduced, which to a greater extent characterizes the state of the unit. The construction of a diagnostic model by using the basic provisions of information theory allows representation the unit as a system containing the risk of uncertainty based on the sign of failure, which is distributed over all structural elements of the unit. An algorithm has been developed for the recognition rule for the area of displacement of uncertainty zones, the translation of which into a digital format with the introduction of a recognition criterion will provide for a defect identification (risk-failure). Simulation modeling has been carried out as a practical implementation of the risk-failure recognition mechanism when the state of a technical system changes, using the example of a cardan joint of a technological machine, which is in severe operating conditions during reclamation work. It is proposed to include in the design of technological machines a digital module for monitoring the technical state of the unit, including: a module for recognizing the elemental structure; a software shell with a calculation algorithm for evaluating parametric changes; interface predictive risk alert module; service control module by state.

References

1. Grib VV, Zorin VA, Zhukov RV (2016) Multi-criteria assessment of the technical condition of mechanisms and machines (dynamics and wear). Repair Reconditioning Modernization 6:19–22
2. Apatenko AS (2013) Effect of vehicles service life on their operational reliability and land reclamation work. Mach Equipment Rural Area 10:4–6
3. Apatenko AS (2013) Analysis of processes and reasons for reducing the intensity of operation of technological machines. Vestnik of Federal State Educational Establishment of Higher Professional Education "Moscow State Agroengineering University named after V. P. Goryachkin", vol 3, issue 59, pp 49–51
4. Zorin VA, Baurova NI, Balovnev VI, Grib VV, Kosenko EA (2019) Informational model of state change in a mechanical system. Russ Eng Res 39:680–682
5. Sevryugina N, Kapyrin P (2018) Technological machines, construction resources, efficiency and safety. MATEC Web Conf 178:06017
6. Khalifa M, Dujun TA (2020) Classification and zoning of riverine territories of small towns on the example of the Belgorod region. Bulletin BSTU Named After V.G. Shukhov 8:101–109
7. Sanin SN, Sapelin KI (2019) Automation control of large diameter bent piplines. Bulletin BSTU Named After V.G. Shukhov 2:156–163

8. Kuznetsov AP, Blau P, Koriath H-J, Richter M (2016) Criteria for energy-efficiency of technological processes, technological machines and production engineering. Procedia CIRP 46:340–343
9. Chou H-H, Zolkiewski J (2012) Managing resource interaction as a means to cope with technological change. J Bus Res 65(2):188–195
10. Siltala N, Järvenpää E, Lanz M (2018) An executable capability concept in formal resource descriptions. IFAC-Papers On Line 51(11):102–107
11. Chalaganidze SI, Katsitadze JB, Kutelia GG (2017) The theoretical and experimental study of the ploughs' ploughs hare sin order to increase the reliability of resource-saving technology using similarity and dimensions theory. Ann Agrarian Sci 15(3):329–331
12. Valerie CY, Sun ZL, Sun L, Li X (2010) Single-machine scheduling time-dependent jobs with resource-dependent ready times. Comput Ind Eng 58(1):10–15
13. Berg S, Wustmans M, Bröring S (2019) Identifying first signal so emerging dominance in a technological in novation system: a novel approach based on patents. Technol Forecast Soc Change 146:706–722
14. Sun X, Zhou X, Chen Z, Yang Y (2020) Environmental efficiency of electric power industry, market segmentation and technological innovation: empirical evidence from China. Sci Total Environ 706:135749
15. Huang H-C, Su H-N (2019) The innovative ful crums of technological interdisciplinarity: an analysis of technology fields in patents. Technovation 84–85:59–70
16. Wang D, Tong X, Wang Y (2020) An early risk warning system for out ward foreign direct investment in mineral resource-based enter processing multi-classifiers fusion resources. Policy 66:101593
17. Sevryugina NS, Kapyirin D (2017) The effectiveness of the choice of means for mechanization of construction and special construction works. Mechanization Constr 78 (11):59–64
18. Zorin VA, Baurova NI, Pegachkov AA (2019) Assessment of products risks of mechanical engineering by results of diagnosing. Periodicals Eng Nat Sci 7(1):287–293
19. Prusov A, Sevryugina N, Ovsienko A (2019) Smart house: model of resource management and safety of vertical transport. E3S Web Conf 97:01016
20. Apatenko AS, Sevryugina NS (2019) Methods of recruiting of mobile repair services and maintenance of machines performing reclamation works. IOP Conf Ser Mater Sci Eng 786:012037
21. Sevryugina NS, Apatenko AS (2019) Digital systems and precision control of the process machinery performance in environmental engineering. Mach Equipment Rural Area 7 (265):35–38
22. Kapyrin P, Sevryugina N (2018) The procedural approach to reliability of objects of the raised level of responsibility. IOP Conf Ser Mater Sci Eng 365:042018
23. Sevryugina N, Kapyrin P (2018) Resource verification of construction objects having increased importance level. MATEC Web Conf 251:03019
24. Brik B, Bettayeb B, Sahnoun M, Duval F (2019) Towards predicting system disruption in industry 4.0: machine learning-based approach. Procedia Comput Sci 151:667–674
25. Ari AAA, Gueroui A, Titouna C, Thiare O, Aliouat Z (2019) Resource allocation scheme for 5G C-RAN: a swarm intelligence base dapproach. Comput Netw 165:106957
26. Coccia M, Watts J (2020) A theory of the evolution of technology: technological parasitism and the implications for innovation management. J Eng Technol Manage 55:101552
27. Prenkert F, Hasche N, Linton G (2019) Towards a systematic analytical frame work of resource inter faces. J Bus Res 100:139–149
28. Zorin V, Baurova N (2018) Assessment of means of mechanization of construction risks using catastrophe theory. MATEC Web Conf 251:03008

Prefabricated Decorative Facade Finishing Material

V. S. Lesovik, L. P. Nagruzova, N. A. Ekler, and M. V. Dayronas

Abstract Living in one's own house is more comfortable in terms of the psychological climate, positive emotions, creative mood, mental activity and increasing the life expectancy of a person. In the Russian Federation, the most popular materials for the construction of walls are: timber, bricks, blocks of wood concrete, expanded concrete, foam concrete. The first two materials have a significantly higher cost, and the rest do not allow getting a complete design and require additional facade finishing of the building, the cost of which is about 25% of the total estimated cost of construction. In this regard, the development of decorative small-piece elements for facade finishing with reliability indicators corresponding to the service life of the building, securely fixed on the main wall material and having a low cost becomes an urgent task. The most convenient technological platform for the production of such finishing elements is cement systems, namely highly mobile self-compacting fine-grained concrete, which allows obtaining high-quality surfaces with low porosity. Cement finishing composites have a high degree of affinity with the main types of used wall materials, including gas silicate blocks, the mineral composition of which is also close to cement stone. In this work, the compositions of decorative fine-grained concrete for the production of finishing facade elements are developed. The issues of obtaining a high-quality smooth surface of products, as well as the effect of the addition of microsilica on the brightness of the color and its resistance under climatic conditions are considered.

V. S. Lesovik (✉)
Belgorod State Technological University Named After V.G. Shukhov,
Kostyukov St., 46, Belgorod 308012, Russia

L. P. Nagruzova · N. A. Ekler
Khakass State University Named After N.F. Katanova, Lenin Avenue 92/2,
Abakan, Republic of Khakassia 655017, Russia
e-mail: L_nag@bk.ru

M. V. Dayronas
Belgorod State Technological University Named After V.G. Shukhov
(North Caucasian Branch), 357202 Stavropol Territory, Mineralnye Vody,
Zheleznovodskaya st., 24, Belgorod, Russia

S. V. Klyuev (ed.), *Digital Technologies in Construction Engineering*,
Lecture Notes in Civil Engineering 173, https://doi.org/10.1007/978-3-030-81289-8_12

Keywords Decorative fine-grained concrete · Affinity of structures · Surface quality of products · Microsilica · Intensity and durability of color

1 Introduction

Multistory buildings, that predominate in the modern development of settlements, are inherently less resistant to natural disasters, resource-intensive, expensive to operate and dispose of, than they create big problems for future generations. In order to optimize the system "man-material-environment", in accordance with the principles of Geonics (Geomimetics), it is necessary to introduce the principle of natural similarity of the created materials, buildings and structures [1–3]. Individual housing construction is most in harmony with the environment and can be organically integrated into the natural landscape. Living in one's own house is more comfortable in terms of the psychological climate, positive emotions, creative mood, mental activity and, finally, increasing the life expectancy of a person.

In the Russian Federation, the most popular materials for the construction of detached house walls are: timber, bricks, blocks of wood concrete, expanded concrete, foam concrete. The cost per unit volume of a wall constructed from these materials is shown in Fig. 1.

However, the values shown in the diagram are conditional, as almost all the presented materials do not allow obtaining a complete structure and requiring additional facade finishing of the building, the cost of which is on average about 25% of the total estimated cost of construction [4]. The reason for this (Table 1) is the high water absorption of the porous materials, which causes seasonal fluctuations in thermal conductivity and reduced frost resistance, as well as low surface strength, which requires protection from mechanical damage.

To date, notch and curb lego-blocks were developed, which, unlike the wall products listed above, do not have cold bridges due to a well-thought-out notch and curb fixation system. However, these blocks do not have enough experience in development and operation, so they do not really compete with the materials from Table 1 [5].

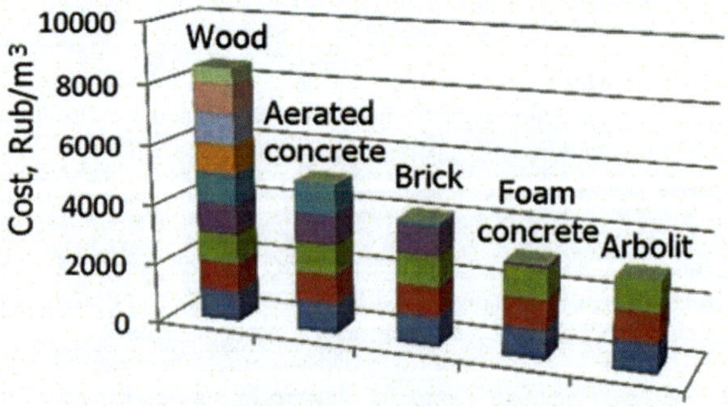

Fig. 1 Average cost of materials for low-rise housing construction

Table 1 Physical, mechanical and performance properties of materials

Material type	Average density (kg/m^3)	Water absorption by mass (%)	Thermal conductivity (W/m K)	Frost resistance (cycles)	Strength (MPa)
Aerated concrete	300–1200	20–25	0.1–0.4	35	0.9–10.0
Foam concrete	300–1200	10–16	0.1–0.4	35	0.5–7.0
Arbolite concrete	500–900	75–85	0.2–0.3	25	2.1–4.5
Expanded clay concrete	300–1500	18	0.15–0.45	25–75	2.5–20.0
Polystyrene concrete	300–600	4	0.1–0.145	70–150	0.7–3.6

The problem of improving the technology of finishing facade elements of the building is devoted to the work of many researchers [5–7]. The quality of finishing largely depends on the texture, color, and durability of the used finishing materials.

At the heart of most of the currently used finishing solutions is the principle of structural independence from the main wall—the so-called suspended facades, in which the finishing material is attached to its own power frame, fixed on the surface to be finished. Such solutions are extremely convenient when reconstructing existing multistory buildings with a large facade area, regardless of its technical condition. However, for finishing low-rise residential buildings, their use leads to a significant increase in price. The small area of the facade, a large number of functional (window openings and doorways, etc.) and decorative (pilasters, wall bends, etc.) complicate the construction of the frame; make many cuts and joints of decorative elements.

Plastering the exterior walls is the most cost-effective finishing solution, but its durability is significantly inferior to the main structures, forcing to repair or re-finish several times during the life cycle of the house. In addition, the quality of finishing largely depends on the skills of the craftsmen and their compliance with the recipes and recommendations of the manufacturer.

In this regard, the development of decorative small-piece elements for facade finishing with reliability indicators corresponding to the service life of the building, securely fixed on the main wall material and having a low cost becomes an urgent task. The most convenient technological platform for the production of such finishing elements are cement systems, in particular highly mobile self-compacting varieties of concrete, which allow obtaining high-quality front surfaces with low porosity. In addition, cement finishing composites have a high degree of affinity with the main types of used wall materials, including gas silicate blocks, the mineral composition of which is also close to cement stone. The high degree of affinity [8, 9] determines the chemical compatibility (including the absence of products dangerous to the other in one of the materials) and the proximity of the thermal expansion coefficients, which is the key to the absence of cracking of the surface of a less massive finishing layer and additional stresses in the mounting solution. The modulus of deformations of the ornamental facade finishing should also be close to the modulus of deformations of the composition of the manufactured block itself.

2 Materials and Methods

For the preparation of fine-grained concrete for finishing compositions, the following materials were used: microsilica (MSU-85) of the ferroalloy plant of Novokuznetsk (Russia); superplasticizer (SP-1) of the Novomoskovsk plant (Russia); light river washed sand, medium, I-class with a fineness module of 2.2; Portland cement of the Topkinsk plant (Russia) CEM II/A-S 32.5 N.

The optimization of the concrete mix compositions was carried out using the method of mathematical planning of the experiment. The quality of the front surface of the product was evaluated by the surface porosity. To do this, measurement zones (squares of 20 × 20 cm in size) were applied on round samples (made of a concrete mixture of the finishing layer) with a diameter of 60 cm, and the number of shells of different diameters was calculated inside each zone. The mobility of concrete mixtures was estimated by the standard cone draft, the determination of physical and mechanical parameters was carried out according to standard methods.

3 Results and Discussion

Facade finishing should provide a number of physical, mechanical and operational requirements: high-quality design of the facade of the building; high weather and frost resistance, the ability to smooth out temperature fluctuations together with the main material [10]. In addition to these conditions, the facade finishing must have a long-term maintenance-free operation.

The primary task is to get a high-quality top surface, smooth, without pores. Experimental studies and the study of literary sources determined the influence of several main factors on the quality of the front surface: the material of the molds, the granulometric composition of the mixture, and the methods of its compaction. Concrete mixes tend to reflect the structure of the mold surface. It is important to use forms that have an ideal surface texture. This is confirmed by experimental studies of concrete samples of facade finishing made in metal (Fig. 2a) and plastic (Fig. 2b) forms.

On the basis of experimental data, the dependence of the surface porosity on the mobility of the concrete mixture is obtained (Fig. 3).

As we can see, high-mobility mixtures contribute to increasing the smoothness of the surface. To obtain self-compacting highly mobile mixtures, it is necessary to

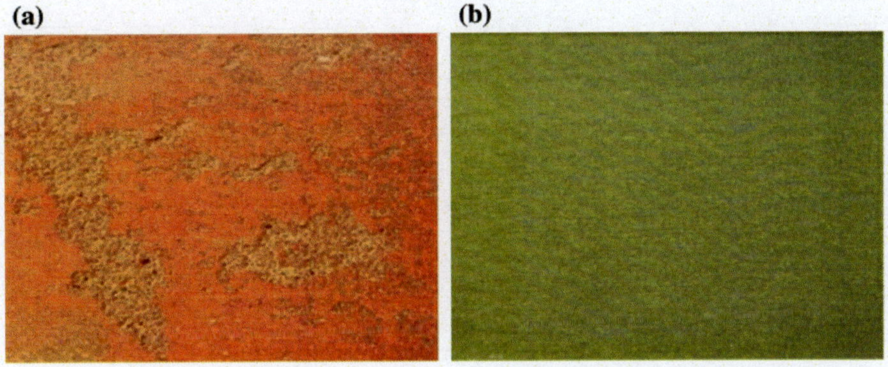

Fig. 2 The quality of the front surface of the samples made: **a** in metal forms; **b** in plastic forms

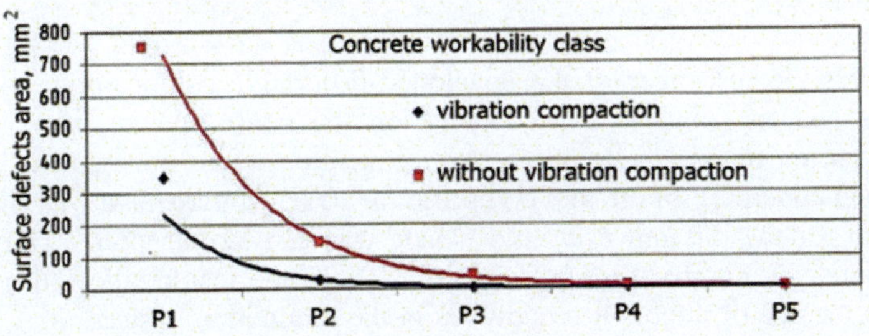

Fig. 3 Dependence of the number of surface defects on the workability of the concrete mix

include highly dispersed fillers in the concrete composition, while simultaneously compensating for the increase in water demand with an increased dosage of a plasticizing additive. When selecting the composition of decorative concrete, an super dispersed additive—microsilica—was used.

The insertion of microsilica leads to an increase in compressive and bending strength, the rate of hardening, adhesion to the aggregate, corrosion and frost resistance, and water resistance. The disadvantage of microsilica, as it was mentioned above, is its high water demand. To reduce the amount of water, a superplasticizer SP-1 was inserted. The deformative properties of concrete (shrinkage, creep) do not deteriorate when replacing up to 25% of cement with microsilica [11]. Frost resistance is provided due to the low permeability and high density of the cement stone.

Due to the application of the mathematical planning method, regression equations were obtained, the strength dependence on the flow rate of water, microsilica and superplasticizer, the values are given per 1 m^3 (Fig. 4). As the development of concrete compositions is carried out for products potentially operated in the zone of intensive humidification (the basement part of buildings), frost resistance tests were conducted. It was found that after 200 cycles of alternating freezing and thawing, the strength decreased by 1%, which meets the regulatory requirements.

Fig. 4 Dependence of strength on the consumption of water, microsilica and superplasticizers

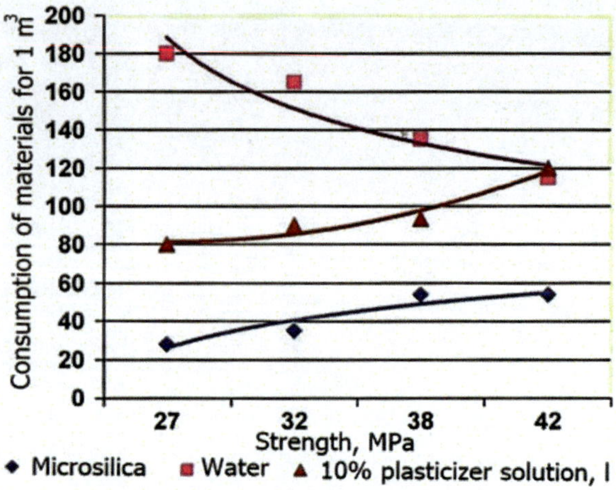

To assess the resistance of the developed decorative compositions to climatic influences and the impact of microsilica on the color scheme, two groups of samples were made (self-compacting concrete with microsilica; ordinary fine-grained concrete). In the first composition, 10% of microsilica was added. The assessment for the difference in color shade was carried out after 10 days, again after 1 month, 6 months (Fig. 5a). The analysis of the results allowed us to conclude that the use of microsilica additives in the production of decorative elements of a wide range of colors: from light shades to more saturated colors. It should be noted that the volume-colored concrete containing microsilica in its composition looked more saturated and expressive.

Testing for resistance to natural factors showed that the samples were on the streets of the city of Abakan (Russia) for 1.5 years, did not change their color (no fading), there is no desalination.

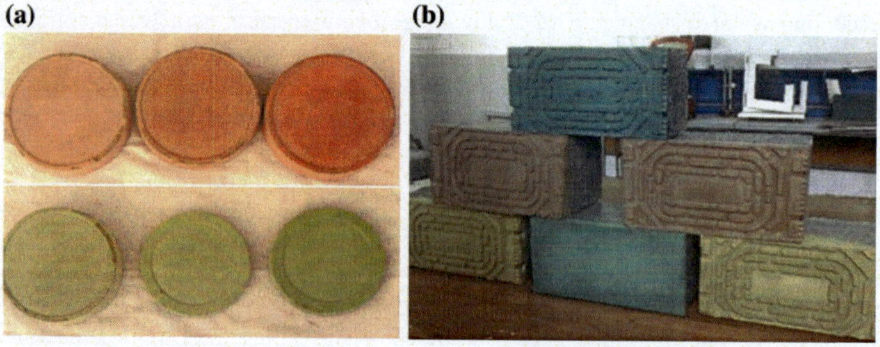

Fig. 5 Samples of facade concrete and products from it: **a** the effect of microsilica on the color range; **b** blocks of different colors and different surface relief

4 Conclusion

The research made it possible to develop small-piece textured finishing products focused on use in low-rise construction. The use of cement binder as the basis for their production makes this technology extremely simple and affordable for small businesses to master. The high quality of the products is achieved through the use of self-compacting concrete mixes, which provide a high-quality surface when using polymer molds, and low porosity of the front layer, which forms the durability and preservation of the aesthetic properties of the products. The operational reliability of the developed finishing products is ensured by implementing the principles of affinity for wall materials made in most cases with the use of Portland cement.

Acknowledgements The work is realized in the framework of the RFBR according to the research project № 18-29-24113. The work was realized using equipment of High Technology Center at BSTU named after V.G. Shukhov.

References

1. Lesovik V, Volodchenko A, Glagolev E, Lashina I, Fischer HB (2019) Geonics (geomimetics) as a theoretical basis for new generation compositing. In: 14th international congress for applied mineralogy (ICAM2019). Springer Proceedings in Earthand Environmental Sciences, pp 344–347
2. Elistratkin MYu, Jamil AN, Galkina AA, Semiokhina VA, Pogorelov VS, Novoselova AA (2019) Application of the provisions of geonics-geommimetics in the development of finishing materials. In: Science-intensive technologies and innovations, international scientific and practical conference dedicated to the 65th anniversary of BSTU named after V.G. Shukhov. BSTU, Belgorod, pp 116–120
3. Lessowik WS (2015) Geonik. Geomimetik als grundlage für die synthese von intelligent bauverbundwerkstoffen. In: 19 Internationale baustofftagung. Ibausil, pp 183–189
4. Elistratkin MY, Lesovik VS, Zagorodnjuk LH, Pospelova EA, Shatalova SV (2018) New point of view on materials development. IOP Conf Ser Mater Sci Eng 032020
5. Sychev SA, Badin GM (2013) Advanced technologies for construction and reconstruction of buildings. BHV-Petersburg, SPb
6. Nagruzova L, Repina Y (2019) Monitoring of unified panels on a wooden frame with polystyrene cement insulation. Bull Constr Equip 3:64–72
7. Tarasenko VN, Samoilova AYu, Mosienko AS (2020) Acoustic comfort of a room as one of the factors in the formation of environmental space. Bull BSTU V.G. Shukhov 12:65–73
8. Lesovik VS, Zagorodnyuk LKh, Chulkova IL (2014) The law of affinity of structures in materials science. Basic Res 3(2):267–271
9. Murtazaev SAYu, Uspanova AS, Khadzhiev MR, Khadisov VKh (2020) Increasing the adhesion strength of the plaster coating with the base. Constr Mater Prod 3(6):17–26
10. Zagorodnyuk LH, Koryakina AA, Sevostyanova KI, Khaheleva AA (2018) Heat insulating composite mixtures with technogenic materials. IOP Conf Ser J Phys 133–137
11. Kaprielov SS, Sheynfeld AV (2001) Effect of the composition of organomineral modifiers of the MB series on their efficiency. Concr Reinf Concr 5:11–15

Phase Composition of Sewage Sludge Ash Ceramics Modified by Drinking Water Treatment Sludge Filtrate

S. A. Shakhov and N. Yu. Nikolaev

Abstract Industrial wastes are widely involved in the building ceramic production. Sewage sludge ashes (SSA) are promising secondary sources for building ceramics production. According to modern concepts, formation processes of ceramic structure can be controlled by adjusting ceramic mass compositions by additives of nanosized particles. In this study, the filtrate of drinking water treatment sludge obtained at the water treatment facilities of Novosibirsk, which is a silicate sol with nanodispersed particles, was used as an additive. Purpose of the study was to assess the effect of silicate sol from drinking water treatment sludge on sewage sludge ash ceramics phase composition and microstructure. Using the thermal analysis, X-ray phase analysis and electron microscopy, the drinking water treatment sludge filtrate additive effect on SSA-clay samples phase composition and microstructure was established. It was found that ceramic obtained by sintering SSA with drinking water treatment sludge filtrate additive is characterized by a matrix microstructure.

Keywords Clay · Sewage sludge ash · Phase composition · Building ceramics · Anorthite · Matrix microstructure

1 Introduction

More and more industrial wastes are involved in the building ceramic production. Sewage sludge ashes (SSA) are promising secondary sources for building ceramics production.

SSA properties were studied in a number of works [1–3], but the results were not widely used in building ceramic industry, because SSA ceramic products not always have the necessary strength characteristics. It is due to the SSA chemical composition, which affects to the phase composition formation of ceramic.

Ceramics phase composition can be controlled by adjusting the ceramic mass composition using special additives. The most suitable modifiers for clay minerals

S. A. Shakhov · N. Yu. Nikolaev (✉)
Siberian Transport University, Dusi Kovalchuk, 191, Novosibirsk 630049, Russia

S. V. Klyuev (ed.), *Digital Technologies in Construction Engineering*,
Lecture Notes in Civil Engineering 173, https://doi.org/10.1007/978-3-030-81289-8_13

structures are nanosized oxides of silicon and aluminum, which are realized in hydrosols [4].

Earlier in [5], it was shown that the drinking water treatment sludge filtrate, obtained after high-speed filters in the Novosibirsk water utility filtering station, is a sol of a silicate composition with a nanoscale particle size.

The purpose of study was to assess the drinking water treatment sludge filtrate effect on the SSA-clay ceramics phase composition and structure formation.

2 Materials and Methods of Research

Kamenskoye deposit (Novosibirsk region) montmorillonite-hydromica clay was used in this study.

Sewage sludge was obtained from Novosibirsk wastewater treatment plant (map No. 39). The sludge was burned in laboratory furnace at temperatures of 850 °C. Drinking water treatment sludge filtrate, obtained after high-speed filters in the Novosibirsk water utility filtering station and separated from the coarse fraction, was used as a modifying additive in the study.

Ceramic mixtures compositions are presented in Table 1.

Weighed portions of the mixtures with different additives amounts were mixed with water or drinking water treatment sludge filtrate. Samples were dried at 110 °C. Then the SSA-clay samples were sintered at 1100 °C with holding at a maximum temperature for 1 h.

Differential thermal analysis (DTA) was performed to identify the phase formation features of the SSA-clay samples. X-ray phase analysis (XPA), which was performed on a Bruker D8 Advance diffractometer using Cu-Kα radiation, was used to determine the phase composition of SSA-clay ceramic samples. The PDF2 database with the Search-Match shell was used to identify the obtained diffraction patterns.

Element maps and microstructure images of SSA-clay ceramic samples were obtained using Hitachi TM 3000 electron microscope with an energy dispersive analyzer.

Table 1 Ceramic mixtures compositions

Compositions №	Content of components (%)		
	SSA (%)	Clay (%)	Drinking water treatment sludge filtrate (%) (in terms of dry matter; over 100%)
1	50	50	0
2	50	50	0.05
3	50	50	0.1
4	50	50	0.25

3 Results and Discussion

The endothermic effect at 50–150 °C temperature range is due to the sorbed water loss by clay minerals. In this case, there is a loss of mass—4.5%. The weight loss occurs at a much lower rate with further mixture heating. Heat absorption is caused by the release of chemically bound water by clay minerals in the 300–500 °C range. Kaolinite is initially dehydrate, turning into metakaolinite, and then decompose into oxides at 750–900 °C [6] (Fig. 1).

The endothermic effect at 700 °C is associated with the montmorillonite crystal structure destruction. According [7], calcium carbonate dissociates in the 770–1010 °C temperature range.

Exothermic effects on the heating curve are due to new crystalline phases formation. The total weight loss for the test sample was 10.8%.

According to the XPA data (Fig. 2), for all SSA-clay samples after sintering at 1100 °C, the main identified phases are anorthite (3.10; 4.05 A0), quartz (3.35 A0) and hematite (2.7 A0).

The anorthite ($CaO \cdot Al_2O_3 \cdot SiO_2$) formation is due to solid-phase reaction between amorphous γAl_2O_3 formed in SSA from aluminum oxychloride and calcium silicate.

Another possible anorthite formation mechanism may be associated with the reaction between amorphous silica and calcium oxide. According study [8], kaolinite dehydration calcium carbonate dissociation are promoted by the anorthite crystallization in clay. The reaction equation has the form (Eq. 1):

$$Al_2O_3 \cdot 2SiO_2 + CaCO_3 = CaO \cdot Al_2O_3 \cdot 2SiO_2 + CO_2 \quad (1)$$

The phase composition results (Table 2) indicate to increase anorthite content with adding drinking water treatment sludge filtrate.

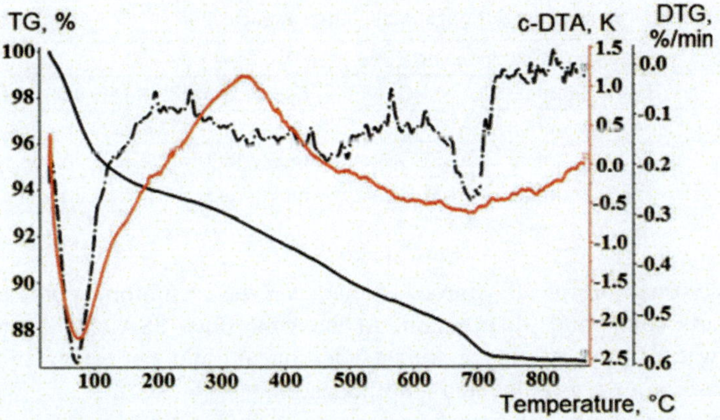

Fig. 1 DTA results (composition № 4)

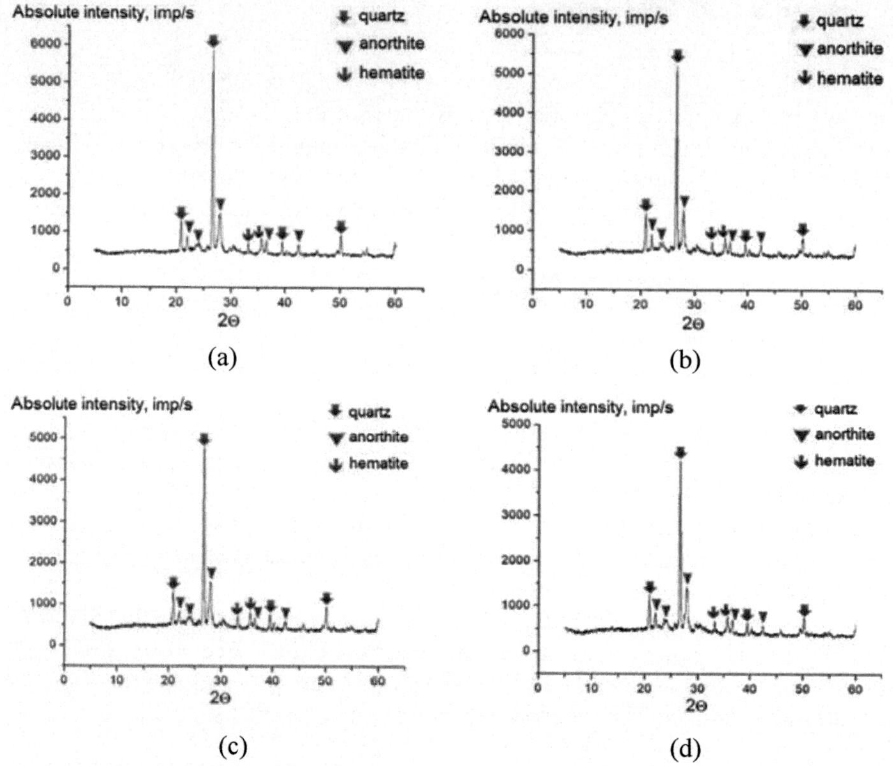

Fig. 2 XPA diffraction patterns of ceramic samples: **a**—50% SSA: 50% clay: 0% drinking water treatment sludge filtrate; **b**—50% SSA: 50% clay: 0.05% drinking water treatment sludge filtrate; **c**—50% SSA: 50% clay: 0.1% drinking water treatment sludge filtrate; **d**—50% SSA: 50% clay: 0.25% drinking water treatment sludge filtrate

Table 2 SSA-clay ceramic samples quantitative phase composition

Sample	Content of components (%)
1	64% anorthite $CaAl_2Si_2O_8$ + 34% quartz SiO_2 + 2% hematite Fe_2O_3
2	67% anorthite $CaAl_2Si_2O_8$ + 31% quartz SiO_2 + 2% hematite Fe_2O_3
3	69% anorthite $CaAl_2Si_2O_8$ + 29% quartz SiO_2 + 2% hematite Fe_2O_3
4	68% anorthite $CaAl_2Si_2O_8$ + 29% quartz SiO_2 + 3% hematite Fe_2O_3

The microstructure of all samples is characterized by uniform pores and crystalline quartz compounds distribution. At the same time, SSA-clay samples with drinking water treatment sludge filtrate addition are still noticeably different in relative position of crystalline and amorphous phases.

The SSA-clay with drinking water treatment sludge filtrate additive samples microstructure feature is matrix microstructure. The spatially arranged matrix microstructure has phase composition differences between filler and binder, which confirmed by the electron microscopy data (Fig. 3d, f). Matrix formed from clay particles is a “binder”. “Filler” formed from quartz grains contained in sewage sludge ash. It is important to note that matrix include amorphous and crystalline anorthite phase. Anorthite crystals reinforce matrix SSA-clay samples microstructure, which provides higher strength and frost resistance.

SSA-clay samples with drinking water treatment sludge filtrate additive are characterized by increased density from 2.15 to 2.21 g/cm^3 and compressive strength from 27.60 to 37.15 MPa (Table 3), which may be due to their matrix microstructure and increased anorthite phase content (from 64 to 68%).

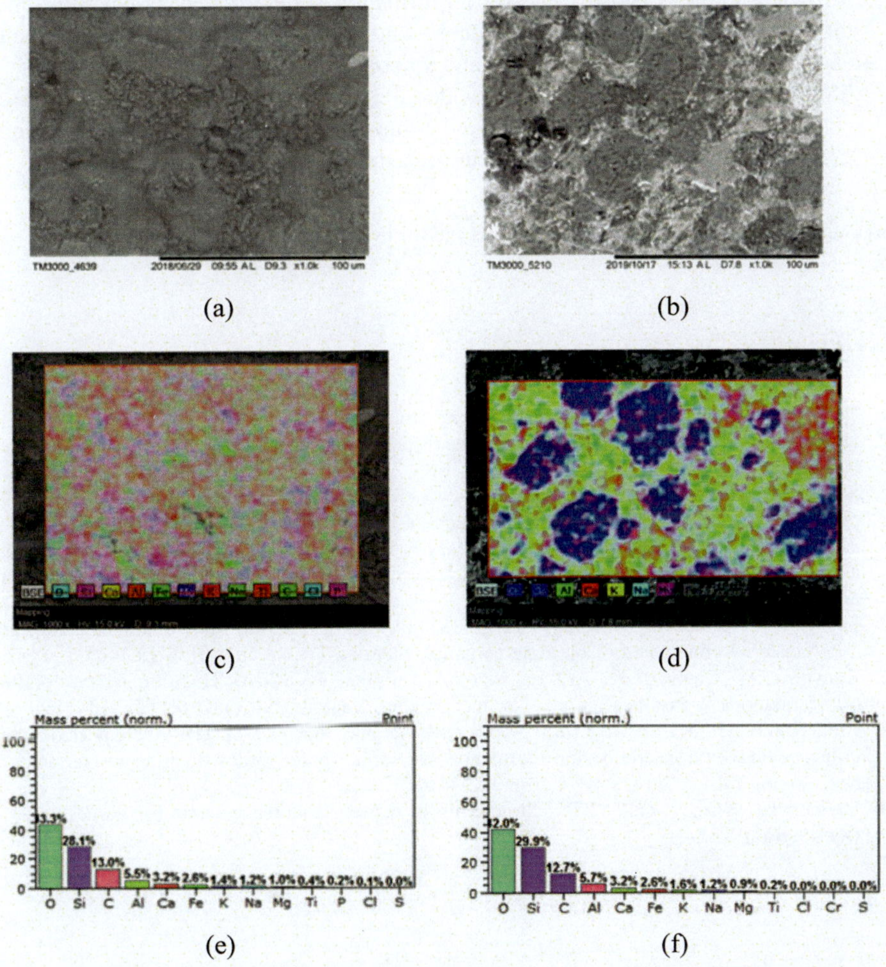

Fig. 3 Microstructure and elemental composition of SSA-clay ceramic samples: **a**, **c**, **e**—composition 1, **b**, **d**, **f**—composition 4

Table 3 SSA-clay ceramic samples water absorption (W, %), density (ρ, g/cm^3), compressive strength (R, MPa)

Compositions №	Water absorption (%)	Density (g/cm^3)	R (MPa)
1	7.79 ± 0.25	2.15 ± 0.05	27.60 ± 3.68
4	7.04 ± 0.05	2.21 ± 0.01	37.15 ± 4.86

4 Conclusions

DTA, XPA and electron microscopy results allow us to conclude that:

1. SSA-clay samples phase composition formation is determined by calcium ions diffusion into the metakaolinite structure and anorthite crystallization from melt enriched with aluminum, calcium and silicon oxides.
2. SSA-clay samples with drinking water treatment sludge filtrate additive are characterized by increased density and compressive strength, which may be due to their matrix microstructure and increased anorthite phase content.

Acknowledgements The reported study was funded by RFBR, project number 20-33-90197.

References

1. Turovskij IS (1988) Sewage sludge treatment. Construction publishing
2. Hakimov FI (1999) Recommendations for the disposal of urban treatment facilities sludge. The Russian State Committee of Ecology
3. Shakhov SA, Klyuchnikova NS, Kozhemyachenko AS (2014) Composition and technological properties of sewage sludge and sewage sludge ashes. Univ News Constr 11(671):103–113
4. Zhenzhurist IA, Zaripova VM, Mubarakshina LF, Hozin VG (2010) The effect of silicon and aluminum oxides hydrosols nanosized particles on the structure formation of clay minerals in an aqueous medium. Glass Ceram 7:28–32
5. Shahov SA, Nikolaev NYu, Rudaya TL (2018) Potential of drinking water treatment sludge as a raw material component of building ceramics. Siberian Transp Univ Bull 1:61–67
6. Kanygina ON, Chetverikova AG, Lazarev DA, Sal'nikov EV (2010) High-temperature phase transformations in iron-bearing clays of the Orenburg region. Bull OSU 6(112):113–118
7. Buruchenko AE, Haruk GN, Musharapova SI, Sergeev AA (2018) The calcium carbonate influence on the phase composition formation of ceramics based on low-melting and refractory clays during firing. Univ News Constr 2:21–28
8. Levickij IA, Klimov YuA (2005) Structuring of densely sintered ceramic for domestic use. Glass Ceram 6:32–36

Specifics of Experimental Studies and Features of Variation Formulations of Numerical Analysis of Composite Roofs Made of Panels-Shells

N. A. Mityakina, G. V. Korenkova, and Yu. V. Denisova

Abstract Modern trends in the use of spatial structures in the roofs of buildings for various purposes are considered. The paper analyzes the specifics of the development of improved types of composite roofs made of reinforced concrete panels-shells per span and the features of the method of their calculation based on experimental and theoretical studies of the deformation and crack resistance of structures. Some results of tests of a large-scale model of a composite panel-shell are presented, aimed at checking the basic physical prerequisites for the development of a method of nonlinear calculation, identifying the types of cracks, patterns of their development and disclosure, and the degree of influence of anisotropy acquired in the process of crack formation in structural elements. Based on the comparison of the values of the longitudinal deformations of concrete in different sections of the shelf of the panel-shell, it is concluded that their distribution over the width of the cross section is uneven, which makes it possible to talk about the feasibility of using the calculation scheme of the variation method with unit functions that take into account the twist of the cross sections of the panel-shell. A variant of constructing the calculation equations is proposed based on the generalization of the V. Z. Vlasov variation method to the class of composite reinforced concrete shells under consideration, with the inclusion in the calculation model of new elements related to the consideration of the contact and support conditions of prefabricated structures, characteristic types of cracks and other features obtained from experimental data. Mixed boundary conditions are formulated that provide a numerical implementation of the proposed version of the calculation equations for the accepted unit functions.

Keywords Spatial structures · Reinforced concrete panels-shells · Experimental and theoretical studies · Large-scale model · Deformation · Crack resistance · V. Z. Vlasov variation method

N. A. Mityakina (✉) · G. V. Korenkova · Yu. V. Denisova
Belgorod State Technological University Named After V.G. Shukhov, Belgorod, Russia
e-mail: natmit2010@bk.ru

S. V. Klyuev (ed.), *Digital Technologies in Construction Engineering*,
Lecture Notes in Civil Engineering 173, https://doi.org/10.1007/978-3-030-81289-8_14

1 Introduction

Spatial systems of building roofs are one of the most expressive tools of forming in modern architecture [1].

Until quite recently, many original solutions to spatial roofs could not be implemented either due to very complex calculations, or due to the impossibility of their implementation because of technological problems. However, today the development of computer-aided design tools and new technological solutions in the field of construction of complex systems [2] open up almost limitless possibilities for the use of thin-walled spatial structures [3] in the construction of buildings and structures for various purposes. The use of these structures in the roofs of buildings, due to the effect of their spatial work, which is realized here to the greatest extent, leads to significant savings in material resources. One of the varieties of spatial structures are various types of panels-shells on the span of various types [1], formed by a thin spatial shell or a flat slab [4] and contour (reinforcing) elements or edges in the form of beams, trusses, etc.

Taking into account the current state of the base of the construction industry, the spread of such structures in most cases is associated with the use of structural schemes of buildings, in which a large-sized spatial covering system is assembled on a construction site from separate relatively small elements. The structure obtained in this way acquires the properties of a composite structure [5].

The main problems here are related to the fact that for the calculation of composite structures made of panels-shells, it is not always permissible to use physical models and normalized physical and mechanical parameters [6, 7] obtained in relation to solid (monolithic) elements, as in some cases this leads to significant distortions in the estimation of limit states. Therefore, the task was set to develop improved types of composite roofs made of reinforced concrete panels-shells per span and methods for their calculation based on experimental and theoretical studies of deformation and crack resistance of structures [8].

2 Methods and Materials

The model of the composite panel-shell of the CPS (see Fig. 1) is a steel-reinforced concrete structure, the flange of which is formed from two cylindrical shells-inserts with dimensions of $1.04 \times 2.0 \times 0.016$ m, supported by two steel-reinforced concrete diaphragms with a span of 4.0 m. The diaphragm tightens and suspensions are made of Class A-III reinforcing bars with a diameter of 10 and 6 mm, respectively. The upper diaphragm chord is designed with a variable width of 20–40 mm and a height of 100 mm. The diaphragms are supported by solid reinforced concrete sections with a length of 500 mm and a thickness of 40 mm.

The composite panel-shell model was tested on a test bench consisting of two supports mounted on reinforced concrete slabs of the power floor (see Fig. 2).

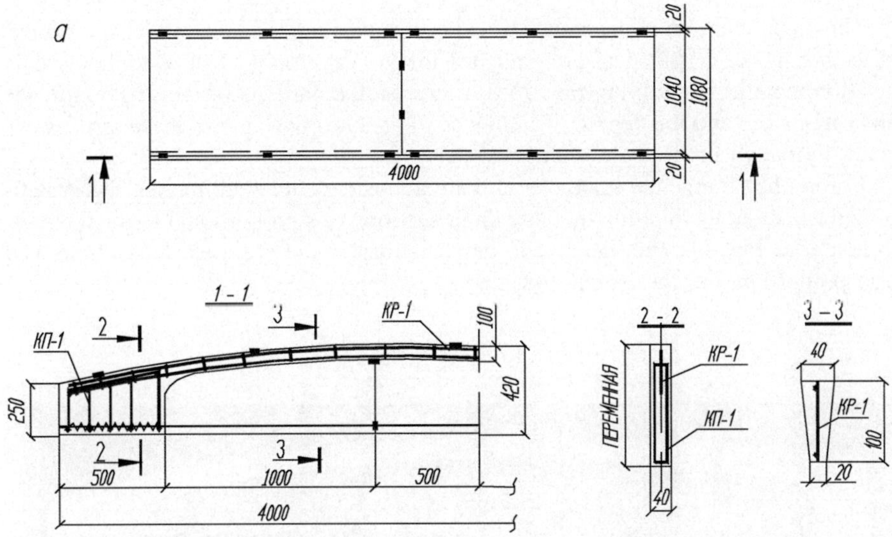

Fig. 1 Design of composite panel-shell model

Fig. 2 General view of the test of the panel-shell model

The main attention in the course of experimental studies of the model was aimed at testing the basic physical prerequisites for the development of a nonlinear calculation method, identifying the types of cracks, the patterns of their development and disclosure, and the degree of influence of anisotropy acquired in the process of crack formation in structural elements.

Before the emergence of cracks in the composite panel-shell model, the growth of deflections in the longitudinal and cross sections was proportional to the increase in load (see Fig. 3). The increase in deformations in the specified load range was also proportional to the increase in load.

a

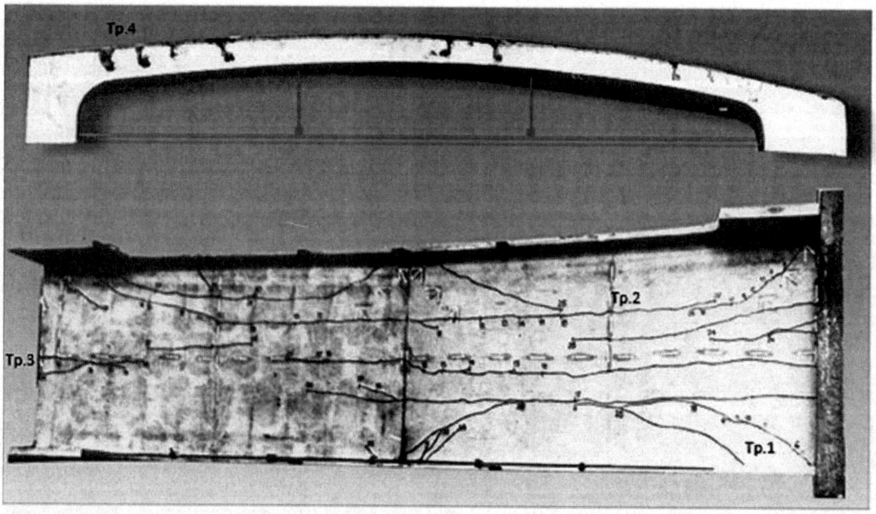

б

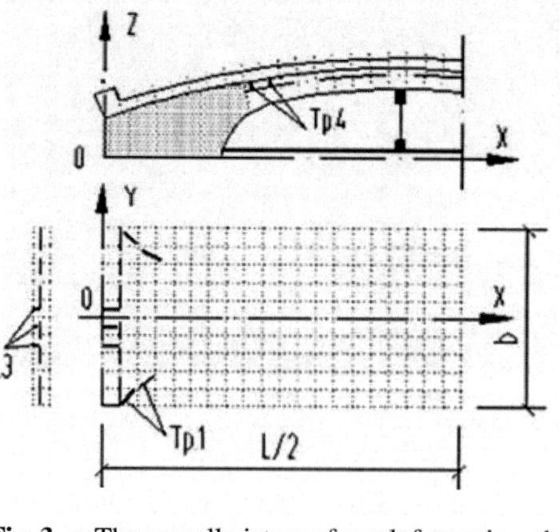

q,кПа	a_{crc}, мм			
	Тр. 1	Тр. 2	Тр. 3	Тр. 4
1.88	0.05			
2.63	0.05 / 0.09	0.05	0.05 / 0.07	
3.38	0.10	0.12 / 0.16	0.05	0.05
3.63	0.11 / 0.14	0.19	0.15 / 0.17	0.08 / 0.11
4.13	0.12	0.30	0.25	0.10
4.88	0.30	0.40	0.30	0.15
5.63	0.40	0.50	0.35	0.18
6.38	0.50	0.50	0.45	0.30

Числитель - эксперимент: знаменатель - расчет

Fig. 3 **a** The overall picture of crack formation, the design scheme of cracks at a load of 2.63 kPa, **b** the width of the opening of cracks of various types during loading in the model of the CPS shell

A comparison of the values of the longitudinal deformations of concrete in different sections of the flange of the panel-shell showed that the most uneven nature of their distribution over the width of the cross section was observed in the support zones. Here, even the sign of these deformations changed along the width of the panel flange: compression at the edges and stretching in the middle part. This allows talking about the feasibility of using the calculation scheme of the variation method with unit functions that take into account the twist of the cross-sections of the panel-shell.

The emergence, development and opening of cracks in the composite panel-shell during loading were observed in the following order (see Fig. 3).

The first through cracks (Cr. 1) were found at the eighth stage of loading at a load of 1.88 kPa. The formation of these cracks began in the angular support zones of the flange at the places where it interfaced with the supporting edge-thickening. Then, as the load increased, they developed longitudinally. The angle of inclination of the cracks to the longitudinal x axis decreased during their development. The described through cracks of the first type (Cr. 1), as they approached the middle zone of the shell, were connected to the cracks formed on the lower surface of the flange of the panel-shell (Cr. 2). Cr. 4 in the support zone of the upper diaphragm chord were fixed at a load of 3.38 kPa. Their formation began at the upper edge of the diaphragm chord in the area of its interface with the supporting solid section.

The formation of transverse cracks in the end ribs of the composite panel-shell (Cr. 3) was recorded at the same loading stage as Cr. 1 at a load of 2.0 kPa. Numerical values of the crack opening width of various types in the elements of the composite panel-shell are presented in the table to Fig. 3 (theoretical values are also shown here).

The exhaustion of the load-bearing capacity of the structure was accompanied by a significant increase in the deflections of the flange, which were more than five times higher than their values at the control load in terms of rigidity. At a load of 6.88 kPa, a sharp increase in the opening of cracks of all types was observed. Their maximum opening in the flange of the panel-shell at the specified load was 0.6 mm. The destruction of the model was characterized by an intense increase in deflections in the middle of the span of the panel flange. At the same time, it was possible to talk no longer about the curvature, but rather about the fracture of the flange of the panel-shell.

In the diaphragms of the panel-shell, there was also a significant increase in cracks in the upper chords (Cr. 4), the opening of which reached 0.6–0.7 mm before destruction. However, at the time of exhaustion of the load-bearing capacity of the flange, the concrete fragmentation of the compressed zone in the diaphragm chords was not observed. The stresses in the tightening of the shell panel also did not reach the fluidity strength. At the stage preceding the exhaustion of the load-bearing capacity, they amounted to 310 MPa.

3 Results and Discussion

The experimental design of the CPS composite panel-shell model in a nonlinear formulation was calculated with the V. Z. Vlasov variation method in combination with the method of sequential loads [9]. At the same time, a two-level calculation scheme was used for the CPS panel-shell. The design scheme of the first level included all the elements of the composite structure as a single spatial system. The design scheme of the second level was built to clarify the parameters of deformation and crack formation of the elements of the composite flange and included a cylindrical shell with appropriate boundary conditions at the longitudinal and transverse edges. For a more complete analysis of the structural deformation features, a finite element method calculation was also performed in parallel with the variation method calculation.

In general, calculations with the variation method and the finite element method evaluate satisfactorily the performance of structures of the studied class. The close coincidence of the above experimental data with the theoretical data obtained by both methods at loads not exceeding the crack formation loads confirms the conclusion [9] that previously stressed reinforced concrete shells can be calculated with a practical degree of accuracy in a linear-elastic formulation before the emergence of cracks.

The tests confirmed the feasibility of using the anisotropic physical model of reinforced concrete in the calculation of the crack resistance of the supporting corner zones for the structures under consideration. To obtain the calculated equations of the shell of the type under consideration, two groups of integral equilibrium conditions for a transverse strip of width *dx* (see Fig. 4) are compiled in the form of the work of all external and internal forces applied to this strip on possible displacements. The final variation equations of equilibrium in displacements are obtained after the expression in the integral equations of the equilibrium of forces through displacements.

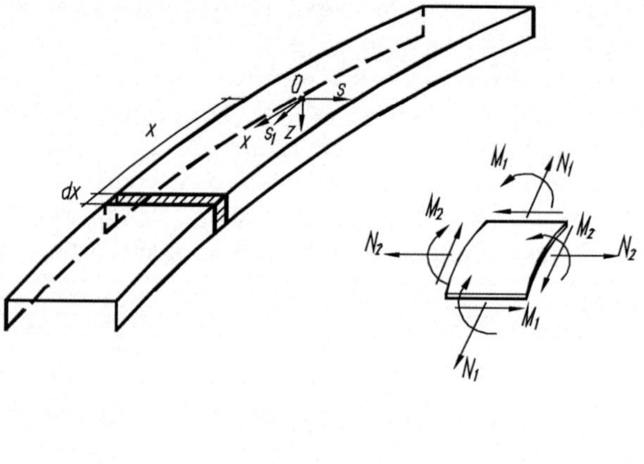

Fig. 4 Direction of the coordinate axes and internal forces in the panel-shell

The four groups of variation equilibrium equations in matrix form are written as:

$$[L_{ji}] \cdot [\overline{U}_1\ \overline{U}_2\ \overline{U}_3\ V_1\ V_2\ V_3]^T = [0\ 0\ 0\ q_1\ q_2\ q_3]^T, \tag{1}$$

where $\overline{U}_1, \overline{U}_2, \ldots, V$—components of decompositions of the desired displacements; $[L_{ji}]$—differential operator [8].

The calculated equations are derived from the general system of variation equations of the displacement method and allow the exclusion of displacement components in accordance with the accepted geometric and static hypotheses [10].

With the introduction of these hypotheses, the general system (1) is reduced to the following form:

$$\begin{gathered}\sum_c \left(\bar{I}_{dc}\overline{U}_c''' - b_{dc}\overline{U}_c\right) - \sum_i \left(I^*_{di}V_i''' + K_1 a_{di}V_i'\right) = \bar{P}^*_d \\ \sum_c \left(I^*_{dc}\overline{U}_c''' + K_1 a_{jc}U_c'\right) + \sum_i \left[I_{ji}V_i^{\prime V} + (K_1 c_{ji} - \rho_{ji})V_i'' + (K_1^2\beta_{ji} + s_{ji})V\right] = P_j' + g_j^*,\end{gathered} \tag{2}$$

According to kinematic hypotheses, system (2) is limited to only six equations with respect to six unknown functions $\overline{U}_1, \overline{U}_2, \overline{U}_3, V_1, V_2, V_3$ and six corresponding coordinate functions $\bar{\xi}_1, \bar{\xi}_2, \bar{\xi}_3, f_1, f_2, f_3$.

Taking into account the conditions (3), a system of basic calculation equations is obtained from the system (2) [8].

$$\begin{aligned} &\int_6 \bar{\xi}_1(s)\,\bar{\xi}_2(s)\,dF = 0, \quad &&\int_6 \xi_1(s)\,\bar{\xi}_2(s)\,dF = 0, \\ &\int_6 \bar{\xi}_1(s)\,\bar{\xi}_3(s)\,dF = 0, \quad &&\int_6 \xi_1(s)\,\bar{\xi}_3(s)\,dF = 0, \\ &\int_6 \bar{\xi}_1(s)\,\bar{\xi}_1(s)\,dF = 0, \quad &&dF = t ds. \end{aligned} \tag{3}$$

4 Conclusion

Thus, on the basis of the generalization of the V. Z. Vlasov variation method to the class of composite reinforced concrete shells under consideration, a variant of the calculation equations for the analysis of their deformation is constructed with the inclusion in the calculation model of new elements concerning the account of the contact and support conditions of prefabricated structures, characteristic types of cracks and other features obtained from experimental data, and mixed boundary conditions are formulated that provide a numerical implementation of the proposed version of the calculation equations for the accepted unit functions.

Acknowledgements This work was realized in the framework of the Program of flagship university development on the base of the Belgorod State Technological University named after V.G. Shukhov, using equipment of High Technology Center at BSTU named after V.G. Shukhov.

References

1. Semenov VS (2010) Modern spatial constructions: synthesis of art, technology and science. KRCU, Bishkek
2. Otsyukov KA (2020) Innovative technologies in construction and their use in organizational and technological events. Constr Mater Prod 3(1):7–13
3. Mileykovsky IE, Trushin SI (1989) Calculation of thin-walled structures. Stroyizdat, Moscow
4. Shugaev VV (1991) Spatial constructions of elements formed on a plane with subsequent bending. Studies of reinforced concrete thin-walled spatial structures. SRIRC, Moscow, pp 5–35
5. Bondarenko VM, Zalesov AS, Serykh RL (1998) Trends in the future development of prefabricated construction. Concr Ferrous Concr 1:2–4
6. Karpenko NI (1997) On modern constructions of general strength criteria for concrete and reinforced concrete elements. Concr Ferrous Concr 3:4–7
7. Poloz MA, Yasser GS, Shevchenko AV (2019) Application of the stepwise and iterative method in calculating the bending of previously stressed prefabricated and monolithic elements taking into account physical nonlinearity. Constr Mater Prod 2(3):12–27
8. Mityakina NA (2014) Experimental and theoretical studies of composite roofs of panels-shells. Publishing House of BSTU named after V.G. Shukhov, Belgorod
9. Mileykovsky IE (1963) Calculation of reinforced concrete cylindrical arches of shells. SCI, Moscow
10. Mityakina NA, Korenkova GV (2019) The main groups of working hypotheses for constructing a computational scheme and resolving equations of composite roofs. Way Sci II 12(70):33–36

Study of the Strength of Flexural Elements Made of Cellular Concrete

L. A. Suleymanova, P. A. Amelin, and Adham A. Hameed

Abstract Reinforced structures made of cellular concrete began to be used relatively recently. Broad experimental data for the development of calculation methods, in contrast to the calculation of structures made of conventional reinforced concrete, used as load-bearing flexural elements, is currently insufficient. However, the acquired experience in the construction and research of the strength and deformability of reinforced structures allows developing practical methods for calculating and designing structures made of cellular concrete. The existing design provisions should be clarified both in terms of checking the boundary values of the design values, and studying them on various types of cellular concrete, which in turn have a wide range of strengths, volume weights and elastic characteristics. The authors consider the nature of the "stress–strain" curve for cellular concrete, and also carry out a calculated and experimental assessment of the breaking moment of foam concrete samples in the framework of a calculated and experimental study of the strength and deformability of flexural elements made of cellular concrete. The refinement of the calculation of the strength of the flexural elements showed a high convergence of the calculated values of the breaking moment and the experimental data, which allows using the calculation made by the authors for the design of structures made of cellular concrete of various physical and mechanical characteristics.

Keywords Cellular concrete · Strength · Reinforcement · Deformability · Breaking moment

1 Introduction

In modern construction, concrete structures and products are most widely used in the construction of residential buildings. A significant role in the further improvement of concrete structures is played by reducing their weight, which

L. A. Suleymanova (✉) · P. A. Amelin · A. A. Hameed
Belgorod State Technological University Named After V.G. Shukhov, Belgorod, Russia

S. V. Klyuev (ed.), *Digital Technologies in Construction Engineering*,
Lecture Notes in Civil Engineering 173, https://doi.org/10.1007/978-3-030-81289-8_15

allows getting a great economic effect and increasing the number of objects being built within the established budget.

As a result, the share of using structures and products using cellular concrete has recently increased, which not only has a lower weight compared to conventional concrete, but also shows good thermal insulation properties, high fire resistance, cellular concrete structures do not contain flammable components, do not support combustion, do not emit toxic substances, which allows it to be used in fire barriers of a high degree of fire resistance [1].

Currently, cellular concrete is most widely used as wall products (blocks, panels, thermal insulation boards), to a lesser extent as flexural elements (roof and floor slabs, lintels) [1, 2].

In the current regulatory documents, due to the lack of knowledge of the properties of cellular concrete structures, there are no provisions for their design. In particular, this applies to reinforced flexural structures, which prevents their practical use.

2 Methods and Materials

To date, Russian codes of practice and European standards use simplified stress–strain compression diagrams when calculating load-bearing structures made of cellular concrete, which are used when calculating structures made of heavy concrete (Fig. 1) [3, 4]. To account for the non-linearity of changes in the stress and strain of cellular concrete, a curved diagram is used in Fig. 2 [5].

Elastic deformations are used when finding the modulus of elasticity of concrete, and are geometrically expressed by a tangent line to the diagram; plastic deformations are expressed by a transversal line to a selected point in the diagram and are used when finding the modulus of concrete deformation. Their sum determines the total deformations of the concrete:

$$\varepsilon_b = \varepsilon_{el} + \varepsilon_{pl}, \tag{1}$$

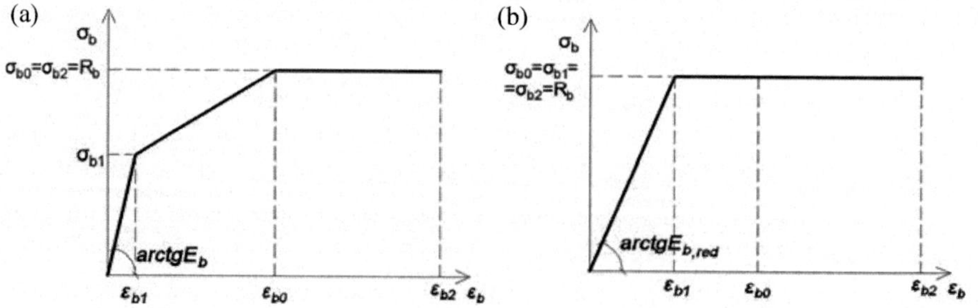

Fig. 1 Diagrams of the state of compressed concrete: a—three-line; b—two-line

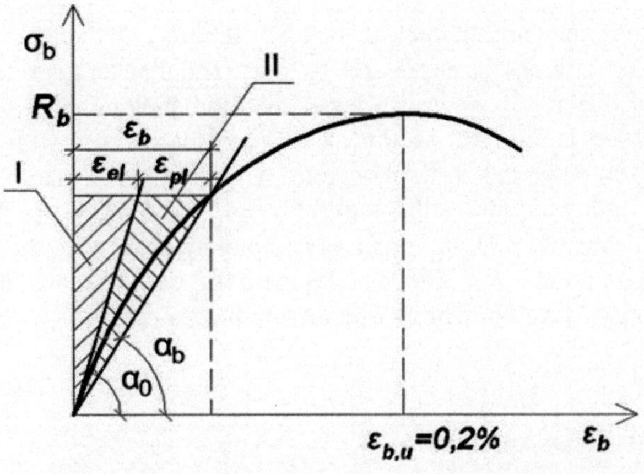

Fig. 2 Curvilinear diagram of concrete compression: I—area of elastic deformations; II—area of plastic deformations; $\varepsilon_{b,u}$—ultimate compressibility; ε_b—total deformations; ε_{el}—elastic component of total deformations under compression; ε_{pl}—plastic deformation

In [6, 7], the nature of cellular concrete destruction under load is described, and the stress–strain diagram is experimentally refined (Fig. 3).

Up to point A, the deformations $\varepsilon_b^{\kappa} = \varepsilon_0^{\kappa}$ are proportional to the stresses σ_0^{κ}. Point A corresponds to the elastic limit of cellular concrete $\sigma_0^{\kappa} = 0.6R_b^n$. At stresses $\sigma_0^{\kappa} \ll 0.6R_b^n$, plastic deformations $\varepsilon_{pl}^{\kappa}$ are practically not observed. As the concrete stress increases, the strain curve rises up to the point B, which corresponds to the concrete stress $0.85R_b^n$. In the AB section, with an increase in stress, the deformations $\varepsilon_{n\pi}^{\kappa}$ increase slightly, which is associated with the development of micro-cracks in cellular concrete. This section of the diagram is characterized by a line whose angle to the abscissa axis is less than the angle of inclination of the OA line $(\alpha_1 < \alpha_0)$.

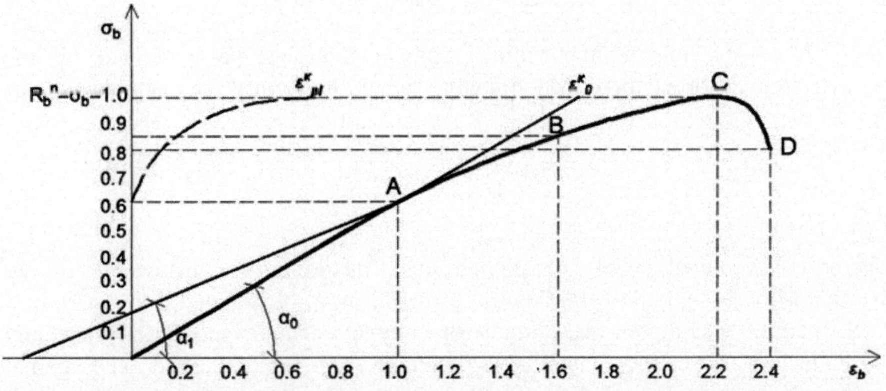

Fig. 3 Experimental stress–strain diagram for cellular concrete

With a further increase in the stress $\sigma_0^{\kappa} > 0.85R_b^{\mathrm{H}}$, the relationship between deformations and stresses is expressed by a curved line convex to the top, the deformations of concrete grow much faster than the stresses. Significant deformations are observed at the BC section, which are associated with the increased development of microcracks in the concrete, which turn into macro-cracks.

The point C corresponds to the highest stress value $\sigma_0^{\kappa} = R_b^n$. At the stresses corresponding to the point D, the prism is completely destroyed, the stresses in the concrete are lower than R_b^n. The development of deformations in this area is associated with the development of irreversible macro-cracks.

3 Results and Discussion

The failure of a flexural reinforced concrete element can occur for two reasons:

- in the case of high concrete strength, when breaking or reaching the fluidity of the reinforcement in the stretched zone;
- from the brittle destruction of the concrete of the compressed zone, with low concrete strength.

Cellular concrete used as wall blocks of buildings (density D300…D700), due to its low density and compressive strength, when used as flexural elements, is destroyed by the second type of destruction, the element is destroyed suddenly.

The destruction of the element according to the first type is viscous, and it is less dangerous, as it develops at a much lower rate and is caused by stresses exceeding the yield strength of the material. Viscous destruction can be achieved by using cellular concrete of a higher class with a density of D900…D1500.

The calculation of the strength of the cross sections of the flexural beam elements depends on the condition:

$$\mathrm{M} \leq \mathrm{M}_{ult}, \tag{2}$$

where M_{ult}—the maximum bending moment that can be perceived by the cross-section of the element; M—effective maximum torque for a double-support beam:

$$M = \frac{qbl_0^2}{8}, \tag{3}$$

where q—calculated evenly distributed load, including the self-weight of the overlap, MPa; b—width of the flexural element, m; l_0—estimated floor span, m.

As part of a comprehensive study of cellular concrete and its physical and mechanical characteristics, the authors conducted computational and experimental studies of the strength and deformability of flexural reinforced elements made of foam concrete [8].

For the calculation and experimental evaluation of the breaking moment of the samples, the authors considered the bending tests of foam concrete beams [9, 10]. The foam concrete mix consisted of Portland cement of the PC-500 brand to the Belgorod Cement Plant (Belgorod, Russia), foaming agent "Penostrom" and quartz sand of natural granulometry M = 1.4. The geometric characteristics of the cross-section and the reinforcement of the elements are shown in Fig. 4. The characteristics of the concrete and reinforcement under consideration are presented in Table 1.

The breaking moment for a rectangular cross section is determined by the formula:

$$M_{ult} = \sqrt{\frac{3}{7}} R_{b1} b h_0^2 \xi \left(1 - \frac{\xi}{3}\right), \tag{4}$$

where R_{b1}—calculated resistance of expanded concrete to compression, kg/m^2. As the test was conducted indoors on dried samples and under prolonged loading, $R_{b1} = R_b \cdot 0.85$. At a mass humidity of 25% or more, the calculated compression resistance $R_{b1} = R_b \cdot 0.7225$, MPa. b—the width of the rectangular cross-section, m; $h_0 = h - a$—the working height of the cross-section, (the distance from the upper face of the cross-section of the element to the axis of the stretched reinforcement), m; a—the protective layer of the reinforcement, m; ξ—the relative height of the compressed zone, determined by formulas (5) and (6) in case of damage to the reinforcement:

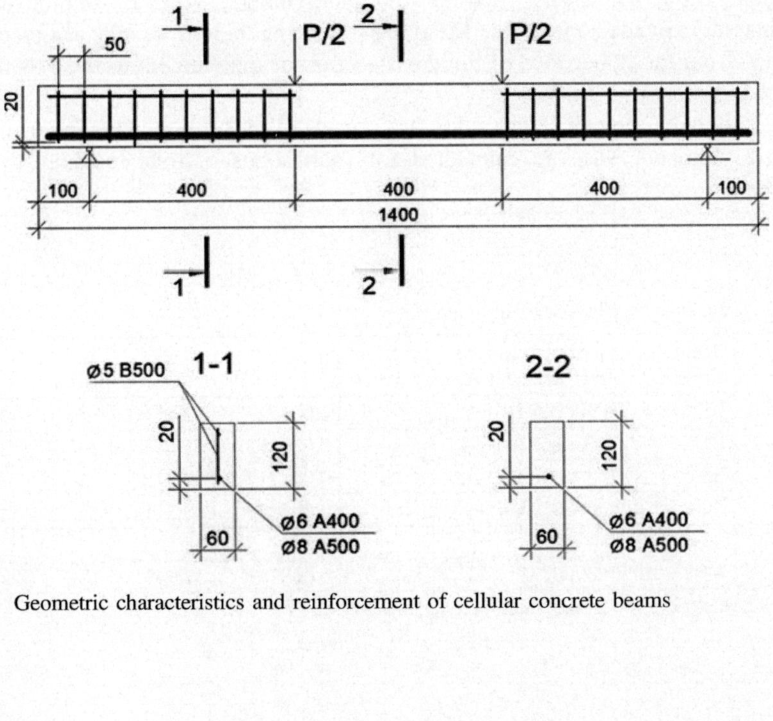

Fig. 4 Geometric characteristics and reinforcement of cellular concrete beams

Table 1 Characteristics of concrete and reinforcement

№ of series	Beam cipher	Characteristics of concrete			Characteristics of reinforcement			
		Strength class	R_b (MPa)	E_b (MPa)	Diameter and class	R_s (MPa)	R_{sc} (MPa)	E_s (MPa)
1	B1-6	B3.5	3.5	3860	Ø6 A400	350	350	202,460
	B1-8				Ø8 A500	435	435	199,230
2	B2-6	B5	4.9	4600	Ø6 A400	350	350	202,460
	B2-8				Ø8 A500	435	435	199,230
3	B3-6	B7.5	6.5	5250	Ø6 A400	350	350	202,460
	B3-8				Ø8 A500	435	435	199,230

$$\xi = \xi_{arm} = \frac{x}{h_0} = \sqrt{\frac{7}{3}\mu \frac{R_s}{R_{b1}}}, \tag{5}$$

when breaking on concrete:

$$\xi = \xi_b = \mu\alpha\left[\sqrt{1 + \frac{2}{\mu\alpha}} - 1\right], \tag{6}$$

where $\alpha = \frac{E_s}{E_b}$; x—height of the compressed area of cellular concrete, m; $\mu = \frac{A_s}{bh_0}$—coefficient of reinforcement of the stretched zone; A_s—cross-sectional area of compressed reinforcement, m^2; R_s—calculated resistance of the stretched reinforcement, MPa; E_s—elastic modulus of reinforcement, MPa; E_b—initial elastic modulus of expanded concrete, MPa; h_0—working height of the cross-section (distance from the upper face of the cross-section of the element to the axis of the stretched reinforcement), m.

The results of the calculated study are presented in Table 2. The values of the breaking moments M_{ult} are calculated for each series of foam concrete beams (Table 3).

Table 2 Calculated studies of flexural elements

№ of series	Beam cipher	Area of stretched reinforcement A_s (m^2)	μ	α	R_{b1} (MPa)	ξ
1	B1-6	0.0000283	0.004717	52.451	3.5	0.61
	B1-8	0.0000503	0.008383	51.614		0.7
2	B2-6	0.0000283	0.004717	44.013	4.9	0.51
	B2-8	0.0000503	0.008383	43.311		0.69
3	B3-6	0.0000283	0.004717	38.564	6.5	0.36
	B3-8	0.0000503	0.008383	37.949		0.62

Note As ξ was taken the lower value of ξ_{arm} andξ_b

Table 3 Calculated and experimental study of the breaking moment of the studied flexural elements

№ of series	Beam cipher	M_{ult} (kNm)	Experimental M_{exp} (kNm)	Deviation (%)
1	B1-6	0.808	0.8	1.05
	B1-8	0.893	0.87	2.61
2	B2-6	0.986	0.98	0.59
	B2-8	1.237	1.25	1.02
3	B3-6	0.979	1.07	9.33
	B3-8	1.520	1.51	0.63

Based on the data presented in Table 3, it can be concluded that the deviation of the calculated values of the breaking moment from the experimental ones is no more than 5%, which indicates a high accuracy of the calculation.

4 Conclusion

The calculation made by the authors demonstrates a high convergence with experimental data, which can be used for the calculation and design of wall panels, floor panels, coverings and lintels made of cellular concrete of various classes in compressive strength and density grades of different thickness and length of spans for residential and public buildings.

To expand the scope of application of this solution, as well as to improve the accuracy of the values by reducing the deviation from the results of experimental studies, it is necessary to test a larger number of flexural elements of various geometric sizes, densities and reinforcement options.

Acknowledgements This work was realized in the framework of the Program of flagship university development on the base of the Belgorod State Technological University named after V.G. Shukhov. The work was realized using equipment of High Technology Center at BSTU named after V.G. Shukhov.

References

1. Suleymanova LA, Kolomatskiy AS, Pogorelova IA, Marushko MV (2017) Improving efficiency of production and use of aerated concrete. Bull BSTU Named After V. G. Shukhov 11:34–42
2. Sari K, Sani A (2017) Applications of foamed lightweight concrete. MATEC Web Conf 97:01097
3. Hedjazi S (2020) Compressive strength of lightweight concrete. Compressive strength of concrete, IntechOpen

4. DIN EN 12602:2016 (2016) Prefabricated reinforced components of autoclaved aerated concrete. German version EN 12602
5. Berezhnoi DV, Laalai F, Sekaeva LR (2020) Numerical modeling of concrete deformation processes during its interaction with reinforcement. IOP conference series: materials science and engineering, vol 709. IOP Publishing, p 044019
6. Nikulin AI, Qasim A-KAF (2019) Improving the methodology for calculating the bearing capacity of eccentrically compressed reinforced concrete elements based on the use of the refined curvilinear deformation diagrams of concrete and reinforcement. Mater Sci Forum 974:570–576
7. Chaipanich A, Chindaprasirt P (2015) The properties and durability of autoclaved aerated concrete masonry blocks. Eco Efficient Masonry Bricks Blocks 215–230
8. Vylegzhanin VP, Pinsker VA (2011) Methodology for calculating reinforced bent products from autoclaved aerated concrete. Hous Constr 12:33–35
9. Novikov MV (2016) Power resistance of normal sections of bent reinforced elements from construction aerated concrete. Bull Civ Eng 56(3):60–66
10. Makhutov NA, Gadenin MM (2020) Unification of calculation methods and tests for strength, resource, and crack resistance. Inorganic materials, vol 56, issue 15. Pleiades Publishing Ltd., pp 1521–1527

Modeling the Mixing Process in Construction Using Edem Solution Software

S. I. Antsiferov, V. V. Tibeykin, Yu. M. Fadin, and A. V. Karachevtseva

Abstract Mixing processes are used in the construction industry. The Edem solution is one of the leading programs using the discrete element method to simulate bulk material processes. Use this program to optimize equipment design, increase productivity and reduce operating costs. The mixing process was simulated on a digital model of a DMPP-1000 twin-shaft mixer. To study the calculation results, the EDEM Analyst postprocessor is used, it allows you to reproduce the mixing process in time, generate videos, perform graphing. On one graph the speed is 1.312 m/s, as at this time it is loaded into the mixer. In the subsequent operating time of the mixer, it is noticeable that the speed of particles is in the insignificant range from 0.987 to 1.173 m/s, which indicates that the average speed of particles in the mixing zone is not significant, which means that there are no stagnant zones, which ensures high homogeneity of the mixture. The graph of the effect of the force of the auger on the material shows that in the first second, when the material touches the working bodies, there is a jump in energy as the material falls, beats against the working bodies.

Keywords Simulation · Edem solution · Mixing process · Digital model · Twin shaft mixer

1 Introduction

The processes of mixing various substances are widely used in many industries, including the building materials industry [1]. The production of dry building mixtures is of great importance in the development of the Russian building materials market, therefore, research in the field of mechanical equipment operating in this industry is focused on the creation of new and further improvement of existing machines [2].

S. I. Antsiferov (✉) · V. V. Tibeykin · Yu.M. Fadin · A. V. Karachevtseva
Belgorod State Technological University Named After V G Shukhov,
Kostyukov St, 46, Belgorod 308012, Russia

S. V. Klyuev (ed.), *Digital Technologies in Construction Engineering*,
Lecture Notes in Civil Engineering 173, https://doi.org/10.1007/978-3-030-81289-8_16

At the moment, Edem Solution is one of the leading programs using the discrete element method to simulate processes associated with the movement of bulk materials [3].

Discrete element method (DEM)—Is a family of numerical methods designed to calculate the motion of a large number of particles, such as molecules, grains of sand, gravel, pebbles and other granular media [4].

2 Materials and Methods

Edem Solution is now being used around the world to optimize equipment design, increase productivity, reduce operating costs, shorten the product development cycle, and stimulate product innovation [5].

EDEM allows you to quickly create models of granular solid systems with specified parameters. The functionality of EDEM [6], allows you to combine mechanical, material and other physical properties in the process of modeling the molecular system of solids [7].

Edem Solution is optimized to run on multi-core computers for lower computing time. But despite this, every day the demand for calculations of larger and more complex models is increasing. Therefore, the developers of Edem Solution are constantly looking for ways to increase the estimated performance of the program [8].

The mixing process will be simulated on a digital model of a DMPP-1000 twin-shaft mixer, previously created in the NX CAD/CAM/CAE system (Fig. 1). For example, the use of the NX CAD/CAM/CAE system allows the design of planetary mixers and its kinematic analysis [9].

The NX system from Siemens PLM Software is a solution for design and technological preparation of production [10]. The use of this product provides team access to the creation of projects from anywhere in the world, has a reliable data storage base and the ability to coordinate projects with employees, customers and

Fig. 1 Digital model of the DMPP-1000 twin-shaft mixer

bosses and make quick decisions. The system is successfully used in the aerospace and automotive industries, in shipbuilding and power engineering, in the production of medical equipment, in the field of machine tools and mechanical engineering [11].

To speed up the mixing simulation process, parts and assembly units that are not involved in the mixing process have been excluded from the digital model, and the geometry of the parts has been simplified, which does not affect the movement of particles. A simplified digital model of a two-shaft mixer has been created, which includes a body and two shafts with working bodies (Fig. 2).

In the EDEM desktop that appears, we name the simulation process "Simulation_Mixing" (Fig. 3).

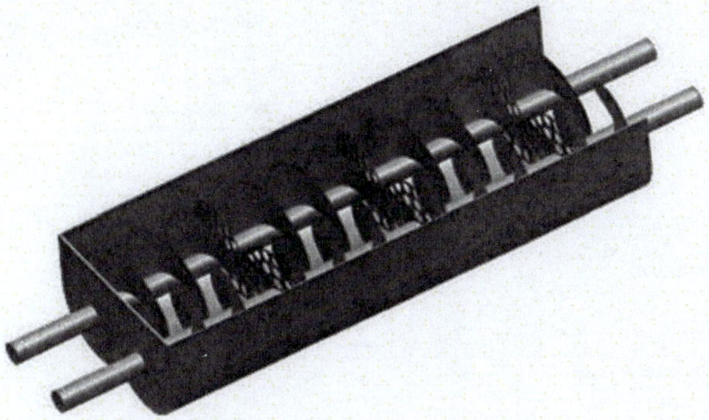

Fig. 2 Simplified digital model of a twin-shaft mixer

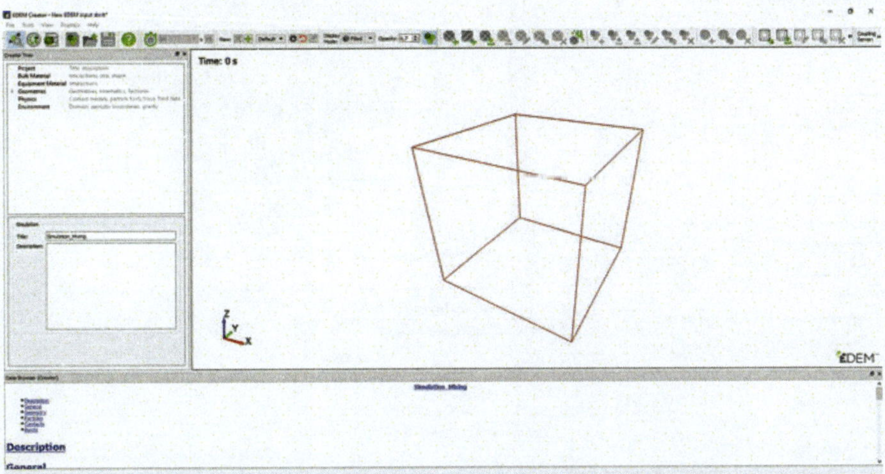

Fig. 3 EDEM workspace

Next, we create the material. We use the Add Bulk Material command. Then add particles using the Add particle command (Fig. 4).

The material will be composed of spherical particles with dimensions of 10 mm. Add material properties and calculate particle properties using the Calculate Properties command (Fig. 5).

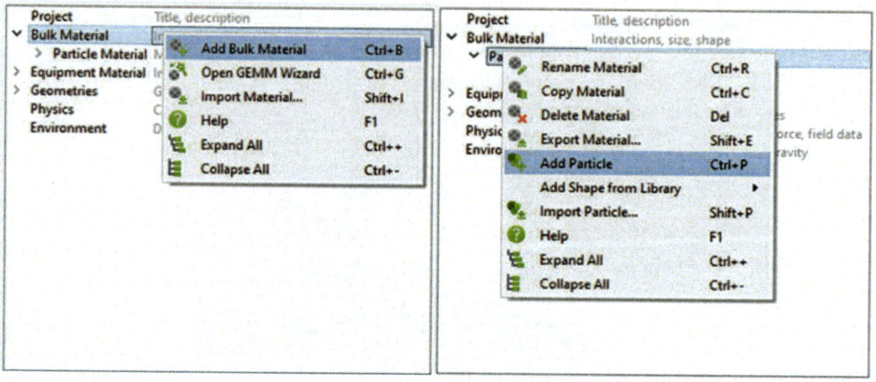

Fig. 4 Material creation

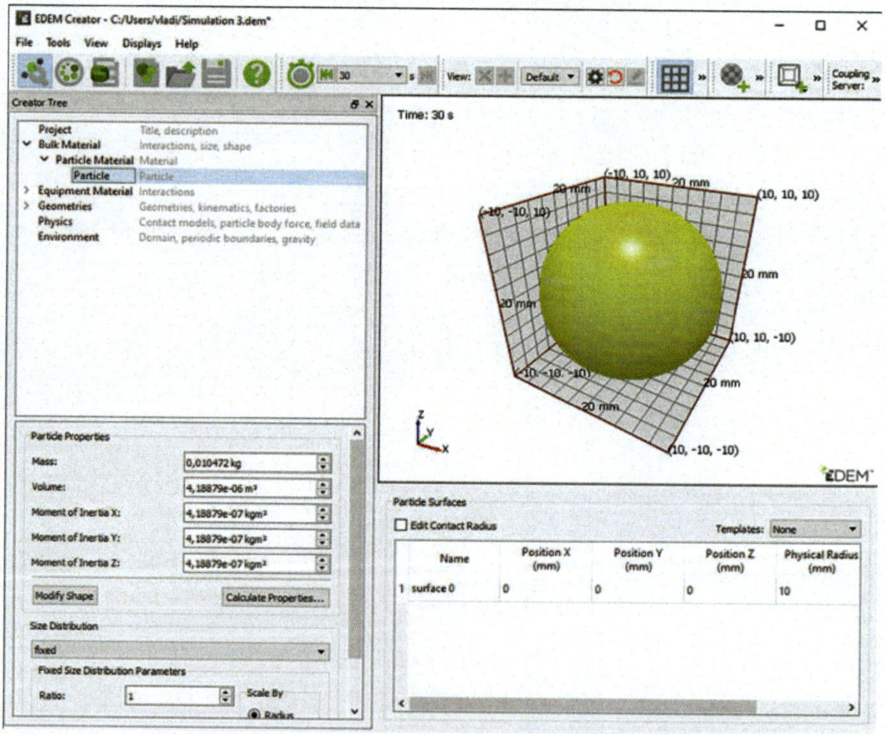

Fig. 5 Setting particle properties

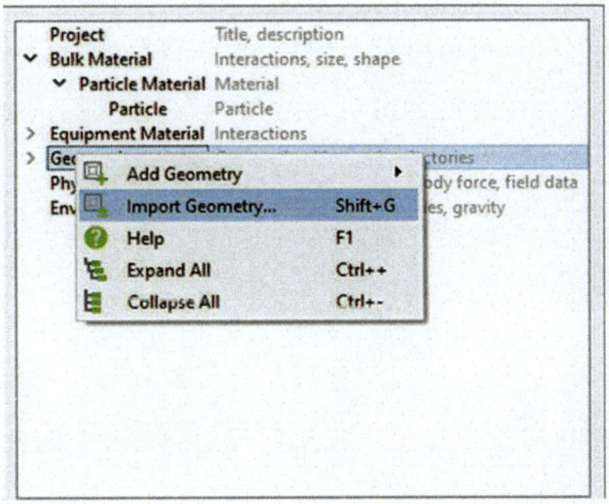

Fig. 6 Importing geometry

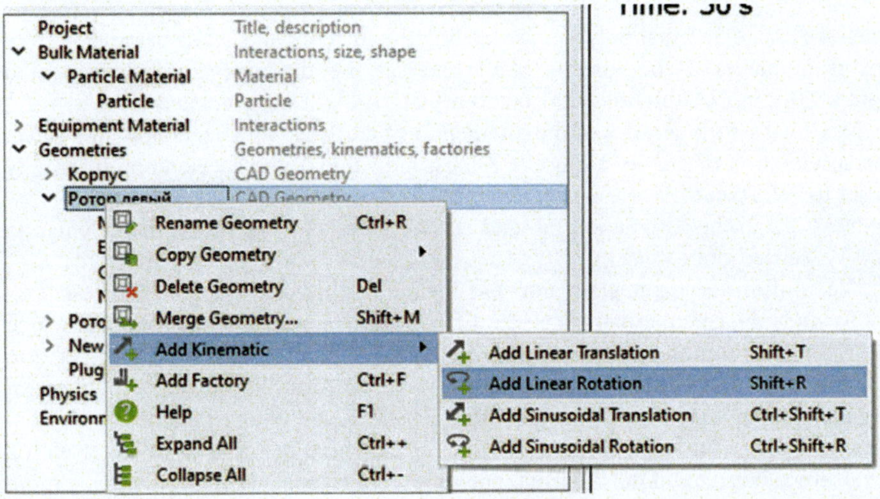

Fig. 7 Adding rotation

Next, we import the previously created simplified 3D model, having previously saved it in the "STEP" format (Fig. 6).

Next, you need to set the kinematics (Fig. 7).

We set the properties of rotation of the shafts. We enter the coordinates for the axes of rotation of the working bodies. We set the speed of rotation of one rotor at 25 rpm, and the other at −25 rpm, so that it rotates in the opposite direction (Fig. 8).

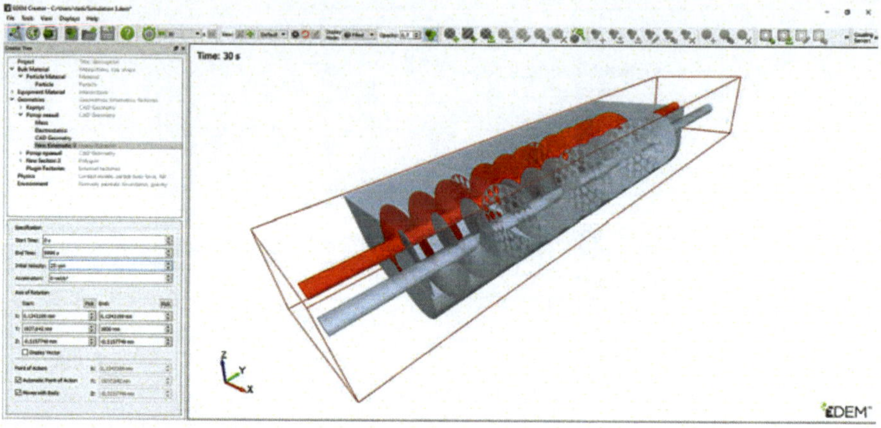

Fig. 8 Rotation parameters

3 Results and Discussions

Then we create a virtual polygon for feeding the material and set its parameters. We put it in place of the loading shaft and set the dimensions of the sides to 840 × 590 mm. A total of 5000 random particles will be generated (Fig. 9).

Next, we set the physics of the interaction of particles with each other and select the acceleration of gravity along the Z axis, the value of the acceleration of gravity is set to -9.81 m/s^2.

Next, go to the Simulation tab and set the parameters at which the calculation will be carried out.

We confirm the parameters and start the simulation of the mixing process.

To study the calculation results, the EDEM Analyst postprocessor is used, which allows you to reproduce the mixing process over time, generate videos, perform graphing and export data for research in universal formats. We select the tab for plotting graphs and select configure the dependencies of the parameters for displaying graphs (Fig. 10) shows that in the areas where the perforated auger with a reverse winding direction is located, back pressure is created. In these areas, a more intense mixing process takes place than in other areas of the mixer.

We select the tab for plotting graphs and select configure the dependencies of the parameters for displaying graphs (Fig. 10) shows that in the areas where the perforated auger with a reverse winding direction is located, back pressure is created. In these areas, a more intense mixing process takes place than in other areas of the mixer.

Analysis of the graph of changes in kinetic energy (Fig. 11), which is due to the forces of resistance to mixing of the material, shows that:

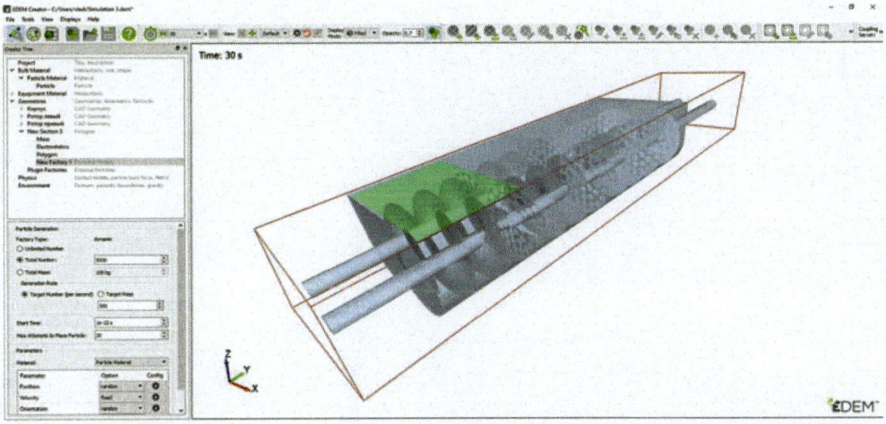

Fig. 9 Polygon parameters

Fig. 10 Trajectories of particle motion at 1, 15, 30 s

- energy surges at the beginning of the mixer operation due to the fact that the material has not yet been evenly distributed over the entire length of the mixer chamber and working body;
- in the subsequent time of operation, the kinetic energy fluctuates in the range of 0.70–0.90 J, this indicates that the same energy acts on each of the particles, which indicates uniform mixing.

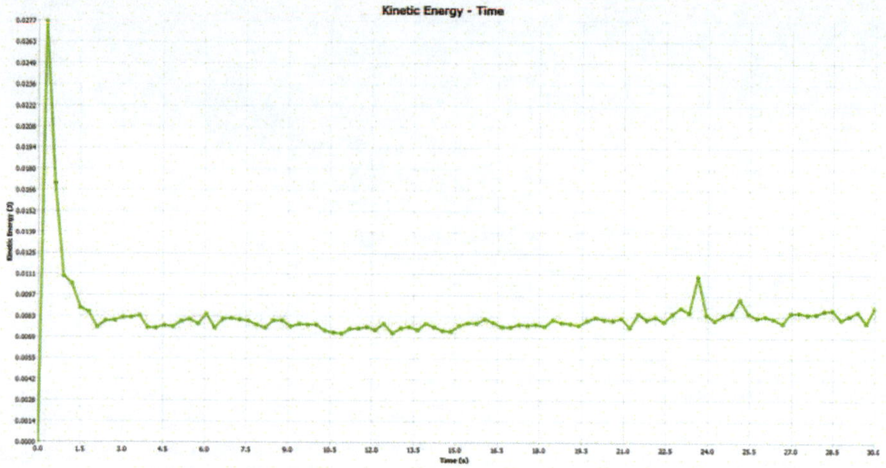

Fig. 11 Plot of kinetic energy versus mixing time

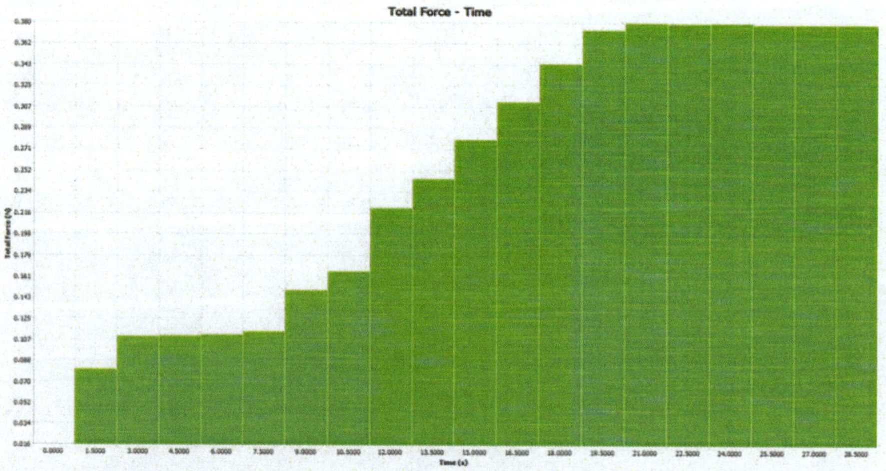

Fig. 12 Graph of the effect of force on the auger

Analysis of the graph of the effect of force on the screw by the material (Fig. 12) shows:

- uniform increase in the force of action of the mixed material on the working bodies. This is due to an increase in the volume of material over time, since the material is fed continuously, the components have not yet been distributed along the entire length of the mixer and the working body, the mixer has not returned to normal operation;

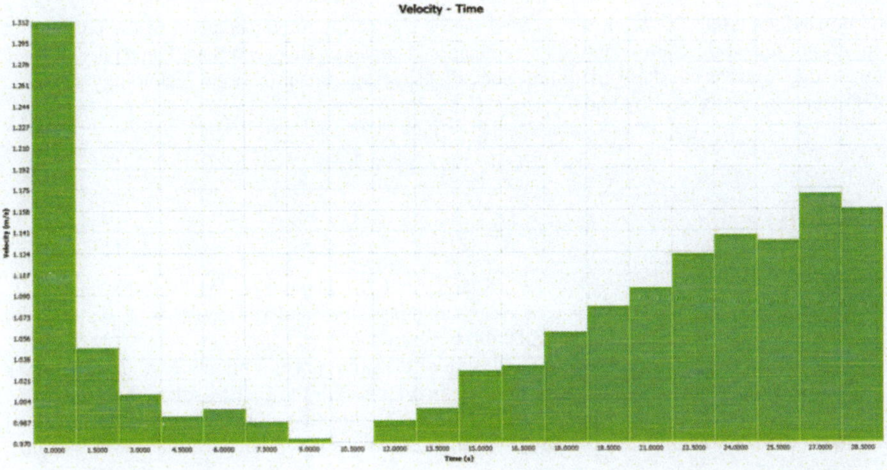

Fig. 13 Graph of particle velocity

- at the end of the graph, the end of growth is observed and the force stabilizes over time, since the mixture fills the mixer body and the material is continuously unloaded. As a result, the volume of the mixture in the mixer body is the same and the force acting on the working elements is also stable, without sudden jumps and drops, which indicates the optimal mode of operation.

Analysis of the graph of the speed of movement of particles (Fig. 13) shows that at the beginning the speed is 1.312 m/s, since at this time the material is loaded into the mixer. During the subsequent operation of the mixer, it is noticeable that the particle speed varies in insignificant limits from 0.987 to 1.173 m/s, which means that the average particle speed in the mixing zone does not differ significantly, which means there are no stagnant zones, which also ensures high homogeneity of the mixture.

4 Conclusion

From the results obtained, the process of modeling the mixing of bulk materials, it can be concluded that the constructive solution of the working body of the twin-shaft mixer is operable. The pitch of the turns of the spiral is selected optimally, as evidenced by the graph of the speed of movement of the particles (Fig. 13). In the areas where the perforated screw with a reverse winding direction is located, sufficient back pressure is created on the incoming mixture components, prompting them to multiple extrusion through the holes in the screw, due to this, the mixing process is intensified and the quality of the resulting mixture is improved.

Acknowledgements The work was carried out as part of the implementation of the program of activities of the world-class scientific and educational center of Belgorod State University, Russia: "Innovative Solutions in the Agricultural Sector", using equipment of High Technology Center at BSTU named after V. G. Shukhov.

References

1. Bogdanov VS, Semikopenko IA, Sharapov RR, Fadin Yu M, Nesmeyanov NP, Gerasimenko VB (2013) Basics of calculating machines and equipment for enterprises of building materials and products. TNT, Stary Oskol
2. Dubrovsky VV (2018) Improving the design of the drum dryer. The future of science, Kursk
3. EDEM Help Manual. Electronic manual for working with the EDEM software product
4. Williams JR, Hocking G, Mustoe GGW (1985) The theoretical basis of the discrete element method NUMETA numerical methods of engineering theory and applications. A A Balkema, Rotterdam
5. Pavlenko AA, Saplinova VV (2019) Use of EDEM SOLUTION in equipment research of the building materials industry energy-saving technological complexes and equipment for the production of building materials. Belgorod
6. DEM Solutions Ltd (Electronic resource). Access mode http://www.dem-solutions.com/
7. Eltsov Yu M, Khakhalev PA, Kharin NP (2016) Investigation of the operating modes of the BWM with various intra-drum devices in the EDEM software. Science-intensive technologies and innovations. Belgorod
8. Williams JR, O'Connor R (1999) Discrete element simulation and the contact problem. Arch Comput Methods Eng 6(4):279–304
9. Bogdanov VS, Chemerichko GI, Anciferov SI, Bogdanov DV, Kulakov LS, Popovich AV, Karachevceva AV (2019) Kinematic analysis of a planetary mixer mechanism. J Phys Conf Series 1353:012030
10. Eltsov Yu M, Kozlov AA, Sedoikin AV, Shirokova Yu L (2011) Designing in NX with teamcenter textbook manual. Publishing house of BSTU, Belgorod
11. Vedmid PA, Sulinov AV (2014) Processing programming in NX CAM. DMK Press, Moscow

Effective Polyfunctional Additive for Composite Materials Based on Cement

N. O. Kopanitsa, O. V. Demyanenko, and A. A. Kulikova

Abstract The article presents the results of studies related to the development of an effective complex polyfunctional additive (PA), consisting of silicon nanodioxide—Ts38, microsilica (MS), thermally modified peat (MT-600) and marble production waste—microcalcite (MCa), to improve the operational properties cement-based composite materials. A complex of physical and chemical studies was carried out: derivatographic, X-ray phase and electron-microscopic analysis of the control and modified cement stone. It was found that the use of a polyfunctional additive affects the hydration processes of cement clinker minerals, improving both the properties of the cement stone and the properties of concrete. The results of X-ray phase analyzes have shown that the use of a polyfunctional additive leads to a decrease in the intensity of $Ca(OH)_2$ peaks and a change in the phase composition of the cement stone, the formation of a cement stone structure mainly from low-basic calcium hydrosilicates. It is shown that the introduction of a polyfunctional additive in an amount of 5%, with a decrease in cement consumption, makes it possible to increase the compressive strength of the cement stone up to 83% and to form a homogeneous dense structure of the cement stone, which will improve the performance properties of composite materials.

Keywords Industrial waste · Microcalcite · Microsilica · Nanosilica · Mt-600 · Cement · Composite materials

1 Introduction

The use of modifying additives, including nanoobjects, which effectively affect the properties and structure of the final material, opens up wide opportunities for improving the technology of composite materials based on cement. There are a lot of studies on the use of modifying additives for various purposes in the production

N. O. Kopanitsa · O. V. Demyanenko (✉) · A. A. Kulikova
Tomsk State University of Architecture and Building, Solyanaya Sq. 2,
634003 Tomsk, Russia

S. V. Klyuev (ed.), *Digital Technologies in Construction Engineering*,
Lecture Notes in Civil Engineering 173, https://doi.org/10.1007/978-3-030-81289-8_17

of concrete and mortars. At the same time, the use of local natural raw materials and production wastes to obtain effective modifying additives in high-quality concretes and mortars is relevant and will significantly expand the raw material base for the production of concretes with increased performance characteristics while maintaining or insignificantly increasing their cost. Thus, on the basis of the set of issues formulated, research on the development of the compositions of complex polyfunctional additives and the assessment of their influence on the properties of cement stone and concrete is relevant. Analysis of literature data [1, 2] shows that the formation of a dense and strong packing of fine-grained concrete components is achieved by building its structure “bottom-up” [3], which implies the introduction of particles into the cement composition, which are smaller in size of cement grains. The use in the concrete mixture of rationally selected and balanced in composition and size fractions of not only small and coarse aggregate, but also filler as well as modifying additives interacting with cement hydration products at nano-, meso- and macro-levels of organization, taking into account the provision of solid components concrete mixture of continuous granulometric range in the entire range of particle sizes will allow to obtain a composite material with improved performance characteristics.

The aim of the presented work is: development of a complex polyfunctional additive based on nanosized silicon dioxide and finely dispersed components of various compositions to regulate the properties of a cement binder. According to the accepted working hypothesis, in order to develop effective polyfunctional additives, it was necessary to select their material and dispersed composition so that each of the components would improve or complement the properties of the previous one.

Earlier, in works [4–8], we substantiated the choice of nanosized and finely dispersed components to obtain an effective polyfunctional additive for composite materials based on cement, providing an increase in their performance characteristics—nano silicon dioxide (Ts38), microsilica, microcalcite, thermally modified peat MT-600. The effectiveness of each of the investigated dispersed components on the properties of cement stone has been sufficiently studied by many authors, but their complex use enhanced by the action of nano-SiO_2 has not been previously studied. Figure 1 shows a model of the formation of the material composition of a complex modifying additive, using the selected components in accordance with the working hypothesis adopted in the work and acquired by cement stone, with the appearance of each subsequent component of the additive, characteristics. Previous studies [4–8] suggest that using the proposed substances in combination with nano-SiO_2 in a polyfunctional additive, it is possible to achieve a synergistic effect on the properties of cement stone due to their physicochemical interaction.

In [9], the results of studies of the effect of each of the established components of the developed additive in combination with nano-SiO_2 on the properties of cement stone are presented. It is shown that the addition of nano-SiO_2 significantly increases the efficiency of dispersed components (MS, MT-600 and MCa), which leads to an increase in the strength of the cement stone due to the activation of the processes of formation of the structure of the cement stone at the nanoscale level.

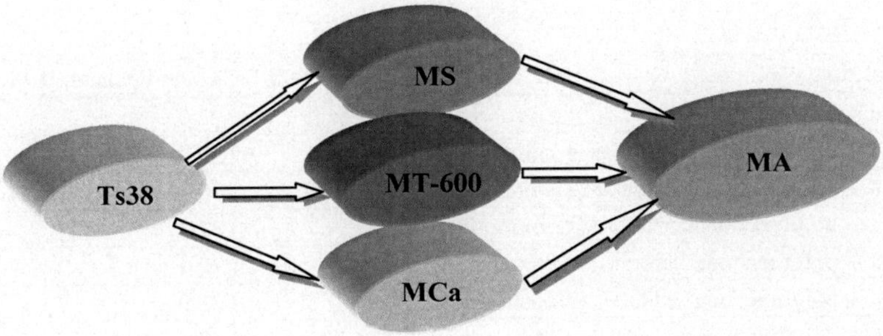

Fig. 1 Model of the formation of the material composition of a complex modified additive for composite materials

To solve this problem, the compositions of the additive were selected in two ways: stoichiometrically and experimentally.

1. Composition—Stoichiometrically: MS—14.4%; MCa—80%; MT-600—5%; Ts38—0.6%;
2. Composition—Experimental: MS—45.2%; MCa—45.1%; MT-600—9.1%; Ts38—0.6%.

2 Materials and Methods

The complex additive was prepared by mixing its components in an intensive mixer with sequential loading of components, starting with the largest fraction: microcalcite, microsilica, MT-600 and Ts38. The total mixing time was 20 min. In the process of mixing, additional grinding of coarse fractions occurs with simultaneous joint activation of all components of the additive and cement. Further, samples of cement stone with the studied additives were made (the amount of additive varied from 5 to 10% of the cement mass), 20 * 20 * 20 mm in size, which hardened under normal conditions and were tested for strength in 3, 7, 28, 120 days of hardening. For each composition, at least 25 samples were prepared (at least 5 for each hardening period). The value of water demand is presented in Table 1.

From the results presented in the table it can be seen that the introduction of a polyfunctional additive increases the water demand of the cement paste from 5.1% to 19.6%, depending on the percentage of the additive.

The strength value was determined as the arithmetic mean value of 5 samples for each hardening period, the coefficient of variation is not more than 5%.

Table 1 The value of water demand

Composition	Water demand, %
Control	27.0
Cement + polyfunctional additive (1 composition, 5%)	29.0
Cement + polyfunctional additive (1 composition, 7.5%)	30.1
Cement + polyfunctional additive (1 composition, 10%)	32.3
Cement + polyfunctional additive (2 composition, 5%)	28.4
Cement + polyfunctional additive (2 composition, 7.5%)	28.9
Cement + polyfunctional additive (2 composition, 10%)	30.0

3 Results and Discussions

Analysis of the data presented in Fig. 2 confirmed that the most effective composition of the complex additive, according to the strength criterion, is: (MS—45.2%; MCa—45.1%; MT-600—9.1%; Ts38—0.6%), in the amount of 5%. The introduction of a polyfunctional additive (PA) increases the strength of the cement stone by 38% (3 days of hardening), in 7 days to 29%, in the late hardening period (28 days) by 83%, in 120 days by 74%, compared with the control sample. The specific surface area of the developed additive $S_S = 2528$ m^2/kg, which is close to the value of the specific surface area of microsilica.

To assess the regularities occurring during the hydration and structure formation of the modified cement stone, a complex of physicochemical studies was carried out. Figure 3 shows a derivatogram of a modified cement stone sample with TG and DSC curves.

On the derivatograms (Fig. 3) of a cement stone with a polyfunctional additive, by 28 days of hardening, endo-effects were recorded at 90–110 °C, which corresponds to the loss of weakly bound water, the endo-effect at 440–487 °C (refers to the decomposition of calcium hydroxide, with the formation of low-basic calcium hydrosilicates type C–S–H(I) and $C_3S_6H_6$, endothermic effect at 581–698 °C, corresponding to dehydration of highly basic hydrosilicates of calcium. The introduction of a polyfunctional additive leads to a change in the phase composition of the cement stone, with the formation of a structure mainly of low-basic

Fig. 2 Kinetics of strength gain during compression of cement stone with the investigated additives

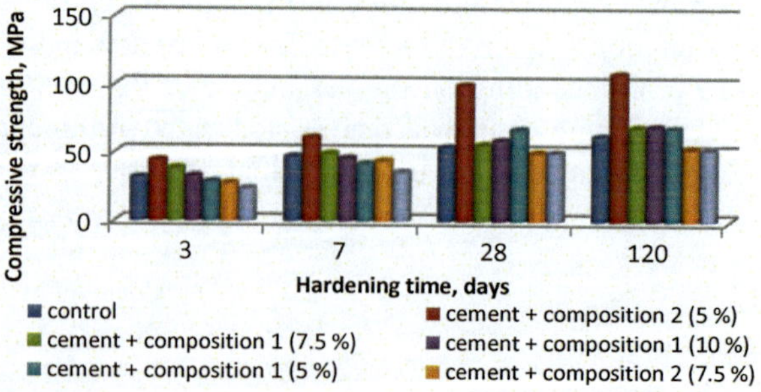

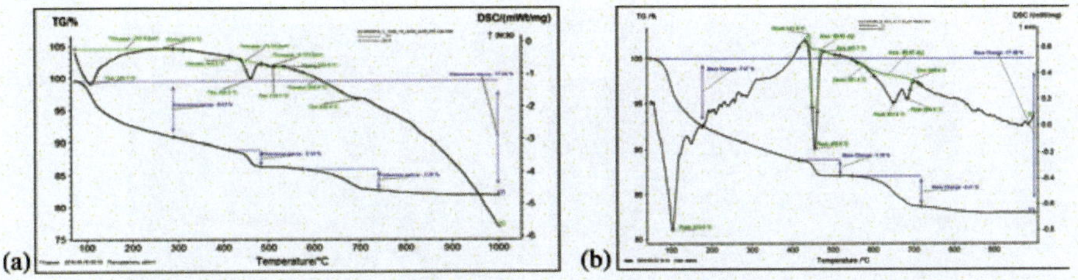

Fig. 3 Derivatogram of cement stone: **a** control, **b** modified by a complex polyfunctional additive

hydrosilicates of calcium (endoeffect at 455 °C), which do not undergo recrystallization processes and increase the strength of the cement stone.

Figure 4 shows an X-ray diffraction pattern of the studied sample of the modified cement stone.

From the X-ray data (Fig. 4) it can be seen that the introduction of a complex polyfunctional additive promotes the formation of low-basic hydrosilicates of calcium, tobermorite and tobermorite-like compounds in the cement stone (d/n = 5.61; 2.62 * 10^{-10} m), 5CaO * $2SiO_2$ * H_2O (d/n = 3.02 * 10^{-10} m), 3CaO * $6SiO_2$ * $6H_2O$ (d/n = 2.77 * 10^{-10} m), CaO * SiO_2 * $2H_2O$ (d/n = 2.17 * 10^{-10} m), C_3S (d/n = 1.92 * 10^{-10} m), $C_5S_6H_5$ (d/n = 3.34 * 10^{-10} m), foshagite (d/n = 1.79 * 10^{-10} m), significantly improving the physical and mechanical characteristics of the cement stone. This "additive-cement" ratio contributes to the amorphization of the structure of the cement stone, which is confirmed by an increased background in the region of small angles and a decrease in the intensity of the peaks of the crystalline phases. According to the results of qualitative X-ray phase analysis, it was found that when a polyfunctional additive is introduced in Portland cement, the intensity of $Ca(OH)_2$ peaks decreases, which may indicate its binding to calcium hydrosilicates, this is confirmed by the identification of additional peaks on X-ray diffraction patterns. To confirm the results obtained on the formation of the structure of a cement stone with a complex polyfunctional additive, an electron microscopic analysis was performed (Figs. 5 and 6).

It was found by scanning electron microscopy that the structure of a cement stone with an additive differs from the control (no additive) by the presence in

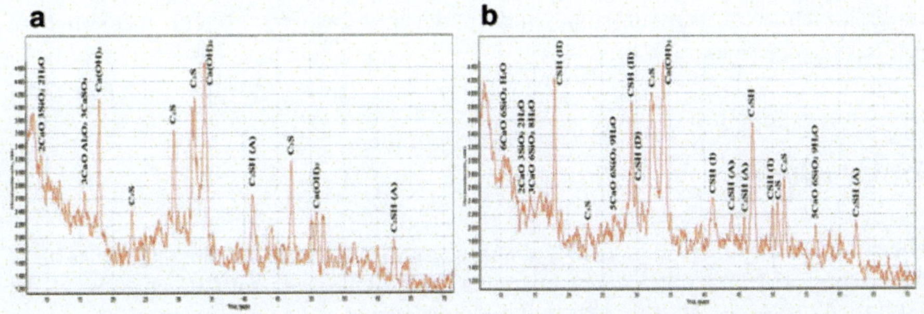

Fig. 4 X-ray of the cement stone: **a** control, **b** modified with a polyfunctional additive

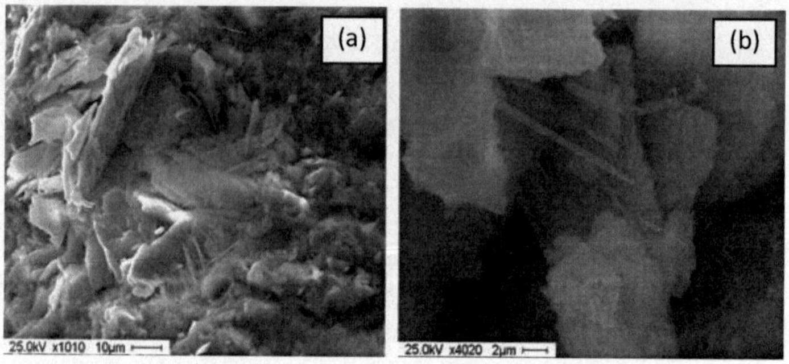

Fig. 5 Electron microscopic analysis of the control cement stone: **a** × 1000, **b** × 4000

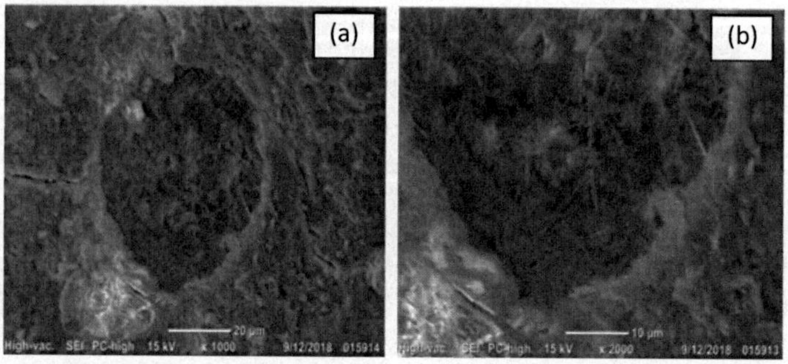

Fig. 6 Electron-microscopic analysis of the modified cement stone with a polyfunctional additive on the cleavage surface: **a** × 10,000, **b** × 2000

cracks and pores of an additional amount of neoplasms in the form of dense accumulations of fibrous crystals, the morphology of which is identical to calcium hydrosilicates of the CSH (B) type. Whereas the control sample of cement stone is characterized by a rather heterogeneous and defective structure with less crystallized pore space, which is due to the uneven distribution and growth of hydration products in the volume of the cement matrix.

Modification of a polyfunctional additive allows obtaining a homogeneous, dense structure of cement stone and increasing the performance properties of composite materials.

4 Conclusion

The completed cycle of experimental studies on the effect of the developed additive on the physical and mechanical characteristics of the cement binder showed the validity and the possibility of regulating the structure and properties of cement stone with modifying additives, based on rationally selected components (nano-SiO_2, microsilica, microcalcite, and MT-600), interacting cement hydration. The mechanisms of the synergistic action of the proposed components of the additives on the structure and properties of the cement stone, consisting in increasing the strength, frost resistance and other performance characteristics, are revealed, which is due to the different mechanism of their action in the formation of the structure and properties of the cement stone. The synergistic action of the additive provides improved performance characteristics of cement compositions, together with a decrease in the cost of the final material, due to the use of components of local resources, or stone crushing waste. It is shown that the introduction of a polyfunctional additive into the composition of the cement paste significantly affects the hydration processes of clinker minerals, changing both the properties of the cement paste and the properties of the hardened cement stone. The polyfunctional additive promotes the binding of portlandite to low-basic calcium hydrosilicates and the formation of additional crystallization centers. The mechanism of action of a polyfunctional additive is complex, which can be explained by a balanced ratio of components both in their chemical composition and in dispersion.

References

1. Belov VV, Smirnov MA (2010) Optimization of the granulometric composition of compositions for the manufacture of non-fired construction conglomerates. Bull Central Reg Branch RAASN 9:65–72
2. Belov VV (2007) Optimization of compositions for the production of pressed fine-grained concrete. Bull Central Reg Branch Russ Acad Archit Constr Sci 6:27–35
3. Drexler KE, Peterson C, Pergamit G (1991) Unbounding the future: the nanotechnology revolution
4. Kopanitsa NO, Sarkisov YS, Demyanenko OV (2017) Peculiarities of silica additives application in building mixes production. Am Inst Phys 020010
5. Kopanitsa NO, Sarkisov YS, Demyanenko OV (2016) The use of nanodispersed silica in the production of building mixtures. Bull Tomsk State University Archit Civil Eng 5:140–150
6. Kosmachev PV, Vlasov VA, Kopanitsa NO, Skripnikova NK, Demyanenko OV (2017) Composite materials based on cement with nanodispersed silicon dioxide. Bull Tomsk State University Archit Civil Eng 4:139–146
7. Kopanitsa NO, Sorokina EA, Demyanenko OV (2018) Influence of the addition of thermally modified peat on the technological properties of building mixtures for 3D printing. Bull Tomsk State University Archit Civil Eng 4:122–134
8. Kopanitsa NO, Sarkisov YS, Demyanenko OV, Gorshkova AV (2017) Cement compositions modified with combined nanodispersed additives. Bull Tomsk State University Archit Civil Eng 4:101–106
9. Demyanenko OV, Kulikova AA, Kopanitsa NO (2020) Assessment of the effect of a complex multifunctional additive on the performance characteristics of cement stone and concrete. Bull Tomsk State University Archit Civil Eng 5:139–152

The Purification of Service Water from Nickel Ions

V. V. Ovchinnikov, A. A. Kulakov, S. A. Maltseva, and I. G. Grigor'eva

Abstract An analysis of the literature data on the extraction of heavy metal ions such as nickel (II), chromium (III) and cadmium (II) from industrial, domestic wastewater was carried out due to the complexes formed by them with polyacrylates, fluoroquinolones (FQ) and other organic substances. The authors of the cited works were judged on the complexity of using the used equipment and, in some cases, on the possible negative biochemical impact on the environment. The investigation relates to organic chemistry—to amino acids: alanine, asparagine and cysteine, which can be used as adsorbent of nickel salts in water in urban water supply systems. The object of the investigation is to improve the efficiency of the water purification method for water supply systems by using amino acids, mainly cysteine, which are capable of adsorbing nickel ions in aqueous solutions. The technical result is achieved by the fact that in the reducing nickel concentration in water, the amino acid cysteine introduced into water binds nickel into a strong chelate complex. This phenomenon has been proved by molecular mechanic calculations. The suggested method of the purification of service or drinking water is not expensive and easy to implement. The purification is carried out with the amino acid cysteine, which is more active than alanine and asparagines, and can be attached in a suitable bag to a water purification industrial column and, when the column is saturated with nickel ions, can be easily replaced with a new bag of cysteine.

Keywords Water · Complex · Dimethyglyoxyme · Amino acid · Alanine · Asparagine · Cysteine

---

V. V. Ovchinnikov (✉) · A. A. Kulakov · S. A. Maltseva · I. G. Grigor'eva
Kazan National Research Technical University Named After A.N. Tupolev-KAI (KNITU-KAI), Kazan, Russia

S. V. Klyuev (ed.), *Digital Technologies in Construction Engineering*,
Lecture Notes in Civil Engineering 173, https://doi.org/10.1007/978-3-030-81289-8_18

1 Introduction

In natural water that has not undergone preliminary purification, there are many different inorganic contaminants and dangerous microorganisms that can cause serious harm to human health. One such natural pollution is nickel, which, at different concentrations, is present in almost many the groundwater of different countries. Necessary to do a clear of service water from heavy metals, especially from nickel salts.

The are some appoarches for this aims between of them can point ultrafiltration using sodium polyacrilate and stability of PAA-Ni complex in the shear field [1], or removal of Ni (II) and Cr (III) by complexation-ultrafiltration using rotating disk membrane and the selective separation by shear induced dissociation [2].

Necessary to note, that mentioned modes are expensive and not easy in the application by the reason of required apparatures.

There is a communication [3, 4] about complexation between of fluoroquinolones (FQs) with metals due to complexation in service water. Nevertheless, in this report is mentioned about antibacterial activity of FQs-metal complexes, which can depend from pH, temperature and type of metal ions, that can be non-favorite reflect on environment [5].

We in our previous paper [6] showed, that such organic substances as amino acids can easy form rather stable complexes with two-valence calcium and magnesium ions that led to reduce of these ions in potable and drinking water. Similar opinion have been stated by author of work [7] at the investigation of complex-formation process of amino acids with metal iones.

This research relates to chemistry of organic compounds, which can be used in water treatment in urban water supply systems. In addition, the purification of any water (technical and drinking) is a necessary existence of a healthy person, since clean water to a lesser extent will cause various diseases [6].

Also the way of receiving ethylidene-diamido-diizopropiliden-O,O′-dinatrium-diphosphonate, in which and ethynediamine $H_2N–(CH_2)_2NH_2$ subject methylacrylic $HOOC–CH_2 = CH(CH_3)$ acid to interaction with acidic phosphoric acidic Na_2HPO_3 sodium at a temperature 70–75 °C in the environment of inert solvent within 1.5–2.5 h is known [8]. In the composition of this organic compound there are amine and acid functional groups. The mentioned compound possesses low technospheric safety due to presence of methyl acrylic acid fragments, sodium phosphoric acid and ethylenediamine fragment, which can cause formation of toxic products in urban water supply system at temperature 70–75 °C. At the same time, the quality of water treatment for water supply systems remains low.

Our research is aimed at improving the quality of technical and drinking water and consists in the fact, that the amino acid is added to technical water. Novelty is that for purification of process water from nickel ions it is new that amino acid cysteine is used as organic product.

2 Methods and Materials

Water purification is carried out as follows [9]:

- at first prepare the standard solutions of nickel chloride with volume of 20 ml with concentration of 0.006, 0.012, 0.018, 0.024 and respectively 0.030 mol·l^{-1}, from which take aliquots on 10 ml, to which add 0.5 ml of 0.1% spirit solution of a dimethylglyoxyme of $C_4H_8N_2O_2$ (Chugaev's reactive) [10]. The solutions are thoroughly mixed and examined after 1.0–1.5 min on a spectrophotometer CFC-2 at a wavelength of λ = 440 nm, using cuvettes with an optical layer thickness of 5 mm with respect to distilled water;
- the correspondence between the mass concentration of nickel in mol·l^{-1} and the corresponding values of optical density (D^o, Table 1) is calculated from the experimental data (Fig. 1) [9].

3 Results and Discussion

After the calibration relationship was established, the optical density of the nickel (II) solutions with amino acids was determined. In three samples of a 20 ml Ni^{2+} ion concentration solution, 128 mg of alanine, 216 mg of asparagine and 245 mg of cysteine were added. The stable complexes of Ni(amino acid)$_2$ are formed in water. Then we took aliquotes by the 10 ml from the obtained solutions and add in it 0.5 ml of a 0.1% alcohol solution of dimethylglyoxime and measured the optical density on a spectrophotometer, followed by calculating their concentrations according to Fig. 1 and the Eq. (1), which has been calculated on the base of represented in Table 1 experimental data. The results are shown in Table 2.

$$D_o = (0.1 \pm 0.05) + (11.5 \pm 3.0)C_{(\text{mol.l})^{-1}} \quad (1)$$

The characteristic of correlation dependence: r 0.936, S_o 5.4, n 4, where r is the correlation coefficient, S_o is the standard deviation, n is the number of standard solutions.

The use of the dependence between the concentration of Ni^{2+} and optical dencity (Fig. 1) allowed to establish in spectrophotometrically investigated solutions the of

Table 1 Experimental data for calculation of the calibration dependence

Standard Solution Number	Concentration of Ni^{2+} ions, mol·l^{-1}	Optical density D^o, A
1	0.006	0.13
2	0.012	0.27
3	0.018	0.36
4	0.030	0.42

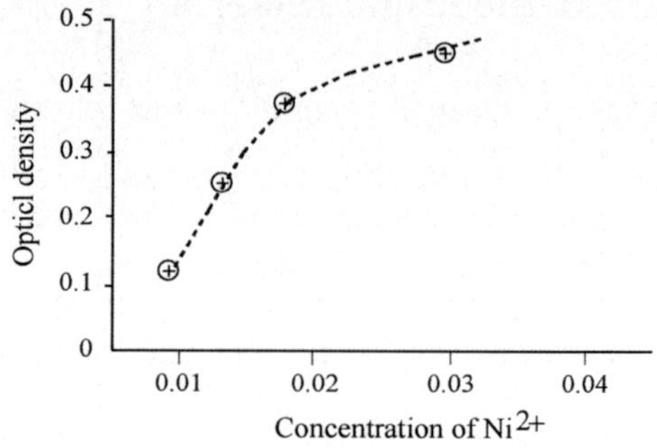

Fig. 1 The dependence of optical density (A) and the concentration of Ni^{2+} ions ($mol \cdot l^{-1}$)

Table 2 Data for calculation of nickel (II) concentrations in service water after addition of amino acids to it

Amino acid	Concentration of amino acid, $mol \cdot l^{-1}$	Optical density D^{o}, A
Alanine	0.07	0.42
Asparagine	0.07	0.42
Cysteine	0.07	0.005

non-bonded Ni^{2+}concntration, which are shown in Table 2 as 0.03 $mol \cdot l^{-1}$ for alanine and asparagine and as about 0.005 $mol \cdot l^{-1}$ for cysteine.

Thus, the addition of a little more than twice the concentration of alanine and asparagine does not give a decrease in the optical density of the dimetilglioksim-nickel complex, but the addition of cysteine leads to a decrease of optical density almost completely, on 84%, which gives a decrease in concentration about on 92%. The ratio of the cysteine to nickel forming complex is 2.35: 1, that can be seen from Figs. 1 and 2. The efficiency of purification of water from nickel salts due to the introduction of alanine and asparagine additives has little effect on its purity, but the addition of cysteine cleanses technical water almost exclusively (~98%).

Because of the obtained by us results show that amino acid clycine extracts the nickel iones from service water more better then dimethylglyoxyme, which is known as Chugaev's reactive, we decided to revise such conclusion through molecular-mechanic calculation, using PCMODEL-MM*X*-1.0 program [10]. The results of calculations are represented on Figs. 3 and 4.

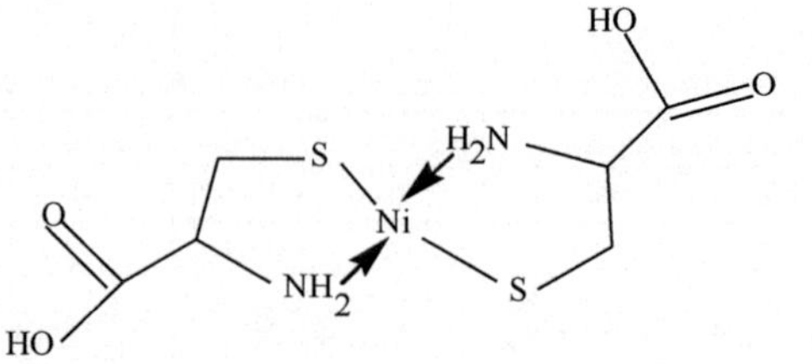

Fig. 2 The scheme of the formation of complex of Ni^{2+} ion with two molecules of cysteine

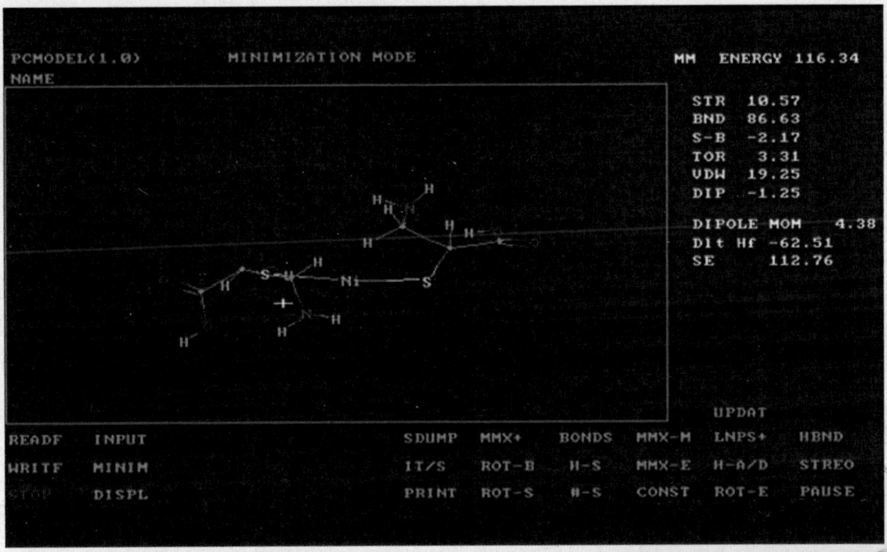

Fig. 3 The structure and calculated thermochemical (in kcal·mol^{-1}) for gas phase data via PCMODEL-MMX-1.0 program of Ni(cysteine)$_2$ complex

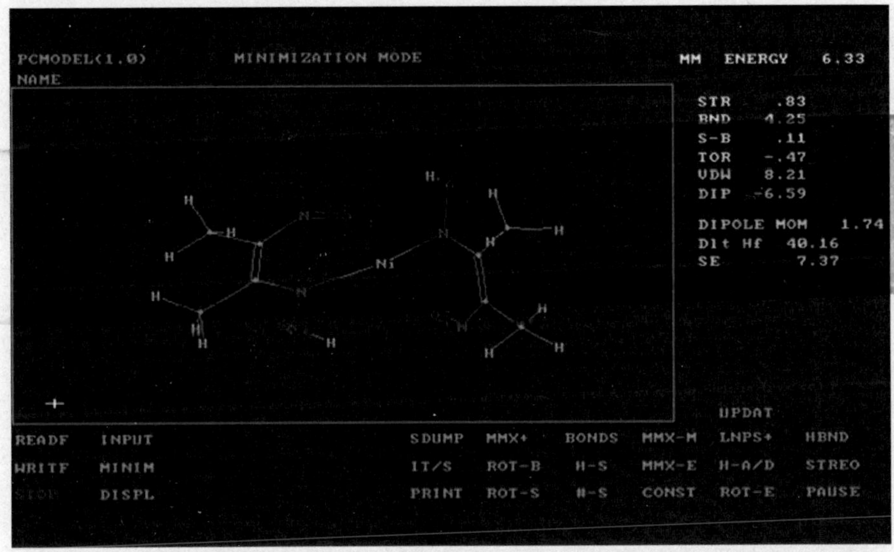

Fig. 4 The structure and calculated thermochemical (in kcal·mol^{-1}) for gas phase data via PCMODEL-MMX-1.0 program of Ni(dimethylglyoxyme)$_2$ complex

In spite of the calculated results have been made in gas phase (Figs. 3 and 4) and are given in kcal.mol^{-1} instead of kJ, it can be seen that the heat of formation of Ni (cysteine)$_2$ complex has a negative value, but the same Ni (dimethylglyoxyme)$_2$ complex has a positive magnitude. This fact can be explained, that the substitution of Chugaev's reactive on cysteine in structure of complex is real possibility.

4 Conclusion

The investigation relates to organic chemistry—to amino acids: alanine, asparagine and cysteine, which can be used as adsorbent of nickel salts in water in urban water supply systems. The object of the investigation is to improve the efficiency of the water purification method for water supply systems by using amino acids, mainly cysteine, which are capable of adsorbing nickel ions in aqueous solutions. The technical result is achieved by the fact that in the reducing nickel concentration in water, the amino acid cysteine introduced into water binds nickel into a strong chelate complex. This phenomenon has been proved by molecular mechanic calculations. The suggested method of the purification of service or drinking water is not expensive and easy to implement. The purification is carried out with the amino acid cysteine, which is more active than alanine and asparagine and can be attached in a suitable bag to a water purification industrial column and, when the column is saturated with nickel ions, can be easily replaced with a new bag of cysteine.

Acknowledgements This work is realized in the framework of scientific research of Kazan National Research Technical University named after A.N. Tupolev-KAI (KNITU-KAI).

References

1. Qiu Y, Hou B, Zhang Q, Ghang X (2018) Treatment of wastewater containing nickel y complexation- ultrafiltration using sodium polyacrylate and the stability of PAA-Ni complex in the shear field. Chem Eng J 334(15):1878–1885. https://doi.org/10.1016/j.cej.2017.11.087
2. Zhang Q, Gao J, Qiu Y-R (2019) Removal of Ni (II) and Cr (III) by complexation-ultrafiltration using rotating disk membrane and the selective separation by shear induced dissociation. Chem Eng Process—Process Intensification 135(1):236–244. https://doi.org/10.1016/j.cep.2018.12.005
3. Chen L, Qiu Y (2019) Removal of Cd (II) from dilute aqueous solutions by complexation–ultrafiltration using rotating disk membrane and the shear stability of PAA–Cd complex. Chin J Chem Eng 27(3):519–527. https://doi.org/10.1016/j.cjche.2018.06.026
4. Cuprys A, Pulicharla R, Brar S-K, Drogui P, Verma M, Surampalli Rao Y (2018) Fluoroquinolones metal complexation and itsenvironmental impacts. Coordination Chem Revies 376(1):46–61. https://doi.org/10.1016/j.ccr.2018.05.019
5. Adusei-Gyamfi J, Ouddane B, Rietveld L, Cornard J-P, Criquet J (2019) Natural organic matter-cations complexation and its impact on water treatment: a critical review. Water Res 160(1):130–147. https://doi.org/10.1016/j.watres.2019.05.064

6. Ovchinnikov VV, Kulakov AA (2018) Quality increasing of technical and potable water due to .complexation between amino acids Ca and Mg Salts. Izvestiya vysshikh uchebnykh zavedeniy. Khimia and khimicheskaya tekhnologiya 61(6):109–112. https://doi.org/10.6060/tcct.201861065.5691
7. Ogorodnikova NP (2010) Autoreferat dissertation... candidate of chemical sciences: 02.00.04. Chemical interaction of metals—copper, iron and manganese with α- and β-amino acids in aqueous and organic media. South. federal. un-t. Rostov-on-Don
8. Gurevich PA, Ovchinnikov VV (2020) Synthesis and pharmacological properties of ethylene-diamido-diisopropylene-o, o'-dinatrium-diphosphonate for its use to improve quality of industrial water of urban heat supply systems. Harold Kazan Technol Univ 23(8):16–19
9. Quantitative chemical analysis of water (2013) Mass measurement procedure of nickel concentrations in natural and waste water by photometric method with dimethylglyoxime. IPA F 14.1: 2.46–96. Environmental regulatory document of Federal significance. The methodology is allowed for the purposes of the state environmental control. Edition, Moscow. https://files.stroyinf.ru/Data2/1/4293776/4293776009.htm
10. Clark T (1990) Computer chemistry. Mir, Moscow

Improvement of the Process of Construction of Bored Pile Shells

V. V. Kocherzhenko and L. A. Suleymanova

Abstract The paper discusses the methods of construction of bored pile shells, which consist in creating a thin-walled shell of monolithic reinforced concrete by lifting a round core-shaper with a diameter less than the diameter of the well by 250–300 mm and compacting the concrete mixture. In the known methods of construction of bored pile shells, compaction of the concrete mixture is carried out by pressing the concrete into the ground with an elastic shell, vibrating and vacuuming in combination with vibrating. The developed technology for the construction of pile shells provides for the complex use of vibrating, vacuuming and electroosmosis for compacting the concrete mixture when lifting the core in the well. The complex method of compaction of concrete mix allows reducing significantly the time of compaction of concrete, to increase the lifting speed of the core-shaper and, consequently, reducing by 25–30% the production time of the pile shell. This leads to a reduction in labor intensity, cost, at the same time provides reliable stability of concrete in the formed shell, improves the quality of work and increases the friction force of the soil on the side surface of the pile shell. To implement the proposed method for the construction of bored pile shells, a device was developed, the design of which provides for the use of a complex method of compacting the concrete mixture when forming the pile shell.

Keywords Pile Shell · Vibrating · Vacuuming · Electroosmosis · Concrete · Device · Bored pile

1 Introduction

A significant place in the foundation construction is occupied by pile technologies, which provide for the use of pre-made piles with subsequent driving them under buildings—driven piles and technologies for making piles in the design position on the construction site—bored piles. Bored piles are used in construction near existing

V. V. Kocherzhenko · L. A. Suleymanova (✉)
Belgorod State Technological University Named After V.G. Shukhov, Belgorod, Russia

S. V. Klyuev (ed.), *Digital Technologies in Construction Engineering*,
Lecture Notes in Civil Engineering 173, https://doi.org/10.1007/978-3-030-81289-8_19

buildings, in difficult engineering and geological conditions and under significant loads on the foundation [1–3].

The load-bearing capacity of piles on the material is usually several times greater than on the ground. Therefore, to reduce the consumption of materials for the production of piles, the designs of pile shells, both prefabricated and bored, were developed [4–6].

The aim of this paper is to improve the technology of construction of bored pile shells and to develop a device that allows this technology to be implemented.

To achieve this aim, the following tasks were solved:

- to provide for measures to eliminate the shortcomings of existing technologies for the construction of bored piles;
- to reduce the production time of pile shells by reducing the compaction time of the concrete mixture in the shell and increasing the lifting speed of the core;
- to increase the stability of concrete in the shell wall by applying a complex compaction of the concrete mixture;
- to increase the friction force on the side surface of the pile shells by reducing the moisture content of the adjacent soil;
- to reduce the complexity of the construction of pile shells;
- to improve the manufacturing quality of bored pile shells.

2 Methods and Materials

As it follows from the analytical review, all known technologies for the construction of bored pile shells provide for the formation of wells in the ground with a diameter of 0.8–1.2 m, the installation of a reinforcement frame, the immersion of a hollow core of the shaper in the center of the well with a diameter of 0.25–0.35 m less than the diameter of the well, the supply of concrete mix and the lifting of the core.

According to the USSR Patent for invention No. 649789. 1979, the method of constructing pile shells provides for the attachment of an elastic shell on the outer surface of the core-shaper. Before removing the core-shaper from the well, a liquid or gas is fed into the gap between the core and the elastic shell under pressure, thus compacting the concrete mixture and pressing it into the walls of the well [7, 8]. The essence of the technology is shown in Fig. 1a.

In addition, for the construction of the pile shell, a technology and device were developed that provides for the location of a vibrator inside the core-shaper, which is activated when the core is lifted from a well filled with a concrete mixture [9]. As a result, the concrete mixture is compacted along the perimeter of the core; the device diagram is shown in Fig. 1b.

In the Russian Federation Patent for invention No. 2006551. 1994, a method for constructing bored pile shells using a tubular core with perforated walls embedded in the ground is established. The core is punched through an annular well and filled

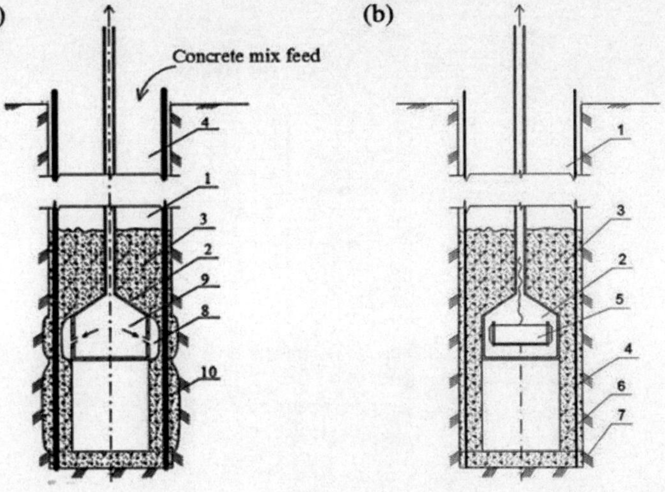

Fig. 1 Construction of a bored pile shell: **a** using an elastic shell and pressing the walls of the well into the ground; **b** using vibrating concrete mixture; 1—well; 2—core-shaper; 3—concrete mixture; 4—reinforcement frame; 5—vibrator installed in the core; 6—bored pile shell; 7—adjacent soil; 8—elastic shell; 9—concrete mixture pressed into the ground of the well wall; 10—compressed gas

with a concrete mixture, after which the radial pressing of the concrete mixture is performed with pressing it into the walls of the well [10].

Also the technology of constructing bored pile shells using vibrating and vacuuming the concrete mixture of the shell when lifting the core, described in the Patent of the Russian Federation for Invention No. 2739531. 2020 is known.

The results of the review of existing technologies and installations for their implementation allowed identifying the main shortcomings of the known technologies for the construction of bored pile shells:

- ensuring the insufficient density of the concrete mixture by known methods of its compaction;
- low resistance of the shell walls;
- insufficient contact of the concrete shell with the ground and, as a result, low ground friction force on the side surface of the pile;
- low speed of lifting the core-shaper from the well and a long time of manufacturing the pile shell.

3 Results and Discussion

The authors developed a device for the construction of pile shells, which allows eliminating the above disadvantages. The scheme of the developed device is shown in Fig. 2.

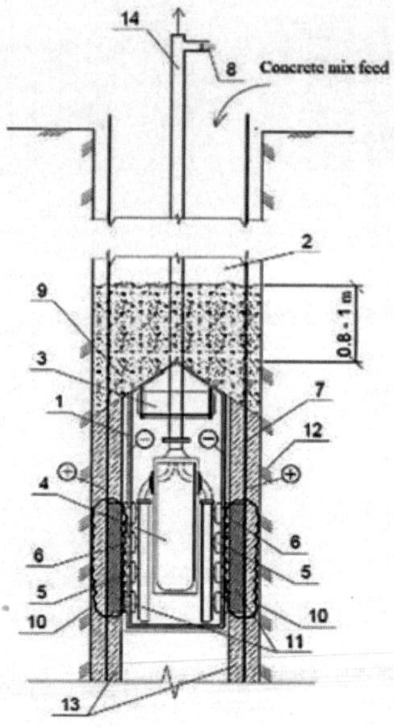

Fig. 2 Diagram of the device for the construction of a bored pile shell: 1—the core-forming body; 2—well; 3—vibrator installed in the core body; 4—container for collecting water removed from the concrete mixture; 5—vacuum shields; 6—holes in the core body; 7—reinforcement frame; 8—electric cable and hoses from the vacuum pump to the shield; 9—concrete mixture; 10—area of the shell concrete exposed only to vibration and vacuuming; 11—shell concrete exposed to vibration, vacuuming and electroosmosis; 12—adjacent soil; 13—concrete pile shell; 14—pipe with a diameter of 70 mm for lowering and feeding the core and passing electric cables and hoses

The design of the developed device (Fig. 4) provides for a circular hollow core-shaper (1) with a diameter of 250–300 mm less than the diameter of the well (2), located at the bottom of the well. At the bottom of the hollow core (1) there is a vibrator (3), vacuum shields (5) and a container for collecting water (4) removed from the concrete pile shell (13). A metal tube with a diameter of 70 mm (14) is rigidly connected to the core from above, inside which electric cables and hoses (8) are passed, connected to a vibrator (3) and to vacuum shields (5). In the core body, opposite the vacuum shields, there are holes for the passage of water removed from the concrete (6).

The diagram of the developed device (Fig. 3) shows the area of concrete of the pile shell exposed only to vibration and vacuuming (10) and the area of concrete exposed to vibration, vacuuming and electroosmosis (11). To create the effect of electroosmosis in the concrete of the pile, a positive potential of direct electric current with a force of 3–15 A/m^2 of the pile surface is applied to the reinforcement frame installed in the well (7), and a negative pole is applied to the pile body [11].

For a comparative analysis of the effectiveness of the developed technology for the construction of pile shells with conventional full-body packed piles, their load-bearing capacity is calculated and the dependence of the specific consumption

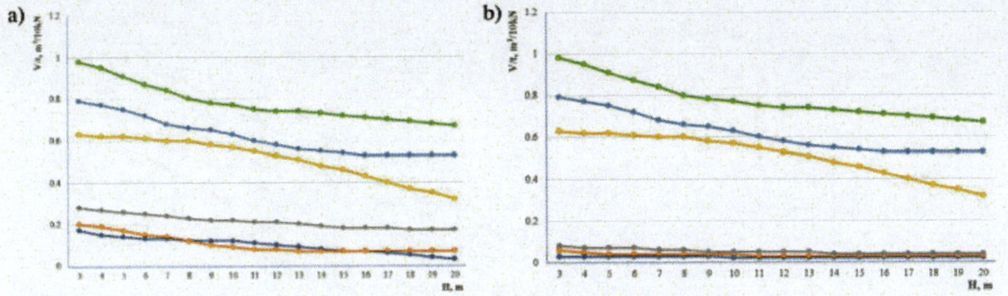

Fig. 3 Dependence of the specific consumption of reinforced concrete of pile shells on the depth of immersion: **a** in clay soils; **b** in sandy soils; ▬—for shell piles with a diameter of 1.0 m; ▬—for shell piles with a diameter of 1.2 m with a shell thickness of 0.25 m; ▬—for shell piles with a diameter of 1.5 m with a shell thickness of 0.25 m; ▬—for a solid pile with a diameter of 1.0 m; ▬—for a solid pile with a diameter of 1.2 m; ▬—for solid piles with a diameter of 1.5 m

of reinforced concrete on the depth is established. At the same time, hanging piles with a diameter of 1000, 1200 and 1500 mm with a shell thickness of 250, 300 and 350 mm, made in clay and sandy (medium size and medium density) soils, are considered.

In the calculations, the resistance under the lower end of the pile is assumed to be the same for the entire pile. The obtained results of the comparative analysis for clay and sandy soils are presented in Fig. 3.

The results obtained allow concluding that the use of the developed structural and technological solutions for pile shells erected with the help of a core-shaper in clay soils reduces the consumption of reinforced concrete in comparison with solid piles by 60–65%; in sandy soils (medium size and medium density)—up to 75%.

The authors established the effect of a complex action (vibration, vacuuming and electroosmosis) on the friction force of the soil on the side surface of the pile shell. Figure 4 shows the dependence of the specific surfacc of the specific friction force on the side surface of piles with a diameter of 1.0, 1.2 and 1.5 m, depending on the depth of immersion in clay soils with different methods of impact on concrete and adjacent soil.

The obtained data show that the combined effect on the concrete and the adjacent soil increases the friction force on the side surface of the pile in the clay soil by 25–27%.

The authors studied the influence of various methods of impact on freshly laid concrete on the duration of the impact of the pile shell, that is, the rate of lifting of the core-shaper along the axis of the well. At the same time, piles of three diameters of 1.0, 1.2 and 1.5 m and a thickness of 250, 300 and 350 mm are also considered. The results obtained are presented in Table 1.

Based on the obtained data, the time for the construction of the pile shell using the developed technology and device, which allows a complex effect on the freshly laid concrete and the adjacent soil, is reduced by 15–20% compared to the use of vibration alone.

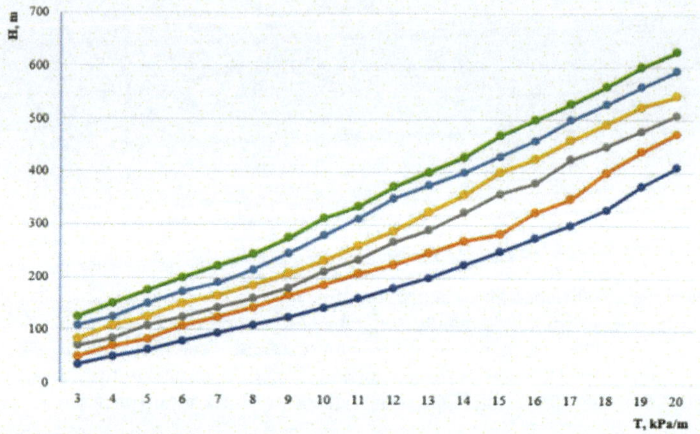

Fig. 4 Dependence of the specific friction force on the side surface of the pile shell in clay soils: —for pile shells with a diameter of 1.0 m when exposed to concrete by vibration; —for pile shells with a diameter of 1.2 m when exposed to concrete by vibration; —for pile shells with a diameter of 1.5 m when exposed to concrete by vibration; —for pile shells with a diameter of 1.0 m with a complex impact; —for pile shells with a diameter of 1.2 m under complex impact; —for pile shells with a diameter of 1.5 m with a complex impact

Table 1 Compaction time of freshly laid concrete under various methods of impact during the formation of a pile shell of different diameter and thickness of the shell

Methods of processing freshly laid concrete into a pile	Pile diameter, mm		
	1000	1200	1500
	Concrete mix processing time (min) at pile thickness (mm)		
	200	250	300
Vibrating	20	25	30
Vibrating and vacuuming	15	20	25
Vibrating, vacuuming and electroosmosis	12	17	22

4 Conclusion

Thus, the device and technology of construction of bored pile shells, based on the use of a complex method of compaction of the concrete mixture, the use of which allows:

- increasing the lifting speed of the core-shaper from 6–7 cm/min to 10–12 cm/min, which reduces the duration of the construction of the bored pile shell by 25–30%;
- increasing the stability of the newly laid concrete mix in the vertical shell by increasing the structural strength of the concrete to 0.4–0.45 MPa and reducing the plastic shrinkage of the concrete;

- reducing the labor intensity of the construction of the bored pile shell and their cost by 20–25%.
- improving the quality of work in the construction of pile shells;
- reducing the consumption of reinforced concrete for the production of piles in comparison with solid bored piles in clay soils by 60–65%, in sandy soils—by 70–75%;
- increasing the load-bearing capacity of pile shells in clay soil by 25%;
- reducing the time for the construction of the pile shell by 15–20%;
- increasing the ground friction force on the side surface of the pile shell.

Acknowledgements This work was realized in the framework of the Program of flagship university development on the base of the Belgorod State Technological University named after V.G. Shukhov. The work was realized using equipment of High Technology Center at BSTU named after V.G. Shukhov.

References

1. Mangushev R, Sotnikov S, Osokin A (2020) Modern technologies of foundation building in the conditions of weak soils of St. Petersburg. E3S Web Conferences. EDP Sci 164:02018
2. Pei HR (2014) Research on foundation stability of building based on foundation bearing capacity. Appl Mech Mater Trans Tech Publications Ltd. 624:661–664
3. Liang RY, Liu HL, Kong GQ (2012) Construction and analysis of cast-in-place tubular piles. American Society of Civil Engineers, GeoCongress
4. Instrumented Large Diameter Bored Piles (2008) Deep foundations on bored and auger piles —BAP V, pp 225–240
5. Leifer D (2015) Advanced building technologies for sustainability. Construction Economics and Building. University of Technology, Sydney (UTS) 12(4):91–92
6. Busch P, Grabe J, Gerressen FW (2013) Influence of the installation process of full displacement bored piles on the subsoil, pp 167–174. Baltic Piling. CRC Press
7. Prekop L (2017) Verification of the vertical bearing capacity of a reinforced concrete pile. Procedia Eng 190:536–539
8. Kocherzhenko VV, Suleymanova LA (2020) Improving the technology of erection of monolithic bored reinforced concrete shell piles. Constr Prod 1:104–108
9. Stepanov M, Dzhabrailova K, Rybak G (2019) The influence of single-layer and double-layer bases vertical soil pressing on the bearing capacity of piles. PNRPU Constr Archit Bull 10 (1):5–16
10. Saenko YV, Nevzorov AL (2015) Deformation and strength characteristics of piling foundation soil. Gornyi Journal 17–22
11. Kocherzhenko VV, Suleymanova LA, Kolesnikov MS (2020) Innovative structural and technological developments of pile foundation engineering. IOP Conf Series: Mater Sci Eng 945(1):012035

Calculation Time Traffic Loading Particles Prior to Reacting with Bi-Directional Rotational Impact

K. A. Yudin and A. N. Degtyar

Abstract Consideration of the mixer with bi-directional rotational action on the material is continued. The rotation of the spherical mixing chamber is carried out by means of a drive using a belt, chain and bevel gears. The mixing of the material in the chamber occurs relative to two mutually perpendicular axes. The resulting complex spatial motion of material particles can be controlled by a frequency converter and the selection of appropriate elements of rotation gears. Preliminary experimental studies have been carried out on the implemented laboratory facility. Earlier, a digital model of the proposed device was created using the well-known graphic products NX and Solid Works, which involves checking the mixer design for patent purity. A laboratory setup has been created. An applied problem is solved mathematically—calculating the time of movement of material particles before colliding with each other. This is one of the tasks describing the interaction of the particles of the mixed material in the mixing chamber of an innovative mixer. The need for regression analysis with the selection of output characteristics, input factors and intervals of their variation is indicated. The analysis of the behavior of particles of a mixed material, implemented using the 3Ds MAX package, is presented. To visualize the behavior of the particles of the mixed material, animation with the indication of time intervals is used. Some screenshots are presented. Blizzard mode shows the movement of spherical and capsule-shaped particles.

Keywords Trajectory of material motion inside the mixing chamber · Mixer · Digital model · Rotation around two the axes

K.A. Yudin · A. N. Degtyar (✉)
Belgorod State Technological University Named After V G Shukhov, 46, Kostukova str, Belgorod 308012, Russia
e-mail: andrey-dandr@yandex.ru

S. V. Klyuev (ed.), *Digital Technologies in Construction Engineering*,
Lecture Notes in Civil Engineering 173, https://doi.org/10.1007/978-3-030-81289-8_20

1 Introduction

Mixing equipment firmly holds its place, being the necessary tools in various areas of the building materials industry, and general construction.

Manufacturers should actively apply various innovative proposals, one of which is a batch mixer, which implements a bi-directional rotational effect on the material in a mixing spherical chamber [1, 2].

The choice of the spherical shape of the mixing chamber is due to the need to increase the intensity of mixing materials, eliminate stagnant loading zones, etc. (Fig. 1).

Bi-directional rotational action on the mixed material in a mixer with a spherical mixing chamber is realized by rotating the carrier with the chamber about the horizontal axis and the chamber itself about the second axis perpendicular to the first.

The operation of a batch mixer involves the optimization of the working cycle, which begins with loading the mixing chamber through the hatch, the implementation of the process of intensive mixing, and then—stopping the chamber, fixing it and subsequent unloading [3]. The design features of the proposed mixer contribute to the appearance of a gyroscopic effect.

Let us consider the mathematical relationships for the interaction of material particles in the mixer. We solve the problem of discretization of the process of interaction of particles of a mixed material in a batch mixer, which implements a bi-directional rotational action on the material [4–7].

2 Methods

2.1 *Calculation of the Time of Movement of Material Particles Before Colliding with Each Other*

Let at the initial moment of time t = 0. The radius vectors $\bar{r}_{i0}$ and $\bar{r}_{j0}$ of the centers of mass of the $\bar{i}$-th $\bar{j}$-th grinding bodies and the vectors of the velocities of the centers of mass $\bar{v}_{i0}$ and $\bar{v}_{j0}$ of these particles are known. We will assume that the

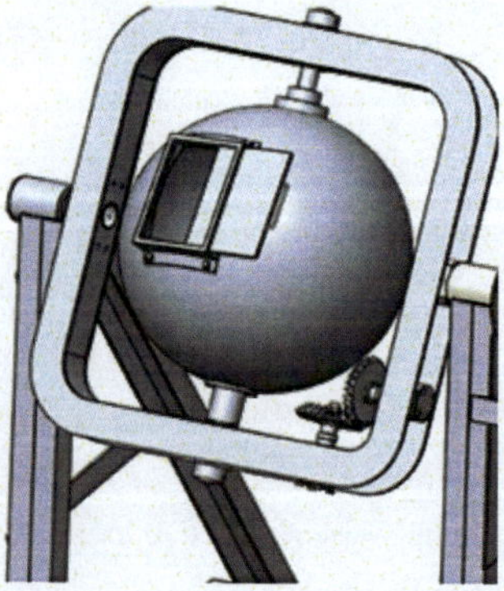

Fig. 1 Spherical rotary mixing chamber

particles move only under the influence of gravity. Writing down the second law of dynamics for each particle and integrating it twice, we get:

$$\begin{aligned} x_i &= x_{io} + v_{xi0}t \qquad & x_j &= x_{j0} + v_{xj0}t \\ y_i &= y_{i0} + v_{yi0}t \qquad & y_j &= y_{j0} + v_{yj0}t \\ z_i &= z_{i0} + v_{zi0}t - \tfrac{gt^2}{2} \qquad & z_j &= z_{j0} + v_{zj0}t - \tfrac{gt^2}{2} \end{aligned} \tag{1}$$

At the moment when the particles touch each other t_{ijs}, their coordinates must satisfy the equation

$$(x_i - x_j)^2 + \left(y_i - y_j\right)^2 + (z_i - z_j)^2 = (2r_s)^2 \tag{2}$$

Substitute (1) into (2) to determine t_{ijs}

$$\begin{aligned} &(x_{io} + v_{xi0}t_{ijs} - x_{j0} - v_{xj0}t_{ijs})^2 + \left(y_{i0} + v_{yi0}t_{ijs} - y_{j0} - v_{yj0}t_{ijs}\right)^2 \\ &\quad + \ (z_{i0} + + v_{zi0}t_{ijs} - \frac{g}{2}t_{ijs}^2 - z_{j0} - v_{zj0}t_{ijs} + \frac{g}{2}t_{ijs}^2)^2 = (2r_s)^2. \end{aligned}$$

Let us group the terms of the equations

$$\begin{aligned} &(x_{io} - x_{j0} + \left[v_{xi0} - v_{yj0}\right]t_{ijs})^2 + \left(y_{i0} - y_{j0} + \left[v_{yi0} - v_{yj0}\right]t_{ijs}\right)^2 \\ &\quad + (z_{io} - z_{j0} + \left[v_{zi0} - v_{zj0}\right]t_{ijs})^2 = (2r_s)^2. \end{aligned}$$

$$\begin{aligned} &(x_{io} - x_{j0})^2 + 2(x_{io} - x_{j0})(v_{xi0} - v_{yj0})t_{ijs} + (v_{xi0} - v_{yj0})^2 t_{ijs}^2 + \left(y_{i0} - y_{j0}\right)^2 \\ &\quad + \ 2\left(y_{i0} - y_{j0}\right)(v_{yi0} - v_{yj0})t_{ijs} + (v_{yi0} - v_{yj0})^2 t_{ijs}^2 \\ &\quad + (z_{io} - z_{j0})^2 + 2(z_{io} - z_{j0})(v_{zi0} - v_{zj0})t_{ijs} + (v_{zi0} - v_{zj0})^2 t_{ijs}^2 = (2r_s)^2 \end{aligned}$$

$$a_{1s}t_{ijs}^2 + 2a_{2s}t_{ijs} + a_{3s} = 0, \tag{3}$$

where $a_{1s} = (v_{xi0} - v_{xj0})^2 + (v_{yi0} - v_{yj0})^2 + (v_{zi0} - v_{zj0})^2;$

$$\begin{aligned} a_{2s} &= (x_{io} - x_{j0})(v_{xi0} - v_{xj0}) + \left(y_{i0} - y_{j0}\right)(v_{yi0} - v_{yj0}) + (z_{io} - z_{j0})(v_{zi0} - v_{zj0}); \\ a_{3s} &= (x_{io} - x_{j0})^2 + \left(y_{i0} - y_{j0}\right)^2 + (z_{io} - z_{j0})^2 - (2r_s)^2. \end{aligned} \tag{4}$$

The solution exists if the discriminant of the equation ≥ 0.

$$(2a_{2s})^2 - 4a_{1s}a_{3s} \geq 0,$$
$$a_{2s}^2 - a_{1s}a_{3s} > 0.$$

Let us introduce the notation

$$\Delta x = x_{io} - x_{j0} \quad \Delta y = y_{i0} - y_{j0} \quad \Delta z = z_{io} - z_{j0}$$
$$\Delta v_x = v_{xi0} - v_{xj0} \quad \Delta v_y = v_{yi0} - v_{yj0} \quad \Delta v_z = v_{zi0} - v_{zj0}$$

Formulas (4) will be written as

$$\begin{aligned} a_{1s} &= \Delta v_x^2 + \Delta v_y^2 + \Delta v_z^2 \\ a_{2s} &= \Delta x \Delta v_x + \Delta y \Delta v_y + \Delta z \Delta v_z \\ a_{3s} &= \Delta x^2 + \Delta y^2 + \Delta z^2 - 4r_s^2. \end{aligned} \tag{5}$$

In this notation, the discriminant is equal to

$$D = (\Delta x \Delta v_x + \Delta y \Delta v_y + \Delta z \Delta v_z)^2 - (\Delta v_x^2 + \Delta v_y^2 + \Delta v_z^2)(\Delta x^2 + \Delta y^2 + \Delta z^2 - 4r_s^2). \tag{6}$$

If $D < 0$ then Eq. (6) s no solutions and the interacting particles in the process of moving along parabolic trajectories do not touch. In this case, it is sufficient to assign t_{ijs} to any sufficiently large value so that when determining the minimum time of movement to the nearest collision, the time does not enter into the selected one.

If $D = 0$ then the equation has one root. This means that the particles will touch each other, but there will be no collision.

If $D > 0$ then Eq. (3) has two roots.

$$(t_{ijs})_{1,2} = \frac{-a_{2s} \pm \sqrt{a_{2s}^2 - a_{1s}a_{3s}}}{a_{1s}}. \tag{7}$$

The lesser of the two roots will be the desired time.

Equation (5) implies

$$a_{1s} < 0.$$

Value $a_{3s} > 0$ as $\Delta x^2 + \Delta y^2 + \Delta z^2 > 4r^2_s$ at the initial moment of time.

Thus, $|a_{2s}| \geq \sqrt{a_{2s}^2 - a_{1s}a_{3s}}$.

Equation (5) implies $a_{2s} = (\Delta\bar{r}, \Delta\bar{v})$, where $\Delta\bar{r} = \bar{r}_i - \bar{r}_j$, and $\Delta\bar{v} = \bar{v}_i - \bar{v}_j$ (Fig. 2).

Since the angle α is obtuse, so $a_{2s} < 0$. It means that in the Eq. (7)—a_{2s} is positive.

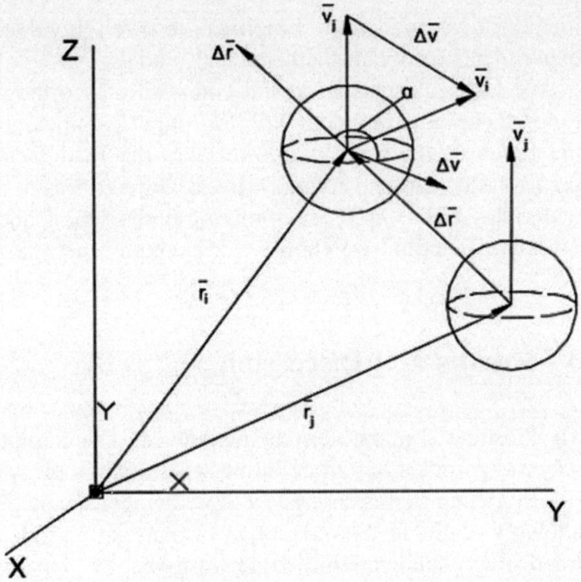

Fig. 2 Interaction of two spherical particles

Thus, the root of Eq. (7) is positive and minimal

$$t_{ijs} = \frac{-a_{2s} - \sqrt{a_{2s}^2 - a_{1s}a_{3s}}}{a_{1s}}.$$

Further, from the arrays of time intervals and (t_{ib}), the minimum is selected. If one of the time intervals for the impact of the particle on the housing (t_{ib}) is found to be minimal, then the number of the particle that hits the housing of the mixing chamber is determined. If one of the time intervals (t_{ijs}) turned out to be the minimum, then the numbers of particles that collide during the movement of the load are determined.

In the future, we turn to the collision of particles with each other and particles with the housing of the mixing chamber.

The proposed mathematical apparatus is necessary for a comprehensive regression analysis [8].

The purpose of the regression analysis is to assess the functional dependence of the average values of the output characteristics on the factor attributes (input factors), expressed as a regression equation. The decisive step in regression analysis is to determine the type of function, since it depends on whether the algebraic equation correctly reflects the essence of the relationship between phenomena. As input factors, one can consider the size of the particles of the mixed material, the filling factor of the mixing chamber, the frequency of rotation of the chamber, the mixing time, etc.

The technical focus of the work—the creation of a patent-friendly innovative device for mixing materials—involves the use of methods of physical and mathematical modeling. Empirical research involves various kinds of machine experiments to obtain mathematical models.

We have chosen the central compositional orthogonal plan (CCOP) of a fractional factorial experiment with four input parameters. As the input parameters and the limits of their variation, we take: the load factor of the mixing chamber—0.25 ÷ 0.3; camera rotation frequency—3 ÷ 8 c^{-1}; particle size of the feed material—0.01 ÷ 0.02 м; t—mixing time of the loaded mixture, 30 ÷ 90 c. As an objective function, we choose, for example, the mixing quality of the mixture.

3 Results and Discussion

To illustrate the problem being solved, calculating the time of movement of material particles before colliding with each other, we use the 3Ds MAX package [9]. Modeling in this case takes place using a special Particle Systems mode, which allows visualizing the movement of material particles inside the mixing chamber. Interesting visualization in Blizzard mode.

We choose the duration of the animation cycle, for example, 50 s. Traditionally, we use standard spherical particles (Spheres) with variations in the parameters reflected in the tabs on the right. The behavior of particles is tracked at specific points in time. The screenshot (Fig. 3) shows the behavior of the particles of the loaded material after a certain time interval. (t = 30 s). The loaded material particles are "emitted" from the rectangular emitter. Shown in Fig. 1 the mixing chamber hatch is made of a rectangular shape. With the help of 3Ds MAX, we can vary the dimensions of the hatch and its location relative to the original plane (localization of the emitter).

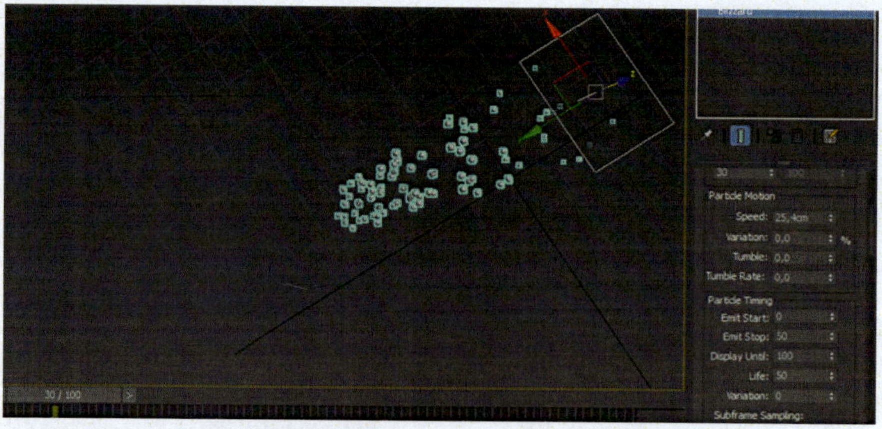

Fig. 3 Behavior of particles of the loaded material at t = 30 s (spheres)

Fig. 4 Behavior of particles of the loaded material at t = 30, 50 s (capsules)

In addition, it is interesting to trace the dynamics of the behavior of elongated particles—the interaction of the so-called Capsules (Fig. 4). In this case, the interacting particles will have large contact zones.

You can observe a decrease in the time before the particles interact. Over time, multiple particle contacts occur. Given the same parameters (number of particles, speed, etc.), at the considered intervals of the animation cycle from 0 to 30 s, from 30 to 50 s, an increase in the number of interactions by 22% is noticeable.

The information obtained in the course of virtual experiments (velocity of movement of material particles, kinetic and potential energy, trajectories of particles, etc.) will optimize the characteristics of the proposed mixer.

4 Conclusion

To confirm the efficiency of the proposed mixer, which implements a bi-directional rotational action on the material, an integrated approach is required. This presupposes the use of an original laboratory setup, carrying out experiments on it and further analysis of the results. Experiments and calculations must be continued with varying parameters to obtain an optimal mixer design suitable for industrial applications.

It is possible to use such mixers in lines for the production of gas silicate products, paint and varnish industry, agriculture and similar low-tonnage industries [10].

Acknowledgements The reported study was funded by the Russian Humanitarian Science Foundation (RHSF 16-16-70006) and RFBR, according to the research project No. 16-38-60146 mol_a_dk.

References

1. Yudin KA, Degtyar AN, Kharin NP (2017) Modelling of bi-directional rotary impact on the material in a spherical mixing chamber. Bulletin of BSTU named after V. G. Shukhov, vol 5, pp 126–132
2. Yudin KA, Degtyar AN, Bogdanov VS, Stovpenko AS (2017) Movement of materials in gyroscopic-type mixer. In: Proceedings of the international conference "actual issues of mechanical engineering" 2017 (AIME 2017), pp 909–913. Novosibirsk State Technical University, Novosibirsk. http://www.atlantis-press.com/proceedings/aime-17/articles
3. Yudin KA, Semikopenko IA, Fadin YM (2017) A device for mixing materials Patent RF 17444
4. Kirkach NF, Balasanian RA (1991) Calculation and design of machine parts. Foundation, Kharkov
5. United States Patent 4326428 (1982) Two degree of freedom rate gyroscope. Date views 27.04. http://www.freepatentsonline.com/4326428.html
6. Uicker JJ, Pennock GR, Shigley JE (2009) Theory of machines and mechanisms. University Press, Oxford
7. Khakhalev PA, Vorobyov ND, Bogdanov VS (2017) The study of the sedimentary core grinding load in a ball drum mill. News of higher educational institutions. Constr Sci Theor J 2:109–115
8. Trusov PV (2012) Introduction to mathematical modeling: a tutorial. Logos, Moscow
9. Autodesk.ru›products/3ds-max/overview
10. Gorlov YP (1989) Technology of heat-insulating and acoustic materials and products. Higher School, Moscow

Mistakes in the Construction of Objects in Constrained Conditions

N. S. Sokolov

Abstract Construction of facilities in constrained conditions requires a special approach associated with the need to develop and implement measures to ensure accident-free operation of buildings of the surrounding development within the geotechnical influence. Often builders neglect the influence of the technology of construction of a new object on the possible negative consequences (cracks appeared on the facades due to uneven precipitation, rolls, etc.) of the buildings in operation. Until now, the concept of "minimum price" is still exaggerated when erecting a part of the building below the zero mark. This approach completely ignores the concept of "technical integrity". At the same time, builders go to any tricks to reduce the cost. Such "irrational" way of constructing may ultimately lead to a significant increase in costs of base part of the building and, as a rule, to increase the duration of erection (approval of a new project replacing it with another geotechnical technology, the new building expertise). This paper examines one negative case from the geotechnical practice of building a 16-storey residential building next to an existing five-storey residential building.

Keywords Geotechnical engineering · Irregular rainfall · Ground anchor · Fence of the pit · Electric discharge technology · Edt piles

1 Introduction

The construction of objects in constrained conditions is the most problematic in terms of the technology of construction production of parts below the zero mark. When designing and constructing, a set of measures should be developed to ensure

N. S. Sokolov (✉)
Chuvash State University Named After I. N. Ulyanov, Moskovskiy prosp., 15, Cheboksary 428015, Russian Federation

N. S. Sokolov
NPF (LLC SPC) "FORST", Ul. Kalinina, 109a, Cheboksary 428000, Russian Federation

S. V. Klyuev (ed.), *Digital Technologies in Construction Engineering*, Lecture Notes in Civil Engineering 173, https://doi.org/10.1007/978-3-030-81289-8_21

accident-free operation of existing buildings. Modern geotechnical technologies allow solving successfully such problems [1–6]. This paper describes a negative example of the construction of a multi-storey residential building next to an existing house.

2 Methods and Materials

Modern geotechnical technologies make it possible to produce drill-injection piles with increased load-bearing capacity. The discharge-pulse technology of the arrangement of piles (EDT piles) is the most adapted for achieving these goals.

The most interesting is the arrangement of EDT piles with multiple widenings. Due to the arrangement of widenings along the pile trunk and under its foot, it is possible to arrange a buried reinforced concrete structure with the possibility of regulating the load-bearing capacity, both on the ground and on the body.

A pile with multiple widenings works as a reinforced concrete buried structure with a number of supports corresponding to the number of widenings. At the initial stage of loading, the upper broadening comes into operation. As the load increases, the underlying extensions are gradually activated, with each extension acting as an additional support.

The practice of manufacturing such piles has shown their high efficiency. The bearing capacity of piles on the ground with one widening is 2.0–2.5 times, and with two–3.0–3.5 times higher than that of piles made without widening.

The advantages of EDT drill-injection piles with multiple widenings are: (1) ease of implementation (added one easy operation); (2) possibility to determine the exact location of the arrangement of widening (determined by the care of fine-grained concrete mixture); (3) the possibility of the arrangement of the required number of widenings for the calculation of the bearing capacity along the length of the pile; (4) to minimize sludge technology; (5) maximum compared to any other technology of arranging drill piles enhance the bearing capacity of piles, as on the ground, and material [7–13].

3 Results and Discussion

The object of the new construction consists of one single monolithic block, which has a rectangular shape in plan. The aboveground part of the construction site consists of 16 floors and a technical floor. The structural scheme is a reinforced concrete monolithic non-nogging frame with self-supporting external walls made of foam concrete blocks supported by inter-floor overlaps. Overlaps and coverings—monolithic reinforced concrete. The foundation solution is provided as a monolithic reinforced concrete foundation plate. The depth of the pit in various areas ranges from 9.3 to 9.6 m. The absolute mark of the pit bottom is 175.30 m of the Baltic System (BS). The pit width in the plan is 27.0 m, and its length is 40.0 m.

In accordance with clause 9.36 of SP 22.13330.2011 “Updated version of SNiP 2.02.01-83x “Foundations of buildings and structures”, the radius of the zone of influence of new construction is equal to $r_{зв}$ = 4 Nc for a pit being developed with a fence made of steel pipes. Thus, the zone of influence was 38.4 m. In the specified zone there is a five-storey two-entrance residential building. The category of the technical condition of the investigated residential building according to the reports on the results of the technical survey is satisfactory. The building is frameless brick with longitudinal load-bearing walls. The spatial rigidity of the building is provided by hard disks of inter-floor overlaps and coverings, blocks of stairwells. The building is five-storey with a basement. In the plan, the building has dimensions of 54. 0 × 12. 75 in height of 17.0 m. The foundations for the building are prefabricated tape blocks of FBS with a thickness of 400–500 mm, laid on a monolithic reinforced concrete belt installed on the foundation plates of the FL. The depth of the foundations is 2.79–3.05 m.

The building is frameless brick with longitudinal load-bearing walls. The spatial rigidity of the building is provided by hard disks of inter-floor overlaps and coverings, blocks of stairwells. The building is five-storey with a basement. In the plan, the building has dimensions of 54.0 × 12.75 in height of 17.0 m. The foundations for the building are prefabricated tape blocks of CFB with a thickness of 400–500 mm, laid on a monolithic reinforced concrete belt installed on the foundation slabs of the FS. The depth of the foundations is 2.79–3.05 m. The walls of the building are made of silicate bricks with a thickness of 510.0 mm on a cement-sand mortar. The internal load-bearing walls with a thickness of 380.0 mm are laid out of silicate bricks, and the inter-floor and attic floors are mounted from reinforced concrete multi-hollow slabs with a thickness of 220.0 mm.

In the original version, the fixing of the pit walls is designed in the form of a loosened interlocking pile wall using two tiers of ground anchors “Atlant” (see Fig. 1a). For this case, for the interlocking pile wall, steel pipes with a cross section of 530 × 8 according to GOST 10,704–91 “Straight-seam electric-welded steel pipes. Grade” with a step of 1.0 m, and along the “W” axis, the pipes are arranged in a step of 0.8 m. In addition, in the immediate vicinity of the building, the pipes are filled with heavy concrete. The mark of the top of the interlocking pile wall of pipes is assumed to be variable—from 184.54 to 184.84 m CS. The bottom of the fence pipes along the axes 1, A and 12 is located at the marks from 171.54 to 171.84 m CS with a length of 13.0 m pipes, along the “W” axis the bottom of the fence pipes is located at 166.84 m CS with a length of 18.0 m pipes. Between the fence pipes, a fence is made of boards with a thickness of 40.0 mm.

At the request of the customer-developer, we (FORST Research and Production Company LLC) were instructed to develop an alternative version of the unfastened interlocking pile wall using steel pipes of the same cross-section dimensions, but ground anchors in three tiers made using electric discharge technology (EDT anchors) (see Fig. 1b). At the same time, the estimated depth of sealing of steel pipes was deeper than the original project. In addition, in the areas adjacent to the

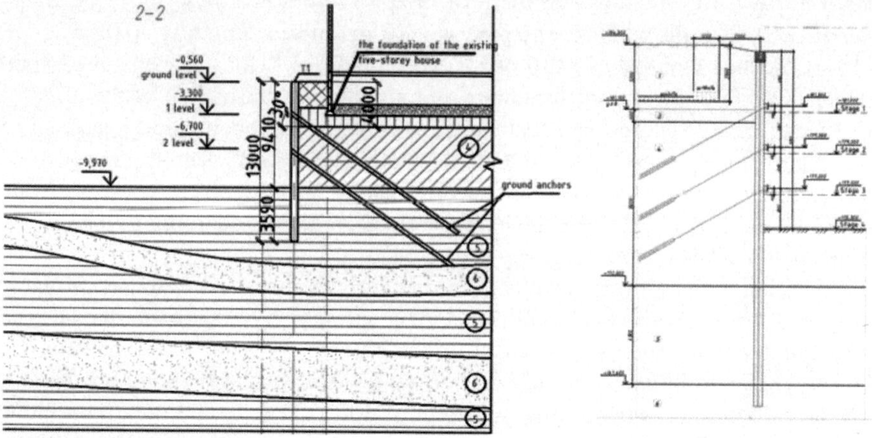

Fig. 1 Variants of loosened pit fences with the help of ground anchors: **a** anchors "Atlant", **b** EDT anchors

existing houses, we designed sheet pile walls made of EDT drill-cut drill-injection piles with a diameter of 350 mm with a monolithic reinforced concrete binding belt on the top of the piles. The projected wall is a solid monolithic buried reinforced concrete structure that would prevent the soil from shedding from under the foot of the foundations of the existing residential building during the excavation of the pit.

The customer, studying both versions of the pit fence, came to a strange conclusion about the need to reduce the cost of the pit fence project. In this case, the meaning of "technical expediency" is completely guaranteed to be ignored, although economic efficiency is acquired. Thus, they ordered the third version of the project of fencing the pit, but without anchoring, using the same as in previous projects, steel pipes and steel shootings, but without the arrangement of solid interlocking pile walls in the areas adjacent to the existing residential building. For the uniform perception of forces from the ground and their transfer to the shootings, a distribution belt of paired channels 50B2 is designed. It is necessary to pay attention to the fact that during the production of works, used interlocking pipes were used as a structural material for the shootings. During the excavation of the pit, they received significant deformations (deflections, crumpling of the cross-section, as well as sinkholes of the soil and asphalt). At the same time, the absence of solid barrier interlocking pile walls in the areas adjacent to the existing residential building during the excavation of the pit led to the soil falling out from under the sole of the foundation. As a result, the existing residential building received instantaneous deformation, as evidenced by the deformation cracks that appeared on the vertical surfaces of the facades, progressing over time. Urgent geotechnical monitoring of vertical movements of sedimentary marks (Figs. 2 and 3) confirmed the worst fears. Part of the residential building on the side of the new construction has sunk and continues to deform (see Fig. 4).

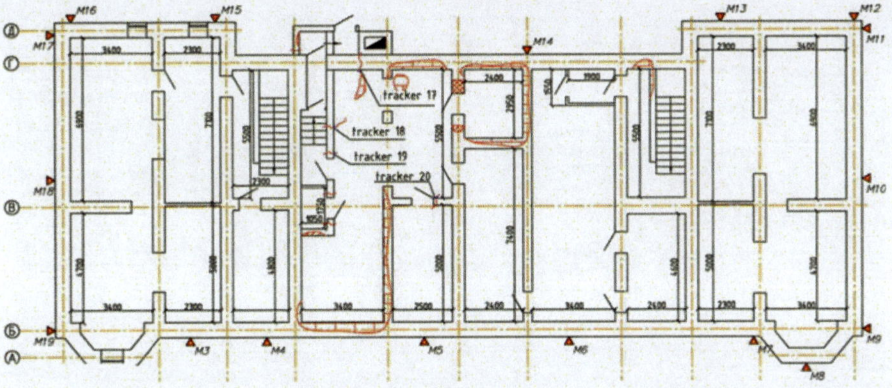

Fig. 2 The plan of the standard floor of the object No. 1

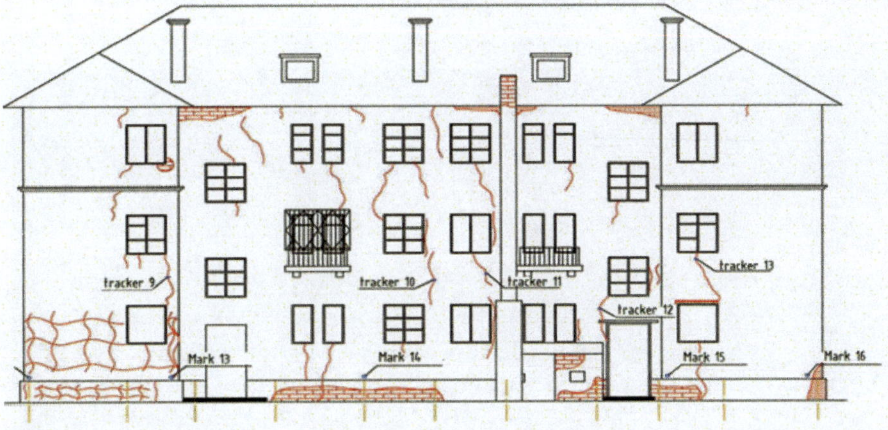

Fig. 3 Facade of object No. 1

The customer urgently decided to develop a project for cementing the deformed part of the base and immediately implement it (see Fig. 4). The results of geotechnical monitoring indicate that even after the geotechnical work on cementing the base, the deformations of the base continued to develop. At the same time, all the maximum permissible deformations have already been exceeded. For example, the most deformed sedimentary mark received a vertical displacement of 52.0 mm (see Fig. 5) at an acceptable value of 20.0 mm.

Descriptions of engineering and geological elements (EGE) according to the technical report on engineering and geological surveys performed at the future construction site are given in Table 1. The construction area belongs to the north-western end of the Volga Upland and is located on the high Oka-Volga right bank. In relation to the Zarechnaya part of the city, the Upland part is raised by 100.0–136.0 m, forming a platform hill, bounded by the Oka and Volga slopes.

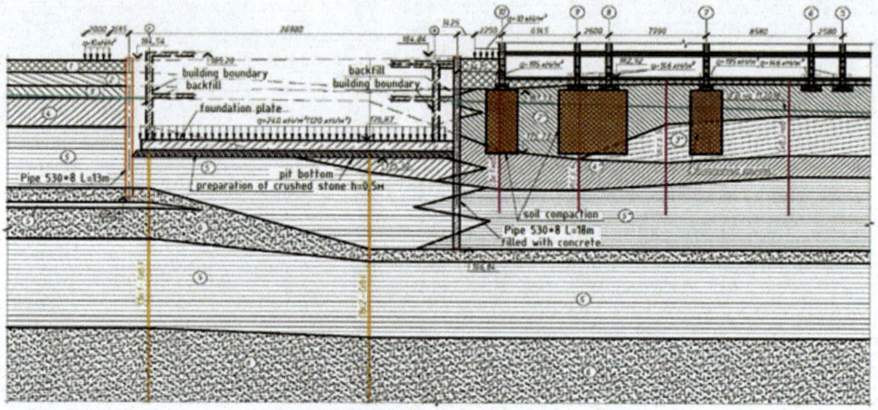

Fig. 4 Vertical binding of the object under construction and cementation fixing of the base of the existing house No. 1 in the engineering-geological section

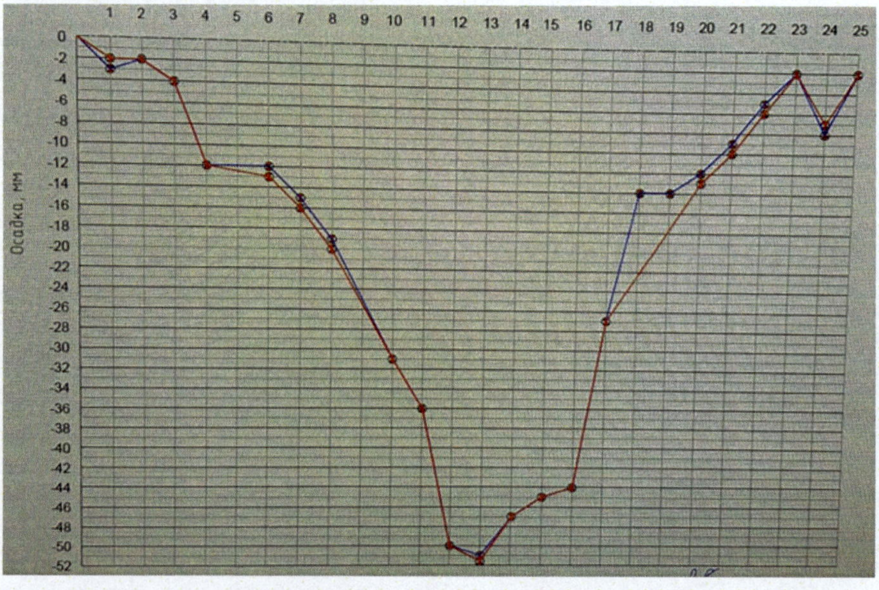

Fig. 5 Graphs of the subsidence deformation marks installed around the existing house No. 1

The terrain of the construction site is planned. The surface level is at 184.3–184.9 m CS (at the mouths of the wells). The planned reference of the workings is made by linear measurements from the solid contours of the situation. The engineering and geological structure of the site up to a depth of 30.0 m is presented: (1) Modern technogenic deposits (tQIV); (2) Middle-upper quaternary deposits of problematic genesis (prQII-III); (3) Middle quaternary lacustrine-alluvial deposits (laQII).

Table 2 shows the descriptions of the EGE with their placement by depth.

Table 1 Geotechnical measures to restore the operational suitability of an existing residential building

№ of projects	Name of geotechnical technology	Measures provided for the protection of the existing building	Carried out activities	Additional geotechnical activities	Results of the work carried out
1	Unfastened retaining wall made of steel pipes and ground anchors "Atlant" in two tiers	no	Project is cancelled	no	
2	Unfastened retaining wall made of steel pipes and ground EDT anchors in three tiers	A solid wall of drill-contact drill-injection EDT piles was designed	Project is cancelled	no	
3	A loose retaining wall made of steel pipes and two shooting belts	A 40 mm thick wooden plank fence is designed	The project was implemented with shootings in one belt	1. Cementation fixing of the base is performed; 2. The foundation of the foundations of the existing house was reinforced	1. Deformations of the house continue; 2. Deformations of the house are stopped

Table 2 Descriptions of EGE by placing them in depth

№ of layer	Age, genesis, description of soils, areal distribution	Power, m
Quaternary system		
Modern deposits (Q_{IV})		
Technogenic deposits		
1	**tQIV- Bulk soil** The loam is brown, dark brown, with layers of sand, road rubble, broken bricks, wood chips. It is filled out in a dry way, classified as a dump of soils; with a non-uniform composition and addition, uneven density and compressibility. Borehole 2 opened the brickwork from a depth of 0.5–3.2 m. Opened by all wells from the surface	0.7–3.2
Undifferentiated Middle-upper Quaternary deposits (*pr*$Q_{II\text{-}III}$)		
2	**pr**$\mathbf{Q_{II\text{-}III}}$ Loess loam is brown, yellowish-brown, with the inclusion of manganese and iron hydroxides, micaceous, sometimes with layers of sandy loam It is opened everywhere under the bulk soil	0.6–2.1

(continued)

Table 2 (continued)

№ of layer	Age, genesis, description of soils, areal distribution	Power, m
Mid-Quaternary eluvial-deluvial deposits (edQ_{II})		
3	**edQ_{II}** The loam is brown, heavily sanded in the roof, layered, with layers of clay	1.4–8.0
Perm system		
Tatar deposits (P_2t)		
4	**P_2t** **Polymictic sand** is brownish-brown, strongly clayey, with layers of clay, in some places with the inclusion of crushed limestone, slightly calcified sandstone. It is opened in the thickness of Tatar clays	opened 0.2–9.2
5	P2t **The clay** is reddish-brown, dark brown, with layers of polymictic sand, with the inclusion of nests and layers of aleurite, aleurolite, sandstone. Opened everywhere in the lower part of the incision	opened 0.5–5.8

4 Conclusion

1. Modern geotechnical methods allow the construction of large and critical structures on weak foundations with minimal subsidence after the commissioning of the object.
2. The most preferred option for each individual site should be chosen based on the feasibility study and the tasks that this design should perform.

References

1. Mangushev RA, Nikiforova NS, Konyushkov VV, Osokin AI (2013) Design and construction of underground structures in open pits. DIA, Moscow
2. Mangushev RA, Veselov AA, Konyushkov VV, Sapin DA (2012) Numerical modeling of technological precipitation of neighboring buildings in the device trench “wall in the ground” II. Bull Civil Eng 5(34):87–98
3. Makovetsky OA, Zuev SS, Khusainov II, Timofeev MA (2014) Ensuring geotechnical safety of the building under construction. Housing Constr 9:34–38
4. Ilichev VA, Konovalov PA, Nikiforova NS, Bulgakov LA (2004) Deformations of the retaining structures upon deep excavations in Moscow. In: Proceedings of fifth international conference on case histories in geotechnical engineering, 3–17 April, pp. 5–24. New York
5. Ilyichev VA, Nikiforova NS, Koreneva EB (2007) Computing the evaluation of deformations of the buildings located near deep foundation tranches. In: Proceedings of the XVIth European conference on soil mechanics and geotechnical engineering. Madrid, Spain, 24–27th September 2007, Geo-technical Engineering in urban Environments, vol 2, pp 581–585. Madrid

6. Nikiforova NS, Vnukov DA (2011) Geotechnical cut-off diaphragms for built-up area protection in urban underground development. In: The pros of the 7th international Sympossium, "Geotechnical aspects of underground construction in soft ground", 16–18 May, 2011, tc28 IS Roma, AGI, 2011.157NIK
7. Nikiforova NS, Vnukov DA (2004) The use of cut off of different types as a protection measure for the existing buildings at the nearby underground pipelines installation. In: Proceedings of international geotechnology conference dedicated to the year of Russia in Kazakhstan, pp 338–342. Almaty, Kazakhstan
8. Petrukhin VP, Shuljatjev OA, Mozgacheva OA (2003) Effect of geotechnical work on settlement of surrounding buildings at underground construction. In: Proceedings of the 13th European conference on soil mechanics and geotechnical engineering, Prague
9. Th. Triantafyllidis, Schafer R (2007) Th. Impact of diaphragm wall construction on the stress state in soft ground and serviceability of adjacent foundations. In: Proceedings of the 14th European conference on soil mechanics and geotechnical engineering, vol 2, pp 683–688. Madrid, Spain
10. Ponomarev AB (2015) Geotechnical monitoring of a house. Housing Constr 9:41–46
11. Sokolov NS, Viktorova SS, Fedorova TG (2014) Piles of increased bearing capacity. In: Materials of the 8th all-Russian (2nd international) conference " New in architecture, design of building structures and reconstruction" (NASKR-2014), pp 411–415. Chuvash state University, Cheboksary
12. Sokolov NS, Rabinow VM (2016) Technology of the piles increased the bearing capacity. Housing Constr 9:11–14
13. Sokolov NS (2016) Technological methods devices bored piles with multiple caps. Housing Constr 10:54–59

Study of the Material Composition of Carbon Black Obtained as a Result of MSW Thermolysis

A. I. Vezentsev, V. S. Sevostianov, A. E. Razdobarin, and R. U. Shamgulov

Abstract This paper presents the results of design and technological developments of resource-saving technologies for processing organic solid municipal waste (MSW), which allow obtaining various types of marketable products. Commercial products were obtained using this technology in the form of: carbon black, liquid hydrocarbon fuel, synthetic hydrocarbon gas. The proposed technology and technical means allow solving a whole range of scientific and technical, research, design, technological and environmental problems. The solutions take into account the conditions of existing production facilities and are aimed at eliminating existing technical and technological limitations in the field of resource and energy saving. The results of experimental studies on the determination of the material composition, structural-morphological and textural characteristics of carbon black, including X-ray phase analysis, laser diffraction, scanning electron microscopy, low-temperature nitrogen adsorption are presented. The results of the energy dispersion analysis indicate the predominance of carbon in the analyzed material, the presence of oxygen, calcium and silicon, and chlorine is also recorded. The product studied in this work was obtained according to the low-temperature thermolysis technology for processing organic solid municipal waste developed and implemented at the company "Ecotrans TC" LLC. Promising areas of application of carbon black obtained during the processing of organic MSW by the method of low-temperature thermolysis are identified.

Keywords Ecology · Solid municipal waste · Carbon black · Material composition · Thermolysis technology · Particle size distribution · Energy dispersive and X-ray phase analysis · Scanning electron microscopy · Low-temperature nitrogen adsorption

---

A. I. Vezentsev · A. E. Razdobarin
Federal State Autonomous Educational Institution of Higher Education «Belgorod National Research University», Pobedy St, 85, Build 12, Belgorod 308015, Russia

V. S. Sevostianov · R. U. Shamgulov (✉)
Belgorod State Technological University Named After V.G. Shukhov, Kostyukov St., 46, Belgorod 308012, Russia

S. V. Klyuev (ed.), *Digital Technologies in Construction Engineering*,
Lecture Notes in Civil Engineering 173, https://doi.org/10.1007/978-3-030-81289-8_22

1 Introduction

The increase in industrial production, as well as the steady increase in energy consumption, continuously increase the anthropogenic and environmental impact on the environment [1].

The main method of solid municipal waste management in Russia is burial [2]. In 2017, the volume of MSW exported to landfill sites amounted to about 50.9 million tons, or 87% of the total volume of MSW exported [3].

The author’s team of researchers of BSTU named after V.G. Shukhov, with the participation of National Research University BelSU and engineering and technical employees of “Ecotrans TC” LLC, conducts scientific and technical developments and the introduction of resource-saving technology and special equipment for low-temperature thermolysis of organic waste into real production [4].

The proposed technology and technical means allow the processing of organic MSW of various morphological and physical composition. The method of low-temperature processing of organic MSW at a temperature of up to 500 °C allows obtaining high-quality products: carbon black, liquid hydrocarbon fuel and synthetic gas.

One of the promising areas of application of carbon black is its use as a sorbent for wastewater treatment.

The most popular industrial sorbent, at present, is absorbent carbon (AC). The consumption of AC in Russia has more than doubled over the past ten years [5].

Product samples were obtained during the thermolysis of organic MSW at a experimental-industrial thermolysis plant [6].

Figure 1 shows the scheme of low-temperature thermolysis with the production of carbon black, liquid hydrocarbon fuel and synthesis gas.

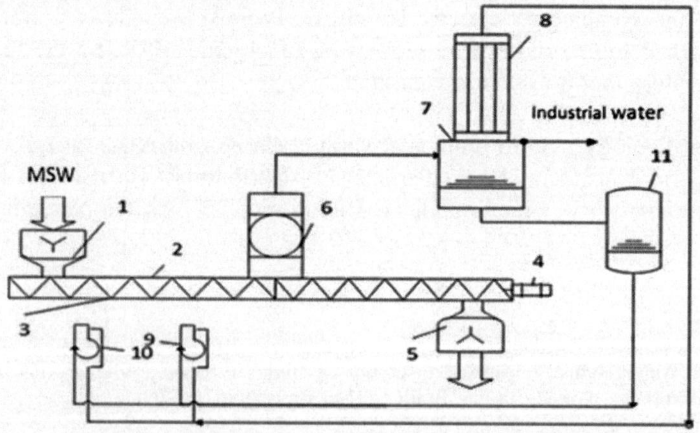

Fig. 1 Scheme of low-temperature thermolysis: 1—feeder; 2—thermolysis reactor; 3—combined conveying body; 4—gear motor; 5—solid product discharge device; 6—filter separator; 7—rectification column; 8—waste gas; 9, 10—gas-liquid burners; 11—liquid hydrocarbon fuel storage tank

The crushed "tails" of MSW or rubber waste are fed into the loading chamber of the feeder 1. Then the mass moves along the reactor 2 with the help of a combined conveying body 3, kinematically connected to the gear motor 4. Under the influence of temperature, the organic mass is decomposed into a solid carbon residue and a vapor-gas mixture. Through the solid product discharge device 5, the solid carbon residue is removed from the reactor. After passing through the filter separator 6, the steam-gas mixture enters the rectification column 7. In the latter, it is divided into industrial water and liquid hydrocarbons. The uncondensed hydrocarbon gas taken from waste gas 8 is burned in gas-liquid burners 9, 10. In the burners, a part of the liquid hydrocarbon fuel taken from the reactor heating tank 11 is also burned.

2 Materials and Methods

Earlier [7], researchers at BSTU named after V.G. Shukhov found that carbon black, a product of thermolysis of MSW, can be recommended as a promising sorbent for the purification of aqueous solutions. However, in the presented and other works, there is no information about the material composition of the obtained material. The purpose of these studies was to determine the chemical, phase, and granulometric compositions, structural, morphological, and textural characteristics of MSW thermolysis products.

2.1 Chemical Composition

During the analysis, an analytical system based on a high-resolution scanning electron microscope Nova NanoSEM was used. The device is equipped with an analytical energy-dispersion spectrometer from EDAX. The analysis was carried out at an accelerating voltage of 30 kV in the immersion mode (immersion in certain artificially formed conditions). The sample was placed in the magnetic field of the objective lens, which reduces aberration and achieves subnanometer resolution. This mode is suitable for the study of non-magnetic samples. The energy-dispersion spectrum of the organic component of the analyzed MSW thermolysis products is shown in Fig. 2.

The chemical composition of the organic component is shown in Table 1.

The chemical composition of the inorganic (lighter) part is shown in Table 2.

According to the results of chemical analysis, it can be assumed that the particles of composition No. 1 are characteristic of the products of thermolysis of organic substances, and composition No. 2—for building materials, possibly cement stone.

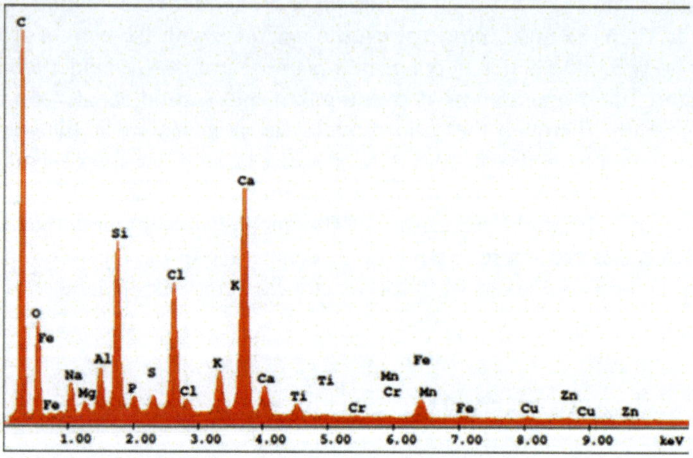

Fig. 2 Energy-dispersion spectrum of the organic part of MSW thermolysis products

Table 1 The elemental composition of the organic component of MSW thermolysis products

Element	Mass, %	Atomic, %	Element	Mass, %	Atomic, %
C	66.84	79.43	K	1.04	0.38
O	14.35	12.8	Ca	6.30	2.25
Na	1.27	0.79	Ti	0.44	0.13
Mg	0.3	0.17	Cr	0.03	0.01
Al	0.97	0.51	Mn	0.05	0.01
Si	3.42	1.74	Fe	1.19	0.31
P	0.33	0.15	Cu	0.23	0.05
S	0.31	0.14	Zn	0.23	0.05
Cl	2.71	1.09			

Table 2 The oxide composition of the light inorganic part of the thermolysis products of solid municipal waste

Element	Mass, %	Atomic, %	Element	Mass, %	Atomic, %
Na_2O	6.29	6.90	CaO	27.73	33.61
MgO	1.91	3.21	TiO_2	2.32	1.97
Al_2O_3	6.90	4.60	Cr_2O_3	0.12	0.06
SiO_2	26.94	30.48	MnO	0.19	0.18
P_2O_5	2.95	1.41	Fe_2O_3	4.90	2.08
SO_3	2.79	2.37	CuO	0.79	0.67
Cl_2O	11.38	8.90	ZnO	0.81	0.67
K_2O	3.99	2.88			

2.2 Phase Composition

The phase composition was determined using the SmartLab (Rigaku) X-ray diffractometer. The X-ray tube is equipped with a Cu-anode, which gives K_a radiation (U = 45 kV, I = 200 mA). The specified diffractometer is equipped with a high-speed D/tex Ultra Hi Pix detector. The Bregg-Brentano pseudoparallel beam focusing is used. The X-ray powder diffraction pattern of MSW thermolysis products is shown in Fig. 3.

The results of the phase composition of the analyzed material are presented in Table 3.

2.3 Granulometric Composition

In this study, we used a laser particle size analyzer “Analysette 22 NanoTec” with a measurement range from 0.01 to 2000 microns, which allows determining the particle size distribution of the analyzed material in suspensions and aerosols. The results of the size distribution of the powder microparticles are shown in Fig. 4.

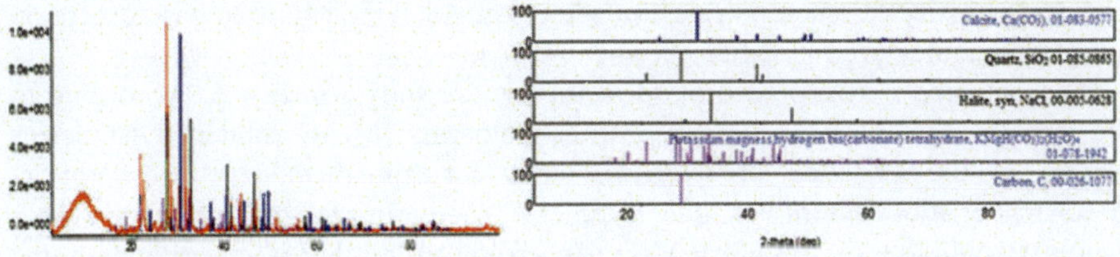

Fig. 3 X-ray powder diffraction pattern of MSW thermolysis products

Table 3 Results of qualitative X-ray phase analysis of MSW thermolysis products

Phase name	Formula	Spatial group	Card number
Amorphous carbon	C	143: P3	00-026-1077
Potassium magnesium dihydrocarbonate	$KMgH(CO_3)_2(H_2O)_4$	2: P-1	01-078-1942
Halite	NaCl	225: Fm-3 m	00-005-0628
Quartz	SiO_2	152: P3121	01-085-0865
Calcite	$Ca(CO_3)$	167: R-3c, hexagon	01-083-0577

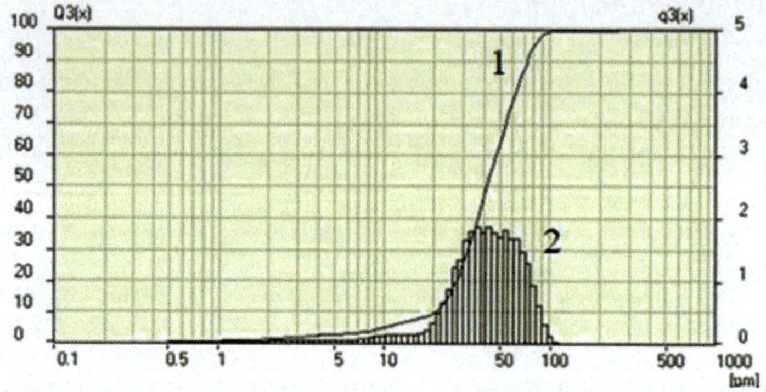

Fig. 4 Particle size distribution of MSW thermolysis products. 1—integral curve in coordinates Q3 (x) = f(μm) (left scale)—each point on the curve indicates how much % of the sample has a particle size less than or equal to this one. 2—histogram in coordinates q3 (x) = f(μm) (right scale)—the number of samples with a given particle size

2.4 Textural Characteristics

We determined the textural characteristics of MSW thermolysis products by the method of low-temperature nitrogen adsorption (BET method). The specific surface area and porosity parameters were evaluated using a TriStar II 3020 gas-adsorption analyzer manufactured by Micromeritics (USA).

The obtained nitrogen adsorption isotherms are shown in Fig. 5. According to the classification of Brunauer adsorption isotherms [8], the obtained isotherms belong to the IV type of isotherms that correspond to physical sorption. A characteristic feature of the type IV isotherm is the presence of a capillary-condensation hysteresis loop. According to the IUPAC classification, the obtained hysteresis loops belong to the H1 type.

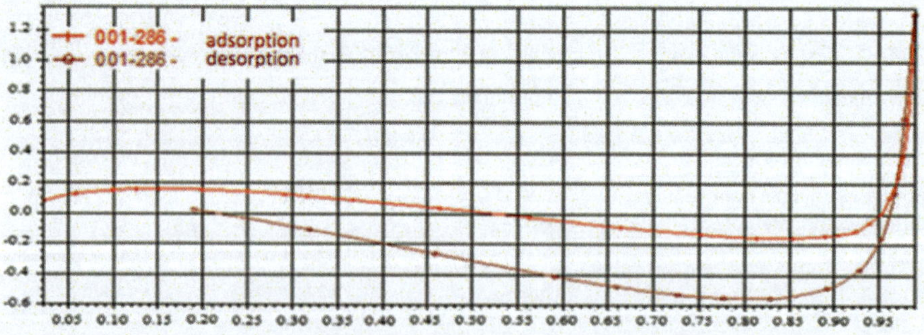

Fig. 5 Isotherms of nitrogen adsorption and desorption by the test sample (The "X" axis is the relative pressure P/Po; the "Y" axis is the amount of adsorbed gas cm^3/g)

Table 4 Textural characteristics of MSW thermolysis products

Specific surface area according to the five-point BET method	0.1271–0.2281 m^2/g
Pore volume at relative pressure P/Po = 0.984479441	0.000288–0.001,148 cm^3/g
Pore size	90.72–165.12 Å

Table 4 shows the textural characteristics of MSW thermolysis products.

3 Results and Discussions

The XRD method revealed low-temperature trigonal quartz in the material under study, which confirms the assumption of the presence of quartz sand in MSW. The predominant ingredient in the MSW thermolysis product is amorphous carbon.

According to the chemical analysis data, it can be assumed that in the initial state during the transportation and storage of MSW, particles of cement stone, quartz sand, chalk, dust and other building materials that can give peaks characteristic of calcium and silicon could get into them.

Based on the energy-dispersion analysis, it is concluded that the analyzed material is dominated by carbon 66.84 wt.%. Oxygen, calcium and silicon are also present in large quantities. Chlorine was recorded at 2.71 wt.%. This amount of chlorine can be explained by the presence of impurity chlorine-containing substances in MSW. A significant content of calcium and silicon oxides was determined. Sodium and aluminum oxides are present.

When studying the granulometric composition, it was found that the average particle size of the analyzed material is 44.07 microns.

Analyzing the graphical dependences of the specific volume of pores on their size, it was found that most of the pores are represented by macropores, which corresponds to a carbon material. The specific surface area of the samples according to the five-point BET method is 0.1271–0.2281 m^2/g. These values correspond to the specific surface area of the coal before activation [9].

The carbon black obtained by low-temperature thermolysis technology is in demand in various areas of industrial production. In particular, it is widely used in paint and varnish, polymer, rubber, electronic, and other industries [10].

4 Conclusion

1. A resource-saving technology and a technological complex for processing organic MSW were developed, and commercial products were obtained: carbon black, liquid hydrocarbon fuel, and synthesis gas.

2. A set of studies to determine the material composition of carbon black, including: X-ray phase analysis, laser diffraction, scanning electron microscopy, low-temperature nitrogen adsorption.
3. The material composition of the MSW thermolysis products, which is mainly represented by carbon (66.84 wt%), was studied. The light component is dominated by calcium oxides (27.73 wt%) and silicon oxides (26.94 wt%).
4. Promising areas of application of carbon black obtained during the processing of organic MSW by low-temperature thermolysis are identified.

Acknowledgements The work was carried out within the framework of project No. 10089447, the program of the REC "Innovative solutions in the agro-industrial complex", the scientific and production platform "Rational use of natural resources".

References

1. Karlovich I, Rumyantseva L (2017) On the problem of global antropogenic impact on the environment. New Sci Experience, Tradit, Innovation 1(3):14–17
2. Wiesmeth H, Starodubets N (2020) The management of municipal solid waste in compliance with circular economy criteria: the case of Russia. Econ Reg 3:725–738
3. Shilkina S (2020) World trends in waste management and analysis of the situation in Russia. Waste Resour 1(5):7
4. Glagolev S, Shein N, Sevostianov V, Obolonsky V, Shamgulov R (2020) Technologies for integrated processing of solid municipal waste. Ecol Ind Russia 12(24):11–15
5. Gorshkova A, Moskova M, Popov A, Shubin I (2018) Sorbent market analysis. Caoutchouc and rubber. LLC "Center for Professional Management" Academy of Business", Saratov, pp 39–42
6. Glagolev S, Sevostianov V, Shein N, et al (2021) Method for low-temperature processing of organic solid municipal waste and installation for its implementation. Pat. 2744225 RU, IPC F23G 5/027, B09B 3/00 (2006.01)
7. Sevostianov V, Obolonsky V, Shamgulov R (2020) Innovative developments in the processing of municipal solid waste. BSTU, Belgorod
8. Greg S, Sing K (1984) Adsorption, specific surface area, porosity. Mir publishers, Moscow
9. Polovneva S, Elshin V, Nosenko A (2015) Specific surface area of active carbons in desorption and reactivation processes. Basic Res 2:1187–1193
10. Kushner V (2008) Materials science schoolbook. OmSTU, Omsk

Analysis of the Influence of Rod Elements on the Mixing Process of Two Components in a Twin-Shaft Paddle Mixer

S. I. Khanin, N. O. Kikin, R. V. Zybin, and E. G. Khanina

Abstract Compulsory mixers have a fairly high performance, the ability to work both in continuous and batch modes, and are easy to maintain. Twin-shaft paddle mixers with horizontal shafts are used to obtain concretes, putties, dry building mixtures. To increase the efficiency of the mixing process of components, a design of a paddle mixer with rod elements remotely installed in front of the working surfaces of the paddles is distance. The research objective of the study was to evaluate the effectiveness of the proposed mixer design for the preparation of dry mixtures. The problems of carrying out in a well-known software product intended for scientific and industrial modeling, simulation of the mixing process of a two-component material on models of paddle mixers with and without rod elements have been solved; research of mixing processes based on the coefficient of heterogeneity of the mixture widely used for these purposes; comparison of mixing processes of components in the considered models of mixers. In the course of the work, the method of simulation was used. As a result of the study, heterogeneity coefficient of the prepared mixtures and the time of their preparation on the considered models of mixers, the efficiency of using the design of the mixer with rod elements were established.

Keywords Twin-shaft paddle mixer · Rod elements · Simulation · Process · Heterogeneity coefficient · Efficiency

1 Introduction

In the production of building mixtures, one of the main processes is the process of mixing components (gypsum, quartz sand, cement, chalk, etc.). Among the wide variety of mixing machines, compulsory mixers should be distinguished, which have a sufficiently high productivity, the ability to work both in continuous and

S. I. Khanin (✉) · N. O. Kikin · R. V. Zybin · E. G. Khanina
Belgorod State Technological University Named After V.G. Shukhov, Kostyukova Street, 46, 308012 Belgorod, Russia

S. V. Klyuev (ed.), *Digital Technologies in Construction Engineering*, Lecture Notes in Civil Engineering 173, https://doi.org/10.1007/978-3-030-81289-8_23

batch modes, and the working bodies of such mixers are characterized by an intense effect on the components being mixed. These machines include a twin-shaft paddle mixer with horizontal shafts [1, 2].

Twin-shaft paddle mixers are used to produce concrete, putty, dry building mixtures. The main mixing elements are paddles mounted on shafts. Modern studies have shown that when a paddle moves in a mixture, a compaction of mixed components is formed in front of its active surface, the speed of movement of particles in which is much lower than at the border of this compaction [3, 4]. As a result, the mass transfer is less intensive, which degrades the quality of the prepared mixture. In work [5], on the basis of experimental data, it is noted that a decrease in the paddle angle leads to a decrease in the seal in front of the paddle. However, reducing the installation angle of the paddle will reduce the productivity of the machine. In this regard, it is necessary to search for new solutions that can solve the problem of compaction of the material in front of the paddle without reducing productivity.

We have proposed a mixer design [6], in which rod elements are installed in front of the paddles, which provide an effect on the compacted material located in front of the paddle when the shafts rotate. At the initial stage of researching the processes occurring in the mixed material, it is advisable to use application programs that allow them to be simulated with significant savings in time and material resources.

2 Materials and Methods

To assess the effectiveness of the proposed design, a simulation of the mixing process of a two-component material was carried out in the EDEM software product intended for scientific and industrial modeling [7–11]. The study was carried out for two mixer models: with rod elements installed in front of the paddles (Fig. 1) and without them. The rod elements were located at a distance of $c = 0.02$ m from the working surfaces of the paddles, and the longitudinal axes of the rod elements were parallel to the lateral edges of the paddle. The characteristics of the mixer models are shown in Table 1.

Two components were simultaneously loaded into the working boxes of the mixers with a coefficient of 0.5: «Component 1» and «Component 2», each of which was loaded into a certain part of the mixer during one second. The paddle shafts were stationary during loading. The characteristics of the components to be mixed are shown in Table 2.

The mixing time of the materials was 66 s, during which the mixing process of the components can be considered to have passed into a steady state. Mixing quality was assessed in two ways. The first method consisted of visual comparison of color images of the mixed components in the mixer body at regular intervals. The time intervals in which the snapshots of the states of the components were taken were $t_c = 5$. This method is auxiliary. For a more accurate and qualitative comparison of

a *b* *c*

Fig. 1 Shell of mixer with paddle shafts: **a** without material; **b** with material, **c** paddle with rod element

Table 1 Characteristics of mixer models

Mixer design	Parameter					
	shell size		Installation angle of paddle, α, *rad*	Shaft speed, n, s^{-1}	The circuncircle diameter of the paddles, D, m	Distance between paddle and rod, m
	Length, l, m	Width, h, m				
Without rod element	0.5	0.288	0.61	1.2	0.155	–
With rod element	0.5	0.288	0.61	1.2	0.155	0.02

Table 2 Particles characteristics of mixing components

Component number	Particles characterization			
	Density, ρ (kg/m^3)	Diameter, d, 1×10^{-3} m	Total mass, m (kg)	Color
1	1900	3	3	
2	2000	6	7	

the efficiency of mixing processes for each model, the heterogeneity coefficient V was determined [12, 13]:

$$V = \frac{100}{\bar{c}} \sqrt{\frac{1}{n-1} \sum_{i=1}^{n} (c_i - \bar{c})^2}, \quad (1)$$

where $\bar{c}$—the arithmetic mean of the concentration of the key component in all n samples of the mixture, %; c_i—concentration of the key component in the ith sample, %; n—number of samples taken, pcs.

To determine the heterogeneity coefficient V, the minimum allowable sample weight G_m was calculated, which depends on the particle diameter of the mixed components [12, 13]:

$$G_M = \frac{1.26 \cdot 10^4 \cdot d^3 \cdot \rho}{C_0}, \qquad (2)$$

where d—the diameter of the particles of the mixed component or the diameter of the lumps from the components of the mixture, m; ρ—particle density or density of particle lumps, kg/m^3; C_0—component particle concentration, %.

For sampling, the shell of the mixers in the longitudinal direction were divided into 5 equal parts (Fig. 2), each of which in the transverse direction, in turn, was divided in the vertical and horizontal directions into 12 equal cells.

Starting from the second second of the mixing process, samples were taken from each cell that exceeded the minimum allowable mass $G_m = 0.172$ kg, determined by expression (2). The EDEM software product allows you to instantly calculate the total mass of the sample, as well as the mass of the key component in each cell of the selected section of the working box of the mixer at any time interval of the mixing process. This method is rather laborious for an operator processing the results of experiments, but it allows one to accurately determine the heterogeneity coefficient of the mixture, which characterizes the quality of its preparation.

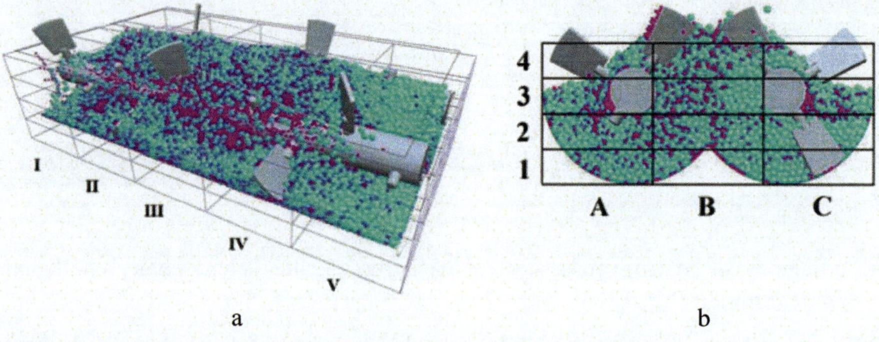

Fig. 2 The scheme of dividing the mixer shell into cells: **a** in the longitudinal direction, **b** in the transverse direction

3 Results and Discussion

Figure 3 shows illustrations of the state of the components of the mixtures at intervals of time in the studied mixer models. Figure 4 shows graphical dependences of the heterogeneity coefficient of mixture on the mixing time.

The most intensive mixing of the components proceeds for the mixer model without rod elements in the first 4 s, and for the model with rod elements—in 3 s. Then, in the time interval from 4 to 30 s in the mixer model without rods, the values of the heterogeneity coefficient decrease to Vb30 = 6.67%. The change in Vb in the interval from tb10 = 10 s to tb30 = 30 s is characterized by pronounced fluctuations in values. Continuation of the process of mixing the components on this model (interval from 31 to 61 s) leads to fluctuations in Vb values from Vb35 = 6.28% (at tb35 = 35 s) to Vb57 = 8.7% (at tb57 = 57 s). The mixing process for the mixer model without rod elements in the considered time interval can be attributed, due to fluctuations in the values of Vb about the average value Vbsr = 7.06%, to a steady-state process.

In the time interval from 4 to 59 s on the mixer model with rod elements, the values of the heterogeneity coefficient decrease to V_{c59} = 3.74%. The change in V_c in the interval from t_{c4} = 4 s to t_{c23} = 23 s is characterized by pronounced fluctuations in values. Then the fluctuations in the V_c values are reduced. In the time interval from t_{c60} = 60 s, up to t_{c66} = 66 s, the mixing process can be attributed to steady-state, where the average value of the heterogeneity coefficient is V_{csr} = 3.85% with the minimum value V_{c64} = 3.68%, attained at t_{c64} = 64 s, and the maximum value V_{c61} = 4.04%, is achieved at t_{c61} = 61 s.

At the initial time interval of 21 s, the mixing process of the components in the mixer model with rod elements proceeds more efficiently than in the model without rod elements. So, for 3 s in the mixer model with rod elements, the value of the heterogeneity coefficient is V_{c03} = 16.15%, in the mixer model without rod elements—V_{b03} = 25.17% (V_{c03} exceeds V_{b03} by 1.55 times). At the tenth second, the values of the heterogeneity coefficient, respectively, are V_{c10} = 12.38% and V_{b10} = 16.03% (V_{b10}/V_{c10} = 1.29 times), and at the twenty-first second, respectively, V_{c21} = 9.46% and V_{b21} = 10, 56% (V_{b21}/V_{c21} = 1.11 times). In the time

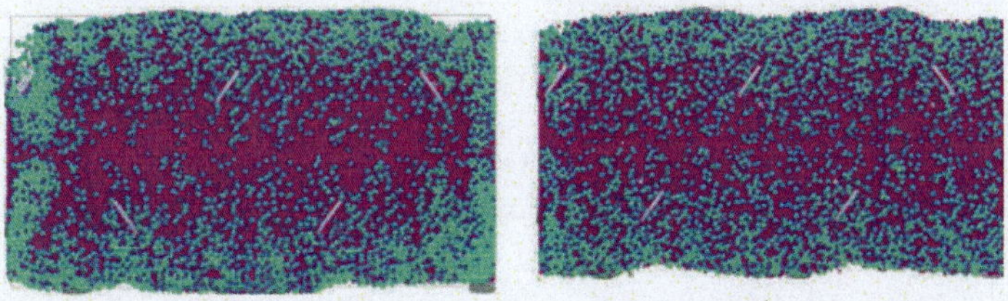

a b

Fig. 3 Illustrations of the state of the mixture obtained: **a** on a mixer model without rod elements at a mixing time t_{b61} = 61 s, **b** on a model of a mixer with rod elements with a mixing time t_{c61} = 61 s

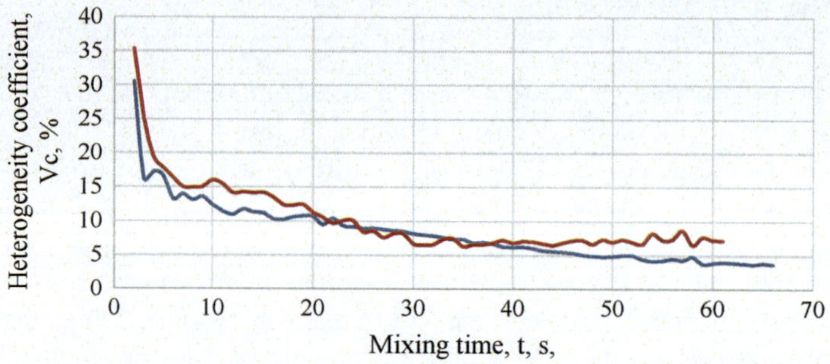

Fig. 4 Dependence of the heterogeneity coefficient on the mixing time in the studied mixer models: — model without rod elements, — model with rod elements

interval from 21 to 38 s, the values of the heterogeneity coefficients for both models are quite close. Thus, the value of the heterogeneity coefficient $V_{c36} = 6.77\%$ obtained on the mixer model with rod elements at $t_{c36} = 36$ s is quite close to the value $V_{b36} = 6.62\%$ obtained on the mixer model without rod elements during the same time. Starting from 39 s, the efficiency of the mixing process on the mixer model with rod elements, in comparison with the mixer model without rod elements, begins to increase. This rise ends at $t_{c60} = 60$. The illustrations of the state of the mixture in Fig. 3 clearly show a more uniform distribution of components obtained on the mixer model with rod elements for the time $t_{c61} = 61$ s, compared to the distribution of components in the mixtures on both models obtained in 38 s, and on the mixer model without rod elements at $t_{b61} = 61$ s.

Comparison of the simulation results shows that the model with rod elements provides a 2.25 times lower value of the heterogeneity coefficient and allows for the preparation of better quality mixtures. The duration of the mixing process of the components in this model is 29 s longer than the duration of the mixing process in the model without rod elements.

4 Conclusion

1. A patent-protected design of a paddle mixer has been developed, which provides an increase in the efficiency of the mixing process of components.
2. In the EDEM software environment, on the mixer models with and without rod elements, virtual experiments were carried out to implement the process of preparing a two-component mixture.
3. Using the results of experiments, a study of mixing processes in the considered models of mixers was carried out, as the main criterion for the efficiency of the process, the heterogeneity coefficient of the mixture, determined by the known expression, was used.

4. The analysis of changes in the models of the state of the mixture over time is carried out. The processes of mixing the components are most intensive in the first 3 … 4 s. The change in the heterogeneity coefficient of the mixture in the mixer model with rod elements V_c in the interval from $t_{c4} = 4$ s to $t_{c23} = 23$ s is characterized by pronounced fluctuations in values, and V_b in the mixer model without rod elements—in the interval from $t_{b10} = 10$ s to $t_{b30} = 30$ s. The steady-state mixing process in the model with rod elements is achieved at $t_{c60} = 60$ s. In the time interval from $t_{c60} = 60$ s to $t_{c66} = 66$ s, the average value of the heterogeneity coefficient is $V_{csr} = 3.85\%$. The steady-state mixing process in the model without rod elements is achieved at $t_{b31} = 31$ s. In the time interval from $t_{b31} = 31$ s to $t_{b61} = 61$ s, the average value of the heterogeneity coefficient is $V_{csr} = 7.06\%$.
5. Comparison of mixing processes of components in the considered models of mixers is carried out. Mixing of components to a heterogeneity coefficient, which reaches the limiting capabilities for a model without rod elements, in both models is carried out for close time intervals. In the model of a mixer with rod elements, a 2, 25 times lower value of the coefficient of inhomogeneity is provided, which characterizes the process as more efficient and allows for the preparation of better quality mixtures. The duration of this process is 29 s longer than the duration of the mixing process in the model without rod elements.

Acknowledgements The study is implemented in the framework of the Flagship University DevelopmentProgram at Belgorod State Technological University named after V.G. Shukhov, using the equipment of High Technology Center at BSTU named after V.G. Shukhov.erences.

References

1. Valigi MC, Logozzo S, Landi L, Braccesi C, Galletti L (2019) Twin-shaft mixers' mechanical behavior numerical simulations of the mix and phases. Machines 7:39–52
2. Yao Y, Feng Z, Chen S, Li BQ, Zhao L, Zhao WA (2016) Double-blade mixer for concrete with improved mixing quality. Arab J Sci Eng 41:4809–4816
3. Demin OV, Pershin VF, Smolin DO (2012) Intensification of mixing bulk materials in a paddle mixer. Ch&ChT 8:108–111
4. Pasko AA, Demin OV (2002) Modeling of the movement of flows of bulk material during flow around a plate. In: Balakirev VS (ed) Mathematical methods in engineering and technology: Sat. XV international scientific conference: In 10 tons. Under the general. Tambov Publishing house of Tamb. state tech. University 10, pp 56–57
5. Pulin VP (1984) Improving the energy efficiency of cyclone forced-action concrete mixers: dis candidate of technical science. Dnepropetrovsk
6. Khanin SI, Kikin NO (2019) Pat. 192657, Russian Federation, IPC B28C 5/14, B01F 7/04 Material mixer. Applicant and patent holder Federal State Budgetary Educational Institution of Higher Education "Belgorod State Technological University named after V G Shukhov". No. 2019119931; declared 06/25/2019; publ. 09/25/2019, Bull. No. 27, 6
7. Bogdanov VS, Khakhalev PA (2014) Methodology for designing energy-exchange linings for ball drum mills. Bulletin of BSTU named after V. G. Shukhov, vol 6, pp 67–72

8. Cai R, Hou Z, Zhao Y (2019) Numerical study on particle mixing in a double-screw conical mixer. J Powder Technol 352:193–208
9. Sinnott MD, Cleary PW (2016) The effect of particle shape on mixing in a high shear mixer. J Comput Part Mech 3:477–504
10. Moon I, Kim J, Roh J, Lee MS (2019) DEM study of a mixer for core manufacturing system. J Comput Aided Chem Eng 46:799–804
11. Khanin SI, Zybin RV, Mordovskaya OS (2020) Increasing the efficiency of the material classification process in the classifying partition of a ball mill. Bulletin of BSTU named after V. G. Shukhov, vol 12, pp 97–107
12. Borshchev VY, Gusev MA, Promtov AS, Timonin AS (2006) Equipment for the processing of bulk materials. Publishing house of mechanical engineering, vol 1, p 208
13. Makarov YI (1973) Apparatus for mixing bulk materials. Mashinostroenie 216

Possible Approach to the Production of Active Silica-Containing Powders

M. V. Morozova, M. V. Akulova, A. M. Ayzenshtadt, and M. A. Frolova

Abstract The assessment of four sand deposits in different regions of Russia on the influence of the dispersion time on the physical and chemical properties of rocks was carried out. It was found that, despite the difference in the chemical, mineral composition and origin of rocks, the size characteristics of the sands at the same grinding time have similar values. The quantitative assessment of the potential energy of the system, which passed after the dispersion of the material into the surface, was carried out taking into account the values of the critical surface tension and the specific surface of the raw material (formed after the mechanical grinding process). For the obtained highly dispersed fractions, the changes in the value of the relative change in the free surface energy ($\Delta E_s/E_{s0}$) were calculated as a function of the sample grinding time. The obtained functional dependences showed a significant increase in $\Delta E_s/E_{s0}$ already at 20 min of rock dispersion. Based on this criterion, it was found that the sands of the Nerekhtskoye deposit in Ivanovo region and the Kenitsa deposit in Arkhangelsk region are most preferable to use as an active fine component for the production of astringent compositions.

Keywords Polymineral sand · Surface tension · Specific surface area · Surface activity

M. V. Morozova (✉) · M. V. Akulova
Federal State Budget Educational Institution of Higher Education, Ivanovo State Polytechnic University, Ivanovo, Russia
e-mail: m.morozova@narfu.ru

M. V. Morozova · A. M. Ayzenshtadt · M. A. Frolova
Federal State Autonomous Educational Institution of Higher Education, Northern (Arctic) Federal University Named After M.V. Lomonosov, Arkhangelsk, Russia

S. V. Klyuev (ed.), *Digital Technologies in Construction Engineering*,
Lecture Notes in Civil Engineering 173, https://doi.org/10.1007/978-3-030-81289-8_24

1 Introduction

Currently, the construction materials industry is very interested in fine-ground mineral powders [1, 2], the introduction of which in the production technology allows to improve the quality of building materials. This approach can significantly improve the quality of composite materials: strength, wear resistance, thermal and chemical resistance, fracture toughness, i.e., increase the operational period and life cycle of the product or structure as a whole [3–6].

The grinding of powders to the nano- and micro-sized level of particles leads to a change in a number of physical and chemical properties of the systems themselves [7–9]. This is due to the fact that a significant proportion of the atoms that make up the material particles are located in the surface layer and have free bonds that determine the value of the free surface energy [10, 11].

When the particle sizes reach nanometers, the specific surface area (S_{sp}) and surface energy increase several times. Therefore, the larger the S_{sp}, the better the interaction of the material particles with the composite matrix becomes (a more uniform distribution of nanoparticles in the material is achieved) [2, 3, 5, 12].

When optimizing the composition of composite materials based on fine-dispersed systems, it is necessary to take into account changes in the properties of the obtained samples depending on their composition correctly. Based on the above, for micro-and nanoscale particles, the most important characteristic is the energy state of the system, estimated by the value of the free surface energy (E_s, J/kg) [2, 10, 13, 14]. In turn, Es is calculated by multiplying the surface tension (a measure of energy storage in a decompressed surface layer) of a highly dispersed system by its specific surface area [10–12]:

$$E_s = \sigma_K \cdot S_{sp} \tag{1}$$

where σ_K—surface tension of highly dispersed material, J/m^2; S_{sp}—specific surface area, m^2/kg.

In [9, 14], a mathematical expression describing the change in the energy state of a composite with a variable composition is presented, estimated by the relative change in the free surface energy, taking into account the surface tension of the system and its specific surface area:

$$\frac{\Delta E_s}{E_{s0}} = (1 - \delta\sigma) \cdot \left(\frac{S_{sp}}{S_{sp0}}\right) - 1 \tag{2}$$

where «0»—index indicating the initial state of the system; δ—the relative change in the value of the surface tension, determined by the formula:

$$\frac{\Delta\sigma}{\sigma_0} = \frac{\sigma - \sigma_0}{\sigma_0} \tag{3}$$

The parameter $\Delta E_s/E_{s0}$ can be used as a criterion for estimating the dispersion interaction of particles in a composite.

Therefore, the aim of this work is to determine the relative change in the surface energy of finely dispersed rocks and to identify the most effective time of their dispersion, accompanied by the maximum energy transition.

The problem of using unclaimed and substandard raw materials to create a hardening filling mass, with the possibility of implementation throughout the territory of the Russian Federation, while ensuring environmental protection measures is relevant.

2 Methods and Materials

2.1 Materials

Four deposits of polymineral construction sands were selected as raw materials: the Krasnoflotsky-Zapad and Kenitsy deposits (Arkhangelsk Region), the Khromtsovskoye deposit (Khromtsovo village, Ivanovo Region) and the Nekhtskoye deposit (Teterinskoye village, Nerekhtsky district).

2.2 Methods

Before starting the tests, the sand was washed and dried to a constant mass at a temperature of 105 °C. Determination of the true sand density was carried out according to GOST 8735-88.

The mineralogical composition of the rocks was determined by recording X-ray diffractograms on the Shimadzu XRD-7000 S X-ray diffractometer (TSC "Arctic", M.V. Lomonosov NArFU). The chemical composition of the sands was determined using the X-ray fluorescence analyzer "MetExpert".

Mechanical activation of sand was carried out by mechanical dry grinding, at a rotor speed of 420 rpm, with large grinding bodies with a diameter of 20 mm on a planetary ball mill Retsch RM100. The size of the obtained highly dispersed particles was determined using the Delsa Nano Series Zeta Potential and Submicron Particle Size Analyzers ("DelsaNano") by photon-correlation spectroscopy based on the principle of dynamic light scattering. The specific surface area of highly dispersed rock systems was determined by the gas sorption method, according to the BET theory, using the Autosorb-iQ-MP analyzer.

To determine the wetting angle at the "Easy Drop" installation at a temperature of 25 ± 1 °C, we used tablet samples made by pressing the ground sand on the "PLG-20" press at a load of 20 kPa into a metal mold with a diameter of 20 mm.

The surface tension of the samples (σ_K) was calculated by the OWRK method (Ounce, Wendt, Rabel, and Kjellble method), based on the measurement of the equilibrium wetting angle (θ) by liquids (distilled water, decane, glycerin, and ethylene glycol) with known values of the surface tension (σ_{liq}) and its dispersion (σ_L^D) and polarization (σ_L^P) components.

At the same time, the obtained functional dependencies of the OWRK method:

$$\frac{\sigma_L \cdot (\cos\theta + 1)}{2 \cdot \sqrt{\sigma_L^D}} = f\left(\frac{\sqrt{\sigma_L^P}}{\sqrt{\sigma_L^D}}\right) \tag{4}$$

are described by linear equations, where θ is the wetting angle of the material under study; σ_L, σ_L^D, and σ_L^P are the total, dispersion, and polarization surface tensions of the working fluids (respectively). And the total surface tension of the system is $\sigma_k = \sigma_S^P + \sigma_S^D$, where σ_S^P and σ_S^D are the polar and dispersion components of the surface tension of the material under study.

3 Results and Discussion

The selected sand deposits do not significantly differ in their true density (ρ_{true}) and have similar values: for the "Krasnoflotsky-Zapad" sand ρ_{true} = 2.71 g/cm^3; for "Kenitsa" sand ρ_{true} = 2.64 g/cm^3; for "Khromtsovskoe" sand ρ_{true} = 2.50 g/cm^3; for "Nerekhtskoe" sand ρ_{true} = 2.60 g/cm^3.

The content of the main elements of the highly dispersed samples in terms of oxides obtained by X-ray fluorescence analysis is given in Table 1.

Table 1 Chemical composition of sands in terms of oxides (%)

Defined component	Sand deposit			
	Krasnoflotsky-Zapad	Kenitsa	Khromtsovskoe	Nerekhtskoe
SiO_2	93.50	90.56	85.23	95.89
Al_2O_3	2.92	5.77	1.97	2.32
MgO	2.92	0.48	0.46	0.16
Fe_2O_3	2.58	0.80	5.75	0.56
CaO	0.67	0.25	3.30	0.20
TiO_2	0.08	0.03	0.05	0.01
K_2O	1.01	0.28	0.01	0.75
SO_3	0.59	0.04	0.68	0.04
P_2O_5	0.05	0.06	0.08	0.02
Na_2O	0.69	1.62	2.41	0.04
MnO	0.01	0.02	0.06	0.01
Cr_2O_3	0.04	–[a]	–	–
SrO	0.01	0.08	–	–

[a]Component was not found

The optimal parameters for the dispersion of rock samples were selected experimentally. The grinding time for all four sand deposits was chosen to be the same: 5, 10, 20, 30 min.

The results of the analysis of the dimensional characteristics of highly dispersed sands after mechanical activation in a planetary ball mill showed that good reproducibility of the results is achieved under the selected grinding modes.

A comparison of the average size characteristics of sand particles obtained with the DelsaNano analyzer showed that the Kenitsa and Nerekhtskoye deposits have a higher grinding capacity than the remaining two deposits (Table 2).

The difference in the values of the specific surface of the analyzed samples with almost equal values of dispersion is explained by the difference in the densities of the materials and, as a result, by the different number of particles in the studied samples. In addition, this difference may be related to the surface roughness of the rock particles.

The surface tension ($\sigma_К$, MJ/m^2) was calculated based on the functional dependence 4, which was characterized by linear equations of the form $y = a \cdot x + b$, where a—characterizes the angle of inclination of the straight line; b—the shift of the regression line. The squares of these values allow calculating the polar and dispersion components of the surface tension. The values of the coefficients of the equation and the critical surface tension for each sand calculated from this dependence are presented in Table 2.

Table 2 Grinding time and dispersion characteristics of sand fractions

Sand deposit	Grinding time (min)	Average particle size (nm)	S_{sp} (m^2/kg)	Linear equation coefficient		R^2	$\sigma_К \cdot 10^3$ (J/m^2)
				a	b		
Krasnoflotsky-Zapad	5	754 ± 6	8241 ± 17	6.90	4.36	0.98	66.54
	10	642 ± 5	10,140 ± 23	6.93	4.35	0.98	67.01
	20	445 ± 3	15,302 ± 15	6.97	4.36	0.98	67.58
	30	387 ± 7	22,231 ± 20	6.98	4.37	0.98	67.81
Kenitsa	5	671 ± 4	7819 ± 13	6.89	4.35	0.98	66.43
	10	559 ± 5	9920 ± 19	6.93	4.36	0.98	67.06
	20	406 ± 3	18,670 ± 12	6.98	4.38	0.98	67.81
	30	329 ± 4	28,603 ± 18	6.99	4.38	0.98	68.00
Khromtsovskoe	5	687 ± 5	910 ± 11	6.97	4.35	0.98	67.51
	10	549 ± 3	1089 ± 13	6.99	4.37	0.98	67.91
	20	467 ± 1	1638 ± 11	7.00	4.35	0.98	67.98
	30	341 ± 2	2319 ± 14	7.05	4.29	0.99	68.11
Nerekhtskoe	5	603 ± 6	778 ± 18	6.76	4.41	0.98	65.22
	10	510 ± 4	887 ± 17	6.83	4.39	0.98	65.84
	20	389 ± 2	2085 ± 15	6.94	4.39	0.98	67.39
	30	313 ± 1	3154 ± 16	6.95	4.38	0.98	67.51

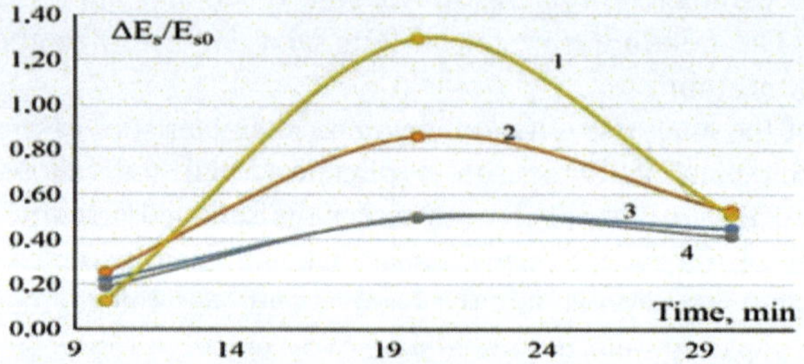

Fig. 1 Dependence of the relative change in the free surface energy of highly dispersed sand samples on the dispersion time: 1—sand of the Nerekhtskoye deposit; 2—sand of the Kenitsa deposit; 3—sand of the Krasnoflotsky-Zapad deposit; 4—sand of the Khromtsovskoye deposit

Figure 1 shows the functional dependences of the relative change in the free surface energy ($\Delta E_s/E_{s0}$) calculated by Eq. 2 and the time of rock dispersion.

4 Conclusion

For highly dispersed samples, within the same region, fairly similar specific surface parameters were obtained with the same grinding time. However, if we compare the values of the sands of different regions, we can see that the sands of Arkhangelsk region have a greater specific surface area than the sands of Ivanovo region (by 9 times more), with the same grinding time.

In this case, the comparison of the free surface energy of the surface unit (surface tension) increases as the grinding time increases. It should be noted that the sands of the studied deposits, despite the difference in the mineral composition and the values of the specific gravity, have similar values of the surface tension of fine-dispersed systems.

Analysis of the dependence of the free surface energy of the systems on the time of sample dispersion showed that at the initial moment (10 min of grinding), the difference in the parameter $\Delta E_s/E_{s0}$ for rock samples was—0.13. With a further increase in the grinding time, this difference increases sharply, reaching—0.80. However, by 30 min of grinding, the values of $\Delta E_s/E_{s0}$ of the sands decreased sharply, and the difference in value was—0.11.

A comparison of the grinding time of sands from different fields showed that the Kenitsa and Nerekhtskoye deposits have a higher grinding capacity than other fields under study. Data on the relative change in the free surface energy of highly dispersed sand samples show that the maximum value of $\Delta E_s/E_{s0}$ for all sands is reached at a grinding time of 20 min. However, the sands of the Nerekhtskoye deposit in Ivanovo Region and the Kenitsa deposit in Arkhangelsk Region have the largest reserves of $\Delta E_s/E_{s0}$. The sands of the Krasnoflotsky-Zapad and Khromtsovskoye deposits showed almost identical parameters for the change in free surface energy over the entire time range of grinding. Optimization of the process of preparation of highly dispersed systems is possible according to the criterion of the relative change in the free surface energy.

Acknowledgements The reported study was funded by RFBR, project number 19-33-60062.

References

1. Tolstoy AD (2020) Fine-grained high-strength concrete Construction. Mater Prod 3(1):39–43
2. Kozhuhova NI, Strokova VV, Chizhov RV, Kozhuhova MI (2019) Chemical reactivity assessment method of nanostructured low calcium aluminosilicates. Constr Mater Prod 2 (3):5–11
3. Klyuev SV, Klyuev AV, Shorstova ES (2020) Fiber concrete for 3-D additive technologies. Constr Mater Prod 2(4):14–20
4. Nelyubova VV, Ogurtsova YN, Sivalneva MN, Strokova VV, Zhao P (2021) Specifics of Structure Formation of Cement Stone with Silica Raw Material - A substrate of photocatalyticcomposite material. MSF 1017:153–162
5. Chernysheva NV, Shatalova SV, Evsyukova AS, Fisher H-B (2018) Features of the selection of the rational structure of the compositional gips binder. Constr Mater Prod 1(2):45–52
6. Basnett A, Sharma P, Krishnaraj L, Ravichandran PT (2020) Characterization of zeolite influence to improve the performance of concrete. IOP Conf Ser Mater Sci Eng912(6):062054
7. Boucedra A, Bederina M, Ghernouti Y (2020) Study of the acoustical and thermo-mechanical properties of dune and river sand concretes containing recycled plastic aggregates. Constr Build Mater 256:119447. https://doi.org/10.1016/j.conbuildmat.2020.119447
8. Gavshina OV, Yashkina SYu, Yashkin AN, Doroganov VA, Moreva IYu (2018) Study of the effect of particulate additives on the setting time and microstructure of high-alumina cement. Constr Mater Prod 1(4):30–37
9. Danilov VE, Ayzenshtadt AM, Frolova MA, Tutygin AS (2018) Dispersion interactions as criterion of optimization of cementless. Compos Binders Inorg Mater Appl Res 9(4):767–771
10. Morozova MV, Akulova MV, Frolova MA (2020) Surface activity of the fine disperse systems on the basis of construction sands innovations and technologies in construction. Lect Notes Civil Eng 95:206–212
11. Glezer AM (2002) Amorphous and nanocrystalline structures: similarities, differences, mutual transitions D.I. Mendeleev. Russ Chem Soc 5:57–63
12. Grishina AN, Korolev EV (2015) Efficiency of modifying cement composites with nanoscale barium hydrosilicates. Constr Mater 2:72–76
13. Binnewies M, Milke E (2002) Thermochemical data of elements and compounds Hannover. Wiley-VCH 928, Germany
14. Korolev EV (2017) Surface tension in the structure formation of materials. Value Calculation Appl Constr Mater 1–2:104–108

Study of the Strength Characteristics of the Support Frame of Horizontal Roller Mill in NX CAE

Yu. M. Fadin, S. S. Latyshev, P. A. Khakhalev, and T. A. Khakhaleva

Abstract The grinding process is one of the main ones in the manufacture of many building materials, such as cement, limestone, gypsum, etc. The design of equipment and machinery for the construction materials industry is intimately associated with the implementation of many types of computations. This paper describes the computation stages to find out the strength characteristics of the support frame of a horizontal roller mill using the finite element method implemented in the CAD/CAM/CAE NX system. The "Advanced Simulation" module of the NX system allows evaluating how the electronic-digital model of the product will act in real use. The collection of loads for analysis, the preparation of the geometry of the electronic-digital model of the support frame of the mill, the generation of finite element grids and the setting of physical and mechanical properties of materials, and the imposition of boundary conditions are considered. Multiple conducting and analysis of preliminary computations using the NX Nastran solver gives the chance to make changes to the design of the support frame quickly and efficiently (the master model principle) to meet the requirements of strength and stiffness of each part of the assembly. This approach to conducting finite element analysis allows correcting the civil engineer's idea, optimizing the amount of material expended on manufacturing, and ensuring the service reliability of the projected product design without attraction of large costs for test trials.

Keywords Horizontal roller mill · NX CAE · "Advanced simulation" module · NX Nastran solver · Structural stresses

---

Yu. M. Fadin · S. S. Latyshev · P. A. Khakhalev (✉) · T. A. Khakhaleva
Department of Mechanical Equipment, Belgorod State Technological University Named After V.G. Shukhov, Belgorod, Russia

S. V. Klyuev (ed.), *Digital Technologies in Construction Engineering*,
Lecture Notes in Civil Engineering 173, https://doi.org/10.1007/978-3-030-81289-8_25

1 Introduction

Modern world industry is moving on, and it is growing and moving to a higher level of technology and productivity along with the development of science and technology every year.

The grinding process is one of the main processes in the production of most building materials. These include cement, limestone, lime, gypsum, various fillers of construction plastics, quartz and marble sand, etc. The great impact of degree of grinding on technological properties of materials and the huge output of them determine the need of research in this area [1–3].

Recently, the most promising method of medium and fine grinding is crushing a layer of material between cylindrical grinding surfaces, there a critical zone is formed, where a compressive force develops that exceeds the ultimate strength of the material. This grinding method is used in a horizontal roll mill.

From what has been said it follows that there is a need to conduct additional studies of the HRM: to find out the relations between the cost of electricity, productivity, characteristics of the grinder, the grinding process and the ultimate product quality [4–7].

Along with improving the efficiency of production, the design departments of mechanical engineering industries are set the problem of conducting reverse engineering of the developing and upgrading structures that includes computations for the strength of machine parts and assemblies. This paper describes the application of the NX CAE software to conduct a strength analysis of the support frame of a horizontal roller mill, taking into account the service conditions and loads arising under the total mass of the structure as well as external forces [8, 9].

The input data for the calculation is a complete electronic-digital model of a horizontal roller mill [10–12], which includes attribute information, such as the physical and mechanical properties of the material of the elements, the conditions for the elements contacting with each other in the assembly.

The boundary conditions were also defined: loads and constraints. The total mass G was calculated by the NX system automatically based on the properties dedicated to the components of materials and amounted to 1183.3170 N. Force F1 = 460 N is involved at the place of attachment of the motor to the frame; force F2 = 2157 N of the weight of the roll 1 is involved on the flanges of the support bracket; force F3 = 823 N of the weight of the roll 2 is operated on the flanges of the stop; force F4 = 2470 N is involved on the axis of the support rollers, taking into account the weight of the cylindrical body with the material. Fixed fastening of the frame base and the hole in the support bracket of the compression unit designed to fix the screw jack is used to limit the calculated structure by movement (Fig. 1).

The next step after the final determination of the loads and boundary conditions acting on the structure is to create idealized (simplified) models for all the elements involved in the calculation. Idealization of the geometry includes the suppression of structural elements that do not have a significant impact on the calculation result—

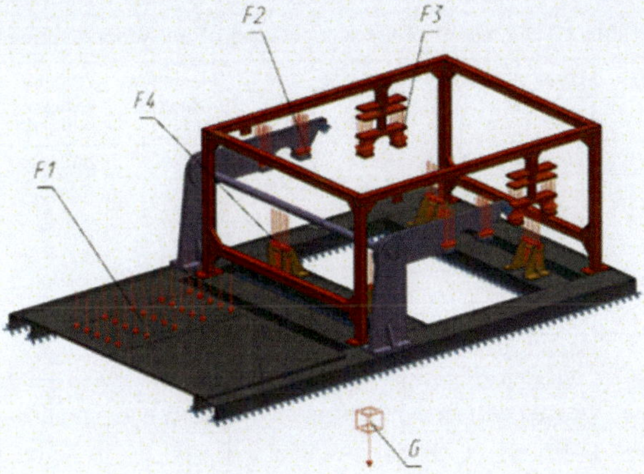

Fig. 1 Design model. G—the total mass of the support frame; F1—the mass of the motor; F2—the mass of the roll 1; F3—the mass of the roll 2; F4—the mass of the roll body with the material

fillets, chamfers, and small holes. Idealization also allows dividing a complex part into several simple volumes.

In the process of creating CAD models that are part of a horizontal roller mill, it is necessary to set the material for each of them. Before this, it is essential to analyze carefully the functional purpose of the designed element and its operating conditions, as the choice of material has a significant impact on the result of calculations. It is assumed that the parts will be made of St3 GOST 380-2005, Steel 45 GOST 1050-88, Steel SHH15 GOST 801-78, more information is given in Table 1.

2 Materials and Methods

To perform the computation of the finite element assembly in NX CAE, you must do the following actions [9]:

1. Create a finite element model (FEM) file for the assembly (.afm).
2. Create finite element models of all the necessary elements included in the assembly, with specific properties for each element or subassembly. Add FEM files to the assembly.
3. Determine the interconnections of the assembly elements. Check and resolve conflicting nodes and grid elements.
4. Create a simulation (computational model) file for a finite element model of the assembly. It determines the boundary conditions and loads.
5. Perform a preliminary calculation.
6. Make design or technological changes.
7. Perform the final computation.

Before calculating the strength of the designed structure for all its elements, you must create finite element models of them. When applying a grid to the element,

Table 1 The results of the computation of the support frame of the horizontal roller mill

№	Name of element	Material	Maximum voltages (MPa)	Working voltages (MPa)	Maximum movements (mm)	Working movements (mm)	Safety factor
1	Longitudinal bar 1	St3 GOST 380-2005	3.449	240	0.002	6	69.58
2	Longitudinal bar 2		6.239		0.003	6	38.46
3	Crossbar 1		2.757		0.001	2.80	87.05
4	Crossbar 1 (2)		3.160		0.0008	2.80	75.95
5	Crossbar 2		4.055		0.0008	2.80	59.19
6	Crossbar 2 (2)		4.062		0.0008	2.80	59.08
7	Sheet		6.469		0.101	3.20	37.10
8	Vertical stand		23.693		0.090	1.64	10.13
9	Vertical stand (2)		33.347		0.069	1.64	7.19
10	Plate		4.187		0.0009	0.90	57.32
11	Plate (2)		5.048		0.0007	0.90	47.54
12	Beam 1		35.464		0.282	3.46	6.76
13	Beam 2		2.275		0.010	2.54	105.49
14	Beam 2 (2)		4.549		0.092	2.54	52.76
15	Beam 2 (3)		5.596		0.019	2.54	42.89
16	Beam 3		3.910		0.081	2.54	61.38
17	Flange		6.169		0.118	0.12	38.90
18	Flange (2)		6.191		0.118	0.12	38.76
19	Stiffening rib		3.898		0.007	0.18	61.57
20	Stiffening rib (2)		13.904		0.087	0.18	17.26
21	Stiffening rib (3)		6.237		0.012	0.18	38.48
22	Stiffening rib (4)		8.705		0.068	0.18	27.57
23	Stiffening rib (5)		2.389		0.034	0.18	100.46
24	Stiffening rib (6)		2.470		0.016	0.18	97.16
25	Stiffening rib (7)		11.202		0.089	0.18	21.42
26	Stiffening rib (8)		6.471		0.088	0.18	37.09
27	Guide pipe		15.359		0.248	0.24	15.63
28	Rod		14.728		0.239	0.40	16.29
29	Plate 1		1.709		0.284	0.60	140.43
30	Plate 2		0.783		0.285	0.60	306.51
31	Stop		14.356		0.302	0.60	16.72
32	Flange		0.781		0.304	0.152	307.30
33	Flange (2)		0.781		0.213	0.152	307.30
34	Stiffening rib		20.846		0.230	0.20	11.51
35	Stiffening rib (2)		20.773		0.230	0.20	11.55
36	Stiffening rib (3)		24.421		0.264	0.20	9.83
37	Left stand		15.036		0.004	0.34	15.96
38	Left stand (2)		15.049		0.004	0.34	15.95
39	Left stand (3)		14.915		0.004	0.34	16.09
40	Left stand (4)		14.927		0.004	0.34	16.08
41	Right stand		15.049		0.004	0.34	15.95
42	Right stand (2)		15.036		0.004	0.34	15.96
43	Right stand (3)		14.927		0.004	0.34	16.08
44	Right stand (4)		14.915		0.004	0.34	16.09

(continued)

Table 1 (continued)

№	Name of element	Material	Maximum voltages (MPa)	Working voltages (MPa)	Maximum movements (mm)	Working movements (mm)	Safety factor
45	Axis	Steel 45 GOST 1050-88	20.755	360	0.007	0.32	17.34
46	Axis (2)		20.749		0.007	0.32	17.35
47	Stiffening rib	St3 GOST 380-2005	2.936	240	0.0009	0.12	81.74
48	Stiffening rib (2)		2.938		0.0010	0.12	81.69
49	Stiffening rib (3)		2.934		0.0010	0.12	81.80
50	Stiffening ribb (4)		2.939		0.0009	0.12	81.66
51	Stand 1		3.033		0.005	1.28	79.13
52	Stand 2		3.023		0.005	1.28	79.39
53	Support-bracket		15.119		0.016	1.92	15.87
54	Axis	Steel 45 GOST 1050-88	15.270	360	0.034	3.44	23.57
55	Flange	St3 GOST 380-2005	2.087	240	0.017	0.168	114.99
56	Flange (2)		1.442		0.015	0.168	166.43
57	Stiffening rib		0.944		0.0004	0.28	254.24
58	Stiffening rib (2)		0.957		0.0004	0.28	250.78
59	Stiffening rib (3)		0.390		0.0003	0.28	615.38
60	Stiffening rib (4)		0.364		0.0003	0.28	659.34
61	Bearing	Steel SHH 15 GOST 801-78	23.997	410	0.003	0.188	17.08
62	Bearing (2)		23.879		0.003	0.188	17.17

first, its parts are built on the outer surface of the product, and only after that the elements are created inside the volume. At the same time, in the process of generating the finite element grid, the system analyzes its quality for the presence of various geometry defects, deviations from the ideal shape, etc.

After creating finite element models for all the elements involved in the computation, we create a set of FE assembly in NX, then find the necessary FE models in the list of files and add them to the assembly. It is extremely necessary to identify and resolve conflicting node and grid numbers; otherwise the computation will not be possible.

The next stage creates data set of assembly simulation; NX Nastran was selected as the solver. The contact conditions of the elements in the assembly were modeled using the surface-to-surface connection, which excludes the displacement of the selected elements.

When working with the computational model (simulation), the loads acting on the assembly were set; the data of them are described above. For the computation, the own weight of the structure was assigned, as well as the forces acting on the frame, the pressure devices and the axes of the support rollers. In addition to the loads, they must set the boundary conditions. A finite element assembly with applied loads and boundary conditions is shown in Fig. 2.

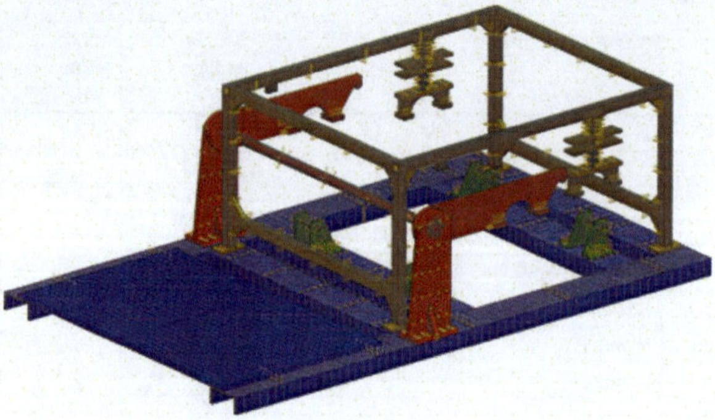

Fig. 2 The final design model of the HRM

3 Results and Discussion

Based on the computation results obtained using the NX Nastran solver, we made a conclusion on the working capability of the computed design under the specified loads and boundary conditions. For this purpose, the strength and rigidity conditions were checked.

The results of the computation showed that the beam 1 of the fixed stop is subjected to the greatest stresses. Therefore, we will check the strength condition on this element.

Checking the strength conditions involves comparing the maximum computational stresses obtained from the CAE analysis of the structure with those allowed for the selected material and determining the safety factor:

$$n = \frac{[\sigma]}{\sigma_{\max}} \geq [n], \tag{1}$$

there

n safety factor;
$[\sigma]$ working stress for the selected material, for steel St3 $[\sigma]$ = 240 MPa;
$\sigma_{\max}$ maximum design voltage, $\sigma_{\max}$ = 35.46 MPa;
$[n]$ working safety factor, $[n]$ = 4.

$$n = \frac{240}{35.46} = 6.77 > 4.$$

Strength condition $n = \frac{[\sigma]}{\sigma_{\max}} \geq [5]$ works.

According to the computation results, the greatest displacements are observed in the spring block flange. Therefore, we can check the stiffness condition for this part. Checking the stiffness condition involves comparing the maximum computation displacements with the working ones:

$$\delta_{\max} \leq [\delta], \tag{2}$$

there

$\delta_{\max}$ maximum computation displacement, $\delta_{\max}$ = 0.34 mm;
$[\delta]$ working movements.

$$[\delta] = \frac{a}{250},$$

there

a overall size of the structure, a = 38 mm.

$$[\delta] = \frac{38}{250} = 0.152\,\text{mm},$$
$$0.34 > 0.152\,\text{mm}.$$

Strength condition $\delta_{\max} \leq [\delta]$ is ruled out.

The computation results are presented in Table 1.

The computation results showed that in the elements Flange, Guide pipe and Stiffening rib that are part of the pressure device, the rigid condition is ruled out, i.e. the maximum displacements are greater than the working ones: δ = 0.304 mm, $[\delta]$ = 0.152 mm; δ = 0.248 mm, $[\delta]$ = 0.240 mm; δ = 0.264 mm, $[\delta]$ = 0.20 mm, respectively. To reduce the load on the flange, we increase its thickness by 3 mm and length by 20 mm. To reduce the load on the guide pipe and stiffening rib we will increase the wall thickness of the vertical stands and beams by 2 mm.

To check the working capability of the structure under the specified loads and constraints, a second test of the strength and rigid conditions was carried out (Fig. 3).

$$n = \frac{240}{22.498} = 10.67 > 4.$$

Strength condition $n = \frac{[\sigma]}{\sigma_{\max}} \geq [4]$ works.

$$[\delta] = \frac{58}{250} = 0.232\,\text{mm} > 0.232.$$

Strength condition $\delta_{\max} \leq [\delta]$ works.

Based on computation structural changes were made [13] to the elements of the horizontal roller mill frame, in particular, the geometry of the flange element was changed, as well as the wall thickness of the stiffening rib and pipe was increased. A re-calculation of the HRM support frame after the made changes showed that the strength and rigid conditions work for all elements of the assembly.

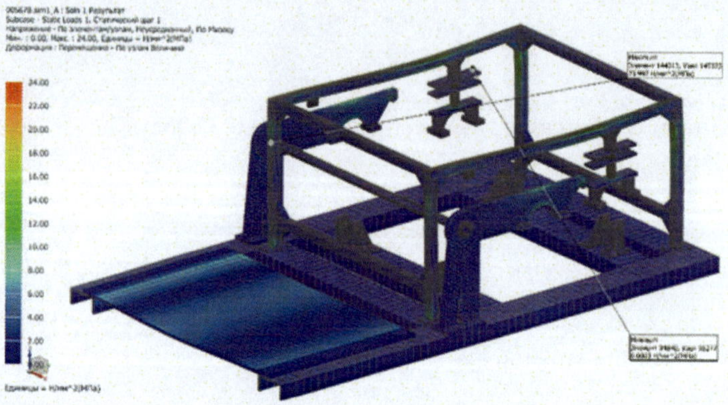

Fig. 3 The results of the stress computation in the support frame of the horizontal roller mill after the implementation of structural stresses

4 Conclusion

The strength computation and analysis of its results showed that the strength and rigid conditions for all elements work. The results of the structural analysis in NX CAE are a color diagram that shows the distribution of stresses and displacements under the influence of their total mass and applied external forces. The color of the diagrams changes from blue to red, where the minimum data of the parameter under study correspond to blue, and the maximum data are in red.

Acknowledgements This work was realized in the framework of the Program of flagship university development on the base of the Belgorod State Technological University named after V. G. Shukhov. The work was realized using equipment of High Technology Center at BSTU named after V.G. Shukhov.

References

1. Harder J (2010) Grinding trends in the cement industry. ZKG Int 4:46–58
2. Schnatz R (2013) Operational experience from India's first MVR vertical roller mill for cement grinding. Cement Int 5:46–53
3. Hoffmann D (2011) The MVR vertical roller mill plus MultiDrive—a successful combination. Cement Int 2:44–49
4. Bauman VA (1981) Mechanical equipment of enterprises of construction materials, products and structures. Mechanical engineering, Moscow
5. Bogdanov VS (2013) Fundamentals of calculation of machines and equipment of enterprises of construction materials and products. TNT Publishing House, Stary Oskol
6. 200 years of industrial progress. The fives FCB Horomill workshop. Case study for grinding plants in the Philippines. CemTech Asia 2014, Kuala Lumpur, http://app.cemtech.com/docs/sessions/. Last accessed 2021/03/18

7. Roll mill type TVM, http://www.tyazhmash.com/products/powerhouse/mill-roller-tvm/. Last accessed 2021/03/18
8. Khakhaleva TA, Bordunov IS, Fadin YM (2019) Review of modern technical solutions in the design of horizontal roller mills. Collection of reports of the XII international scientific and technical conference, vol 1. Gubkin, pp 420–423
9. Yeltsov MY, Khakhalev PA (2014) Fundamentals of calculating the strength of the product in the NX application Extended simulation: textbook: BSTU Publishing House, Belgorod
10. GOST 2.053–2013 Electronic structure of the product. General provisions. Standartinform, Moscow
11. Khakhalev PA, Samsonova PS (2019) Development of an electronic model of an assembly unit in NX under the control of Teamcenter. BSTU Publishing House, Belgorod
12. Yeltsov MY, Kozlov AA, Sedoykin AV, Shirokova LY (2013) Designing in NX under the control of Teamcenter. DMK Press
13. Bogdanov VS, Yeltsov MYu, Shirokova LYu, Khakhalev PA (2016) General principles of the development of equipment for the production of building materials based on various configurations of the product composition. Cem Appl 5:70–73

Disintegrator with Intensive Action on the Ground Material

A. V. Shatalov, A. N. Maslovskaya, V. A. Shatalov, and U. V. Golubeva

Abstract Grinding coarse-grained materials into fine powders is the most complex and energy-intensive technological operation in the production of various materials [1]. The need to obtain products with high dispersion is explained by the fact that their use is technologically and economically more efficient than coarse materials. The fineness and quality of grinding of materials is of great importance for the intensification of various technological processes. Recently, for the production of fine powders, mills of intensive action with a high speed of material loading have found wide application. To increase the efficiency of the grinding process and productivity, a disintegrator with inclined cylindrical spreading nozzles has been developed. Calculations and studies have shown that this design of the disintegrator can significantly increase abrasion and shock loads on the crushed material, and therefore intensify the grinding process and increase productivity. The practical value of the work lies in the creation, on the basis of theoretical developments and experimental studies, of an improved disintegrator design, which increases the efficiency of the grinding process. The results of the work can be used in the building materials industry for the production of various multi-composition mixtures.

Keywords Grinding · Intensification · Disintegrator · Finely dispersed material · Productivity

A. V. Shatalov (✉) · A. N. Maslovskaya · V. A. Shatalov
Belgorod State Technological University Named After V.G. Shukhov, Belgorod, Russia
e-mail: tkmm_bstu@mail.ru

U. V. Golubeva
Belgorod State National Research University, Belgorod 308015, Russia

S. V. Klyuev (ed.), *Digital Technologies in Construction Engineering*,
Lecture Notes in Civil Engineering 173, https://doi.org/10.1007/978-3-030-81289-8_26

1 Introduction

Any modern production of building materials is inextricably linked with the grinding of raw materials and semi-finished products [1]. The need to produce products with high dispersion is explained by the fact that, due to their increased reactivity, technical application is technologically and economically more efficient than coarse materials [2].

At the same time, grinding coarse-grained materials into fine-dispersed powders is one of the most complex and energy-intensive technological operations in the production of building materials, mineral processing, processing of materials in the chemical, fuel and other industries [3]. The dispersion of the resulting product largely determines the quality of the powders obtained and affects the increase in their technological and consumer properties.

Recently, for the production of fine powders, mills of intensive action with a high loading rate of the material have found wide industrial application, providing a complex shock-abrasive effect on material particles [4, 5]. Such grinding plants are disintegrators, which have a relatively high specific productivity, low specific energy consumption, small dimensions, as well as the ability to grind material with natural moisture.

The aim of this research was, based on the analysis of existing structures, to develop a fundamentally new design of the disintegrator, determine the rational modes of the grinding process, make a calculation of its main design and technological parameters.

2 Materials and Methods

To increase the efficiency of the grinding process and productivity, we have developed a disintegrator with obliquely installed cylindrical spreading nozzles, which contribute to the radial acceleration of particles in the direction of the rows of impact elements (Fig. 1) [6, 8–10].

The disintegrator contains a cylindrical body 1 with a tangential discharge pipe 2, an axial loading pipe 3, to which the upper disk 4 is rigidly attached. Concentrically located striking elements 5, 7, 8 are installed on the upper 4 and lower 6 disks. The pitch between the striking elements of the first inner row is greater than the striking elements of the subsequent rows, and this pitch depends on the size of the starting material so that the pieces of the starting material freely pass between the striking elements of the first inner row. The cross-section of the percussion elements is shown in the form of a square.

In the central part of the grinding chamber, at the outlet of the loading nozzle 3, there are cylindrical spreading nozzles 9, bent in the direction opposite to the direction of rotation of the upper disc, at the end of each of the spreading nozzles, a diffuser 10 is fixed, the larger diameter D of which is equal to 0.8 h of the height of

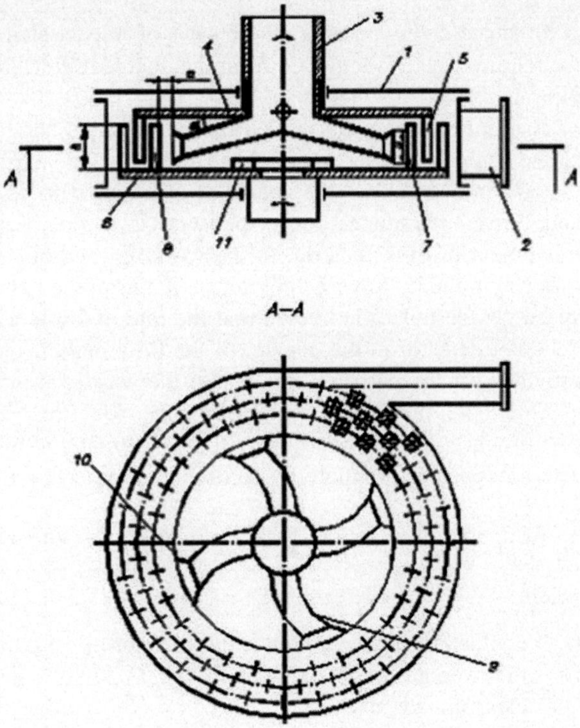

Fig. 1 Disintegrator with obliquely mounted cylindrical spreading nozzles

the shock elements, and the distance a between the ends of the diffusers and the impactors exceeds the maximum size of crushed particles. The angle of inclination of the spreading nozzles is greater than the angle of repose of the crushed material. On the lower disc 6, the blades of the fan wheel 11 are rigidly fixed, creating a ventilation effect in the dead zone under the spreading nozzles. The maximum pitch between the impact elements is 0.1–0.5 of the particle diameter of the initial product, and the minimum pitch is 1.2–1.3 of the particle diameter of the finished product.

The disintegrator works as follows (Fig. 1). The material to be crushed is fed through the loading nozzle 3 and directed to the spreading nozzles 9, from where, under the action of centrifugal forces and wall pressure forces through the diffuser 10, it is thrown into the zone of action of the first inner row of impact elements 7 and receives the speed corresponding to this row. Pieces of material are destroyed and, under the action of the tangential component of the impact force and centrifugal force, are thrown onto the next row of impact elements 5, rotating in the opposite direction.

Striking the impact elements of the second row, the material particles bounce off it, changing the velocity vector, and are ejected from the trajectory of the second row, moving further. The use of the spreading nozzles 9 can significantly increase the speed of the radial emission of the particle to the rows of impact elements.

After successively passing all the rows of impact elements, the material, crushed to the required size, is removed from the disintegrator through the tangential discharge pipe 2.

A number of designs of centrifugal batchers has been developed for the accelerated supply of raw materials to the zone of active action, in which the bulk material moves under the action of centrifugal force. To determine the operating and design parameters of dispensers, it is necessary to know the patterns of movement of bulk material in the working area of the dispenser. The state of the material in the hollow loading nozzle 3 during the expiration is characterized by the speed of free fall. It is known that the rate of fall is a function of the height H and does not depend on the density of the flow particles [1]. The design of the shaft 3 provides for the particles to fall into the acceleration zone from a certain height.

One of the main characteristics of the considered disintegrator is performance. With a continuous supply of material by booster pipes to the zone of active exposure, productivity can be determined by the following formula:

$$Q = \psi \cdot D_i \cdot C \cdot z \cdot Vo \cdot \rho, \tag{1}$$

where

ψ coefficient taking into account the run-up of particles in a centrifugal field;
Di the average particle size, m;
C clear impact size, m;
z number of the striking elements of the first inner row;
Vo speed of material movement along the working surface of the striking element;
ρ bulk density of material, kg/m^3.

3 Results and Discussions

Figure 2 shows the layout of the first and second inner rows of the disintegrator striking elements. Let's introduce the following designations: O—the center of rotation of the rotors; R_1 and R_2—distances from the center of rotation of the rotors to the middle of the striking elements, respectively, of the first and second rows; a—side of the square of the cross-section of the striking element; DE = L_1—the minimum distance between the working elements of the second row, at which the slip of particles between them is possible.

Having determined the angles β_2 и β_1 from the triangles AOB и AOC, the minimum distance between two percussion elements in the second row is obtained:

$$L1 = DE = 2DK = 2R2\sin\frac{\arcsin\frac{(R_1+\frac{a}{2})\cos\alpha}{R_2-\frac{a}{2}} - \arcsin\frac{(R_1+\frac{a}{2})\cos\alpha}{R_2+\frac{a}{2}}}{2} \tag{2}$$

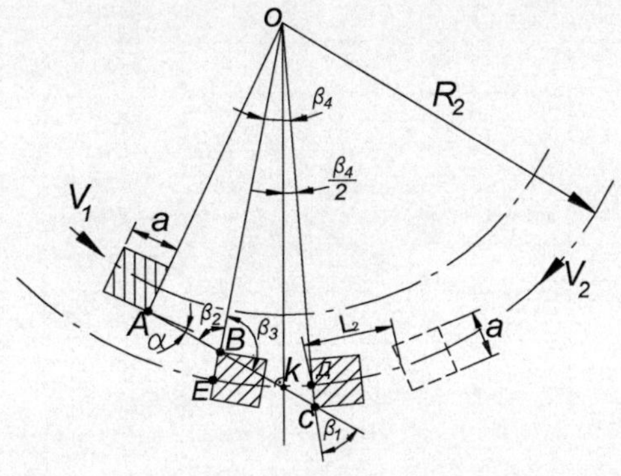

Fig. 2 Scheme for determining the distance between adjacent impactors

In fact, the second row moves towards and L_1 will be larger than in Eq. (2). During the movement of a particle of material with a velocity V on a segment of the BC, the impactor of the second row must pass the path L_2 with an angular velocity of ω_2. The minimum distance L between two adjacent impactors is determined as the sum of L_1 and L_2:

$$L = 2R2\sin\frac{\left[\arcsin\frac{\left(R_1+\frac{a}{2}\right)\cos\alpha}{R_2-\frac{a}{2}} - \arcsin\frac{\left(R_1+\frac{a}{2}\right)\cos\alpha}{R_2+\frac{a}{2}}\right]}{2} + \frac{\omega_2 R_2}{V_1}\left(R_2-\frac{a}{2}\right)\frac{\sin\left[\arcsin\frac{\left(R_1+\frac{a}{2}\right)\cos\alpha}{R_2-\frac{a}{2}} - \arcsin\frac{\left(R_1+\frac{a}{2}\right)\cos\alpha}{R_2+\frac{a}{2}}\right]\cdot\left(R_2+\frac{a}{2}\right)}{\left(R_1+\frac{a}{2}\right)\cos\alpha} \quad (3)$$

Similarly, we determine the distance between adjacent percussion elements on the third row. The number of collisions of material particles with impact elements in the grinding chamber exceeds the number of rows of impact elements due to an increase in the concentration of material particles in the material processing zones (2nd and 3rd rows of impact elements). The grinding degree reaches 500 and more [7, 8].

Thus, in the course of analytical studies, the dynamics of movement of particles of the crushed material in the accelerating units of the disintegrator was established, taking into account its design features. Analytical relationships are obtained for calculating the trajectory of movement of material particles in the grinding chamber of a disintegrator with various impact elements. The use of a disintegrator developed by us [6] with inclined cylindrical spreading nozzles increases the concentration of material particles in the flowing zones of processing of the 2nd and 3rd circles with impact elements, leads to an increase in abrasive and shock loads on the crushed material, makes it possible to significantly intensify the grinding process and increase performance. Experimental studies were carried out on the experimental setup we created. Figure 3 shows the experimental dependence of the

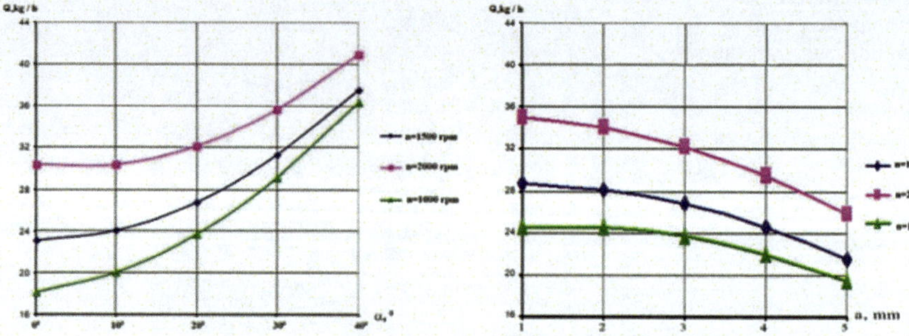

a) b)

Fig. 3 **a** Dependence of productivity on the angle of installation of the plates Q = f(α), **b** Dependence of productivity on the gap between the rows of beats in the last row Q = f(a)

productivity Q on the angle of installation of the plates, α and on the gap between the rows of beats, *a* in the last row. The maximum efficiency is achieved at $\alpha = 40^0$ and a clearance of *a* = 1 mm.

4 Conclusion

1. Based on the analysis of the development of technology and technology for disintegrator grinding, a new design of a disintegrator with horizontal discs is proposed.
2. Analytical relationships were obtained for calculating the trajectory of movement of material particles in the grinding chamber of a disintegrator with various impact elements.
3. Derived analytical dependences to determine: the speed of movement of particles of the material in the disintegrator; the number of striking elements in each row of the disintegrator with a different cross section; productivity and power consumption, taking into account the design features and the mode of the grinding process.
4. Experimental studies have been carried out to determine the optimal parameters of the grinding process.

Acknowledgements This work was realized in the framework of the Program of flagship university development on the base of the Belgorod State Technological University named after V. G. Shukhov. The work was realized using equipment of High Technology Center at BSTU named after V. G. Shukhov.

References

1. Uralsky VI, Sinitsa EV, Uralsky AV, Sazhneva EA (2017) Technological module of the closed circulation cycle. Bulletin of BSTU named after V.G. Shukhov 10:144–148
2. Sevostyanov MV (2016) Resource-saving equipment for complex processing of man-made materials. Bulletin of BSTU named after V.G. Shukhov 4:140–145
3. Shatalov AV, Dubinin NN, Shevlyakov VI (2014) Energy-saving grinding complex. Ecology and rational nature management as a factor of sustainable development: Sat. doc. Intern. scientific-techn. Conf. BSTU Publishing House, Belgorod
4. Glagolev S, Shein N, Sevostianov V, Obolonsky V, Shamgulov R (2020) Technologies for integrated processing of solid municipal waste. Ecol Ind Russ 24(12):11–15
5. Dubinin NN, Evtushenko EI, Nemets II, Nosov OA, Osokin AV (2014) Rotary machines for production of ceramic wall materials. Res J Pharm Biol Chem Sci 1710–1718
6. Maslovskaya AN, Bogdanov VS, Semikopenko IA, Sharapov RR (2007) Patent of the Russian Federation № 2291745, B02C 13/22 Disintegrator
7. Shatalov AV, Mikhailichenko CA (2016) Technological complexes for grinding silica-fine materials. Monograph. BSTU named after V.G. Shukhov, Belgorod
8. Maslovskaya AN, Bogdanov (2009) Disintegrator with high-speed feed unit for crushed material. Bulletin of BSTU named after V.G.Shoukhov 1:101–103
9. Sevostyanov MV, Ilina TN, Uvarov VA, Shinkarev LI (2015) Methods for compacting technological materials and technical tools to implement them. Bulletin of BSTU named after V.G. Shukhov 2:107–111
10. Sevostyanov VS, Shatalov AV, Shatalov VA (2018) Processing of basalt fiber production waste. IOP conference series: materials science and engineering 327

Control of Physical and Chemical Processes at the Phase Boundary in the Formation of Building Composites

A. M. Ayzenshtadt, M. A. Frolova, Yu. V. Sokolova, and T. A. Drozdyuk

Abstract This work aims to demonstrate methodological approaches to solving problems of controlling the transformation processes of matter at the phase boundary of the macro-surface and in dispersed systems, which, in our opinion, can be used as criteria for the quantitative characterization of one of the fundamental laws of the scientific direction "Geonics"—the law of "Affinity of structures". The proposed approaches are based on the fundamental principles of physical and colloidal chemistry, crystal energy. The energy interpretation of the kinetic model of the topochemical interaction of the components and the analog Hamaker constant value are proposed as criteria. The methodological principles of experimental determination of these characteristics are developed. The proposed approaches were tested on the example of various fine-dispersed compositions: glyoxal-bark, basalt-bark, and basalt-polymineral quartz-feldspar sand.

Keywords Structure affinity law · Kinetic model · Surface tension · Rate constant · Hamaker constant · Interfacial interaction · Dispersed systems

1 Introduction

The priority direction of research in the Russian and world scientific practice is the creation of innovative materials based on non-traditional raw materials using the principles of rational nature management, the introduction of new "green" materials and the reuse of natural and technogenic waste. The implementation of this provision contributes to improving materials production efficiency, reducing the burden on the environment and creating a sustainable system of "man-material-habitat" [1, 2]. In this regard, the interest of researchers in new chemical compounds and approaches in the formation of building compositions is understandable [3]. For example, as an unconventional binder, glyoxal, synthesized by the oxidation of

A. M. Ayzenshtadt · M. A. Frolova · Yu. V. Sokolova (✉) · T. A. Drozdyuk
Northern (Arctic) Federal University named after M.V. Lomonosov, Arkhangelsk, Russia
e-mail: yu.sokolova@narfu.ru

S. V. Klyuev (ed.), *Digital Technologies in Construction Engineering*,
Lecture Notes in Civil Engineering 173, https://doi.org/10.1007/978-3-030-81289-8_27

ethylene glycol, is of considerable interest [4, 5]. Glyoxal is able to form spontaneously a polymer structure even at negative temperatures. The product is one of the substitutes for formaldehyde, however, unlike the latter, it belongs to the 3rd hazard class, non-toxic [6]. The effectiveness of the use of this organic compound as a binder for strengthening weak ground bases of highways and engineering communications, as well as a reactive binder component of cementless ground concrete was established [7].

Another example of obtaining environmentally efficient building materials using non-traditional components is thermal insulation composites based on basalt fibers and saponite-containing binder [8]. The use of these raw materials makes it possible to obtain materials that retain their thermophysical characteristics in a wide humidity-temperature range of operation.

The work related to the controlled process of mineralization of the wood matrix [9] made it possible to obtain environmentally friendly building materials from plant raw materials with increased physical and mechanical, thermophysical and fire-technical characteristics, resistant to oxidative destruction.

Achieving a synergistic effect of the interaction of structural elements of different genesis, composition and morphology in the production of compositions with characteristics that exceed the properties of the original components is possible only with the interaction between the structural elements at the phase boundary, which forms strong bonds (physical and/or chemical nature) between elements characterized by physical and chemical affinity.

The main role in these processes is played by the surface of dispersed materials, which differs from the bulk phase by an increased energy potential due to the presence of active centers [10]. Based on this, the control system for the processes of forming the quality of building composites should be based on the functional relationship between the composition of the components, their properties and the characteristics of the contact surface. Therefore, one of the means of optimizing the composition of the developed composite is to comply with the law of "Affinity of structures" [11], the quantitative expression of which can be criteria determined on the basis of the fundamental provisions of physical chemistry.

This work aims to demonstrate several methodological approaches to solving problems of controlling the transformation processes of matter at the phase boundary of the macro-surface and in dispersed systems, which, in our opinion, can be used as such optimization criteria.

2 Methods and Materials

2.1 *Energy Interpretation of the Kinetic Model*

To control the processes of structure formation at the phase boundary of the macro-surface and in dispersed systems and to create composite materials of optimal composition on their basis, a kinetic model of topochemical interaction was proposed, taking into account the diffusion and kinetic processes at the phase boundary. The information criterion for creating such a model, as shown in [12], is the surface tension that occurs at the interface between the dispersed phase—the dispersion medium.

Mathematical expression for the description of the kinetics of topochemical transformations is obtained based on two basic principles: heterogeneous reactions at the phase boundary are first-order reactions [13, 14], and the degree of transformation of the active centers of the surface (α) is proportional to its surface tension (σ). The main kinetic equation of the topochemical process will have the following form:

$$\frac{\mathrm{d}\left(1-\frac{\sigma_{\mathrm{t}}}{\sigma_0}\right)}{\mathrm{d}t}=k\cdot S\cdot\left(1-\frac{\sigma_{\mathrm{t}}}{\sigma_0}\right), \tag{1}$$

where k—the rate constant of the solid-phase process, kg/(m^2 min); S—specific surface area of the solid phase, m^2/kg; t—interaction time, min.

It should be noted that this kinetic equation is applicable for a system in a quasi-equilibrium state. Therefore, when integrating the expression (1), it is necessary to choose the initial time period t_1 (reaching the end time of the induction period of interaction and equalizing the rates of supply and consumption of the substance as a result of the chemical reaction), which will correspond to the parameter σ_1. In integral form, this equation takes the following form:

$$\ln\frac{\sigma_1-\sigma_{\mathrm{t}}}{\sigma_0-\sigma_1}=K(t-t_1)-\ln\frac{\sigma_0}{\sigma_1}. \tag{2}$$

The slope of the functional dependence

$$\ln\frac{\sigma_1-\sigma_{\mathrm{t}}}{\sigma_0-\sigma_1}=f(t-t_1). \tag{3}$$

allows calculating the rate constant of the interaction of reagents on the interface during the formation of the desired composition. Comparison of the absolute values of the constant $k = K/S$ [12] for different reaction conditions (or changes in reacting components) can be used to estimate the affinity of structures.

The proposed energy model was tested on two compositions containing finely dispersed bark of Scots pine as the solid phase [12]. A distinctive feature was the composition of the added liquid phases. In the first case, an aqueous solution of glyoxal (Composition 1) was used, in the second—an aqueous suspension of finely dispersed basalt (Composition 2). The main information parameter (taking into account that the surface of the solid phase participates in the topochemical interaction with its active centers) was the polarization component value of the surface tension ($\sigma_{\mathrm{S}}^{\mathrm{P}}$), determined by the method of "Owens, Wendt, Rabel and Kaelble" (OWRK) [15–17]. The specific surface area of the finely dispersed bark and the "bark-basalt" mixture was measured by the sorption method (BET method).

2.2 The Van Der Waals Interaction Constant

The key element of this composite design method is to determine the value of the van der Waals interaction constant. Thus, the DLFO theory [18, 19] gives a directly proportional relationship between the van der Waals interaction constant of dispersed particles (A_B) and the Hamaker constant (A):

$$A = \frac{\pi^2 N_A^2}{V_m} A_B, \tag{4}$$

where V_m—molar volume; N_A—Avogadro constant.

The practical implementation of such a solution can be associated with the initial stage of the selection of raw materials, which consists of the experimental determination of the Hamaker constant of the composition of the raw mixture, the combination of components in which gives the maximum value A [20]. So, if we denote the Hamaker constant of the composition as A_{1-2}, and a similar parameter for, for example, two assumed raw materials components A_1 and A_2, then, based on [18], this principle will be implemented based on the following mathematical expression:

$$A_{1-2} = \sqrt{A_1 A_2}, \tag{5}$$

In [20], we revealed the methodological features of determining the Hamaker constant for fine powder systems, however, due to the current inability to overcome some instrumental difficulties in determining this parameter, the calculation allows us to talk about the analog value of this constant (A_m).

This methodological approach was tested on a finely dispersed two-component system of basalt-polymineral quartz-feldspar sand.

3 Results and Discussion

3.1 Testing the Kinetic Model

As a result of the conducted experiments, in which the concentration ratios of the components of the interacting mixtures were the variable parameters, the following mathematical expressions were obtained for Eq. (3) (confidence coefficient $R = 0.95 \pm 0.03$):

$$\ln\frac{\sigma_1 - \sigma_t}{\sigma_0 - \sigma_1} = 3.0 \times 10^{-3}(t - t_1) + 0.33, \tag{6}$$

$$\ln\frac{\sigma_1 - \sigma_t}{\sigma_0 - \sigma_1} = 1.2 \times 10^{-2}(t - t_1) + 0.01, \tag{7}$$

Table 1 Experimental and calculated parameters for determining the reaction rate constant

Studied system	Constant term $b = \ln\frac{\sigma_0}{\sigma_1}$ of the dependence $y = Kx + b$	Slope K of the dependence $y = Kx + b$	Specific surface area, S, m^2/kg	Reaction rate constant, k, kg/(m^2 min)
Composition 1	0.33	0.003	111,671	2.69×10^{-8}
Composition 2	0.01	0.012	92,408	2.07×10^{-9}

Expression (6) is obtained for Composition 1, and (7) is obtained for Composition 2. Table 1 shows the experimental and calculated data of the kinetic experiment.

The obtained data show that the rate constant of the topochemical interaction of glyoxal with the active centers of the mechanically activated bark is higher than the rate constant of the interaction in the finely dispersed basalt-bark system. At the same time, from the point of view of the chemical composition and structure of the reacting objects in Composite 1 compared to Composite 2, the affinity of the structures is much higher, which is quantitatively confirmed by the values of the constants determined by the proposed method.

3.2 *Evaluation of the Van Der Waals Interaction*

For a finely dispersed two-component system of basalt-polymineral quartz-feldspar sand, the constants A_{11} (basalt powder) and A_{12}—powder in which a part of the dispersed phase (1—basalt) is replaced by fine particles (2—polymineral sand) were calculated (Table 2).

Table 2 Indicators of the dispersion interaction and the wetting angle of the surface

No	Angle θ_0 for solutions with different concentrations of ethanol (%)				$(A_{11}, A_{12})10^{20}$ (J)
	96	80	70	60	
1	21.0	21.4	21.9	22.4	7.22
2	20.3	21.6	22.0	22.6	8.07
3	20.6	21.8	22.2	22.8	8.00
4	19.8	20.9	21.9	22.2	8.36
5	19.6	20.3	21.4	22.3	8.67
6	19.3	19.9	21.2	22.1	8.76
7	19.1	19.6	20.8	22.0	8.82
8	18.9	19.7	20.7	21.7	8.61
9	18.8	20.3	21.0	21.8	8.67
10	19.7	20.5	21.2	22.2	8.36

The functional relationship of the following values is established: $(A_{11}, A_{12}) = f(C_s)$, which is well described by the equation of a polynomial of the second degree and has the following analytical expression ($R^2 = 0.92$):

$$(A_{11},A_{12}) = -10^{-23}C_s^2 + 8.15 \times 10^{-22}C_s + 7.38 \times 10^{-20}, \quad (8)$$

where C_s—fine sand powder content, %.

Expression (8) allowed calculating the quantitative content of the components in this two-component system, which is characterized by the maximum value of the analog Hamaker constant (the best variant of compliance with the law of "Affinity of structures"). In this case, the ratio of the components is equal to 59% basalt powder and 41% sand powder.

4 Conclusion

The proposed approaches to the identification of the key functions of interfacial interaction in the formation of the properties of building composites are based on the fundamental principles of physical and colloidal chemistry, crystal energy. This fact allows identifying the criteria for the quantitative characteristics of one of the fundamental laws of the scientific direction "Geonics"—the law of "Affinity of structures". It is promising to use the energy interpretation of the kinetic model of the topochemical interaction of components and the value of the analog Hamaker constant as these criteria.

Acknowledgements The research was carried out on the unique scientific equipment "Physical Chemistry of Surfaces of Nano-Dispersed Systems" within the framework of state task No. 0793-2020-0005.

References

1. Lesovik VS (2006) Improving the efficiency of production of building materials, taking into account the genesis of rocks. Associations of construction universities, Moscow
2. Lesovik VS (2012) Geonics. Subject and tasks. BSTU Publishing House, Belgorod
3. Korolev EV (2020) Prospects for the development of construction materials science. Academia. Archit Constr 3:143–159
4. Salaev MA, Krejker AA, Magaev OV, Malkov VS, Knyazev AS, Borisova ES, Khanaev VM, Vodyankina OV, Kurina LN (2011) Ethylene glycol oxidation over supported catalyst in tubular reactor. Chem Eng J 172(1):399–409. https://doi.org/10.1016/j.cej.2011.05.079
5. Mamontov GV, Knyazev AS, Paukshtis EA, Vodyankina OV (2013) Adsorption and conversion of ethylene glycol on the surface of Ag-containing catalyst modified with phosphate. Kinet Catal 54(6):735–743. https://doi.org/10.1134/S0023158413060086
6. Kolychev NM, Arzhakov VN, Arzhakov PV, Serikbaev RE, Kuchkina MA (2013) Glyoxal—a broad-spectrum antimicrobial disinfectant. Sci J KubSAU 87(03):1–10

7. Sokolova YV, Ayzenshtadt AM, Malkov VS, Strokova VV (2018) Glyoxal and its derivatives for stabilization of aluminium silicate colloid-dispersed systems. J Phy: Conf Ser 1135:012109. https://doi.org/10.1088/1742-6596/1135/1/012109
8. Drozdyuk T, Ayzenshtadt A, Frolova M, Nosulya A (2018) Thermal insulation composite with the use of mining waste. IOP Conference Series: Materials Science and Engineering 365032062. https://doi.org/10.1088/1757-899X/365/3/032062
9. Ayzenshtadt A, Lesovik V, Frolova M, Tutygin A, Danilov V (2015) Nanostructured wood mineral composite. Proc Eng 117:45–51
10. Cao G, Ying W (2012) Nanostructures and nanomaterials. Synthesis, properties and applications. Scientific World Publishing house, Moscow
11. Lesovik VS, Zagorodnyuk LKh, Chulkova IL (2014) The law of affinity of structures in materials science. Fundam Res 3–2:267–271
12. Ayzenshtadt AM, Sokolova YV, Danilov VE, Frolova MA (2020) Energy interpretation of a kinetic model of topochemical reactions. Solid State Technol 63(6):2530–2541
13. Mihajlovic I, Strbac N, Zivkovic Z (2004) Kinetic modelling of chalcocite particle oxidation. Scand J Metall 33(6):316–321
14. Stromberg AG, Semchenko DP (2006) Physical chemistry, 6th edn. Higher School, Moscow
15. Ariawan D, Mohd Ishak ZA, Salim MS, Mat Taib R, Ahmad Thirmizir MZ (2017) Wettability and interfacial characterization of alkaline treated kenaf fiber-unsaturated polyester composites fabricated by resin transfer molding. Polym Compos 38(3):507–515
16. Encinas N, Pantoja M, Abenojar J, Martínez MA (2010) Control of wettability of polymers by surface roughness modification. J Adhes Sci Technol 24(11–12):1869–1883
17. Karl CW, Klüppel M (2011) Characterization of elastomers by wetting: Roughness and chemical heterogeneity. Chem Listy 105(15):275–276
18. Deryagin BV, Churaev NV, Muller VM (1985) Surface forces. Science, Moscow
19. Deryagin BV, Abrikosov EM, Lifshits EM (2015) Molecular attraction of condensed bodies. Phys Chem Success 185(9):982–1001
20. Ayzenshtadt AM, Shinkaruk AA, Frolova MA (2021) Possible criterion for evaluating the compatibility of components in the building mixtures. Lect Notes Civil Eng 95:280–286

Removal of Cu^{2+} Ions from Aqueous Media by Modified Biomass of *Pisum sativum* Pods

K. I. Shaikhieva, R. Z. Galimova, L. V. Denisova, and E. S. Antyufeeva

Abstract The possibility of using native and chemically modified pea pod shells biomass to remove the Cu^{2+} ions from model aqueous solutions was investigated. Pea pods were modified by treating their surface with 1, 2 and 3% solutions of sulfuric acid and sodium hydroxide. The sorption properties of native and modified pea pods with respect to copper (II) ions were studied. It was found that the sorption capacity of pea pods treated with 2% sulfuric acid solution with respect to Cu^{2+} ions was 3 times as high as the sorption properties of native *Pisum sativum* pods (0.97 mmol/g and 0.33 mmol/g, respectively). When pea pods were modified with sodium hydroxide, an increase in the sorption capacity of the materials was observed with an increase in the concentration of sodium hydroxide solution—0.33, 0.36, 0.66, and 0.96 mmol/g for native pea pods and those modified with 1, 2, and 3% sodium hydroxide solution. Using the resulting constants of the Langmuir and Dubinin-Radushkevich equations for the processes under study, the thermodynamic parameters of Cu^{2+} ions adsorption from aqueous media by native and acid-treated pea pods were calculated. It is obvious that the chemically modified sorption material has a higher sorption capacity, which makes it possible to extract Cu^{2+} ions from aqueous solutions more efficiently, which will reduce the anthropogenic load on the natural environment.

Keywords Pea pods · Adsorption isotherms · Gibbs energy · Sorption energy

K. I. Shaikhieva (✉) · R. Z. Galimova
Kazan National Research Technological University, Kazan, Russia
e-mail: shaykhievak@gmail.ru

L. V. Denisova · E. S. Antyufeeva
Belgorod State Technological University Named After V.G. Shukhov, Belgorod, Russia

S. V. Klyuev (ed.), *Digital Technologies in Construction Engineering*,
Lecture Notes in Civil Engineering 173, https://doi.org/10.1007/978-3-030-81289-8_28

1 Introduction

Currently, the world community is rapidly developing a new innovative environmental protection area—the use of agricultural and wood biomass wastes as reagents for removing pollutants from water environments. The advantages of such reagents are their high tonnage, low cost, annual production and often high efficiency compared to commercial samples. Numerous reviews show the possibility of using agricultural waste biomass for the removal of heavy metal ions [1–10], dyes [9, 11–15], oil and petrochemical products [16–19], pesticides [20], and other pollutants.

The scope of research is so large that recently review articles are regularly published on the use of various agricultural wastes as sorption materials (SM).

In the temperate zone and southern regions of European Russia, peas (*Pisum sativum*) are grown for the production of human food. After processing the biomass of this plant, the waste is, at best, fed to farm animals and at worst it rots in piles, worsening the environmental situation in the processing areas.

The world literature provides information about the use of by-products of the legume crops (*Fabáceae*), dicotyledonous plants of the *Fabales* order, as SM.

In particular, information is provided on the use of the shells of bean pods (*Phaseolus vulgaris*) [21, 22], chickpeas (*Cicer arietinum*) [23], pigeon peas (*Cajanus cajan*) [24], cowpeas (*Vigna unguiculata*) [25], and other legumes as SM to remove heavy metal ions from aqueous media.

Based on the above, the adsorption of Cu (II) ions by native and chemically treated pea (*Pisum sativum*) pods was studied.

2 Materials and Methods

The surface morphology of sorption materials was studied using an Innova (Bruker) scanning probe microscope in the intermittent contact mode. RTESP (Veeco)rectangular cantilevers with silicone probes were used for scanning. The resonant frequency of these cantilevers is in the range of 250–350 kHz, and the radius of curvature of the probe is 10–13 nm. To eliminate the distortions related to the "jitter" of the microscope affected by external noise, SG0508 anti-vibration system was used, which is able to smooth out vibrations with a frequency of up to 0.5 Hz (lower limit). The results of the study of the surface morphology of native and modified pea pods are shown as histograms of the sorption materials surface roughness distribution.

Cu (II) ions were absorbed with native and modified pea (*Pisum sativum*) pods at 20 °C (the room temperature). To this end, the suspended sorption material weighing 1 g was placed in flat-bottomed 250 cm^3 flasks. Then 100 cm^3 of copper (II) sulfate solution was added, the concentration of Cu(II) ions in which varied from 100 to 1500 mg/dm^3.

The flasks with weighed portions of SM were tightly closed with stoppers and shaken for 5 h. After that, the solution was filtered through Blue Ribbon paper filter and the equilibrium concentrations of Si(II) ions in the filtrates were determined by trilonometric titration in the presence of murexide indicator.

Based on the obtained data, the sorption capacity of pea (*Pisum sativum*) pods (A) was calculated using Formula 1 and sorption isotherms were constructed shown in Fig. 1.

$$A = (C_s - C_e) \cdot V/m, \quad (1)$$

where

A is the sorption capacity for heavy metal ions (mmol/g),
S_s is the initial concentration of heavy metal ions (mmol/dm^3),
C_e is the concentration of heavy metal ions after sorption (mmol/dm^3),
V is the solution volume (dm^3), m is the weight of the sorption material (g).

3 Results and Discussion

The histograms of the surface roughness distribution of native pea pods and those modified with 3% solution of sulfuric acid and 3% solution of sodium hydroxide are shown in Fig. 1a–c, respectively.

It was found that the modification of pea pod surface with sulfuric acid and sodium hydroxide solutions resulted in the formation of a more rough surface. It is known that sorption materials with a more developed surface show better sorption properties compared to materials with a uniform surface.

The isotherms of Cu^{2+} ions adsorption from aqueous media by native and modified pea pods with sulfuric acid and sodium hydroxide solutions of different concentrations are shown in Figs. 2 and 3, respectively.

Figure 2 shows that the sorption capacity of pea pods modified with 2% sulfuric acid solution for copper (II) ions is three times as high (0.97 mmol/g) as native pea

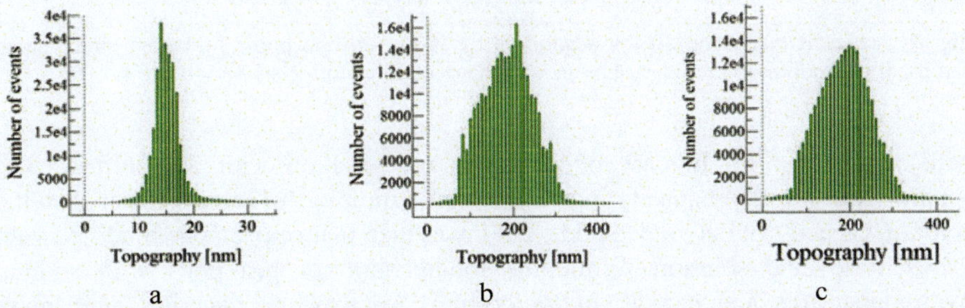

Fig. 1 Histograms of the surface roughness distribution of pea pods: **a** before modification and after modification with 3% solutions of, **b** sulfuric acid, **c** sodium hydroxide

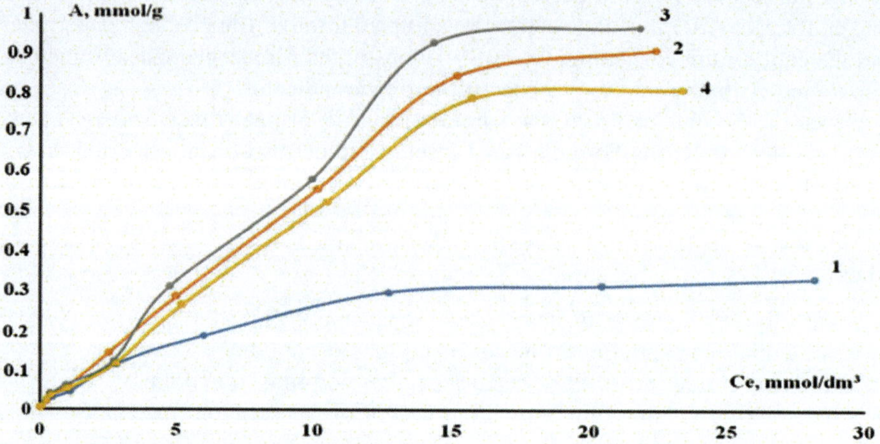

Fig. 2 Isotherms of copper (II) ion adsorption by *Pisum sativum* pods: 1—before modification and modified sulfuric acid solutions with concentrations of: 2—1%, 3—2%, 4—3%

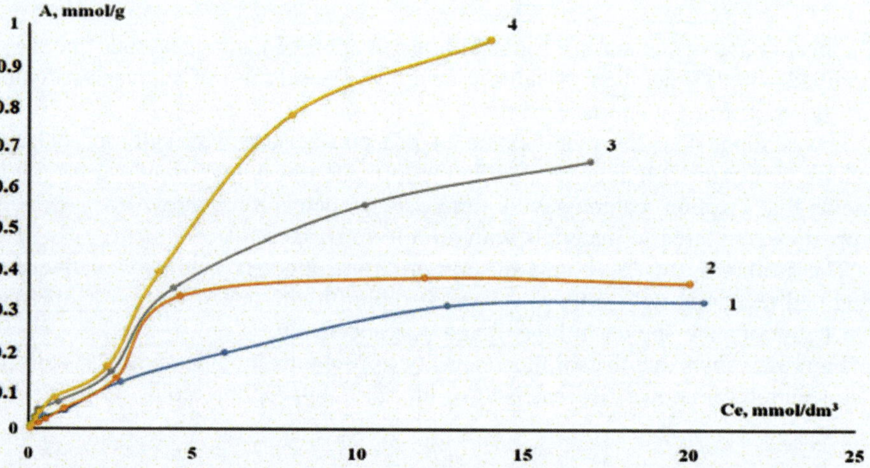

Fig. 3 Isotherms of copper (II) ion adsorption by *Pisum sativum* pods: 1—before modification and modified sodium hydroxide solutions with concentrations of: 2—1%, 3—2%, 4—3%

pods (0.33 mmol/g). In case of pea pods modification with 3% sulfuric acid solution, the sorption capacity increases to 0.81 mmol/g, which is greater than the sorption capacity of native pea pods, but lower than that of pea pods modified with 2% sulfuric acid solution. While the modification of pea pods with sodium hydroxide results in an increase in the sorption capacity of the materials, an increase in the concentration of sodium hydroxide solution—0.33, 0.36, 0.66, and 0.96 mmol/g for native pea pods and those modified with 1, 2, and 3% sodium hydroxide solution.

The results of processing the isotherms of copper (II) ion adsorption with native and modified pea pods in the framework of the most common monomolecular sorption models: Langmuir, Freundlich, and Dubinin-Radushkevich are presented in Table 1.

Table 2 shows that almost all the processes considered are best described by the Langmuir model, that is, adsorption occurs on a uniform surface of the SM where all active centers have the same sorption energy and the sorption process occurs uniformly throughout the material surface. In some cases (for pea pods modified with 1% solution of sulfuric acid and sodium hydroxide), there is a better similarity with the Freundlich model. This phenomenon can be accounted for by the fact that the modification of the material in this case did not proceed evenly, which resulted in the formation of active centers with a higher sorption energy and the uneven process of copper (II) ions adsorption.

Table 1 Equations and values of approximation coefficients of isotherms of Cu (II) ions sorption by pea pods

	The adsorption model		
	Langmuir	Freundlich	Dubinin-Radushkevich
Before modification	$y = 6.72x + 9.24$ $R^2 = 0.99$	$y = 0.71x - 1.32$ $R^2 = 0.99$	$y = -7.88x - 1.24$ $R^2 = 0.88$
1% H_2SO_4	$y = 6.39x + 10.57$ $R^2 = 0.95$	$y = 0.83x - 1.14$ $R^2 = 0.99$	$y = -5.58x - 0.88$ $R^2 = 0.67$
2% H_2SO_4	$y = 4.18x + 8.71$ $R^2 = 0.98$	$y = 0.26x + 0.25$ $R^2 = 0.63$	$y = -2.99x - 1.00$ $R^2 = 0.67$
3% H_2SO_4	$y = 6.09x + 13.20$ $R^2 = 0.98$	$y = 0.80x - 1.15$ $R^2 = 0.99$	$y = -5.44x - 0.75$ $R^2 = 0.75$
1% NaOH	$y = 27.03x - 11.53$ $R^2 = 0.88$	$y = 0.99x - 1.33$ $R^2 = 0.97$	$y = 1.35x - 3.52$ $R^2 = 0.01$
2% NaOH	$y = 6.03x + 2.50$ $R^2 = 0.99$	$y = 0.76x - 1.05$ $R^2 = 0.98$	$y = -3.99x - 0.41$ $R^2 = 0.76$
3% NaOH	$y = 8.34x + 1.01$ $R^2 = 0.99$	$y = 0.91x - 1.02$ $R^2 = 0.99$	$y = -9.74x + 2.19$ $R^2 = 0.56$

Table 2 Thermodynamic constants of Cu^{2+} ions sorption processes

Sorbent	E (kJ/mol)	ΔG^0 (kJ/mol)
Before modification	0.882	−1.245
1% H_2SO_4	4.432	−5.103
2% H_2SO_4	5.432	−6.620
3% H_2SO_4	3.185	−3.711
1% NaOH	1.431	−1.280
2% NaOH	2.094	−3.076
3% NaOH	4.793	−5.235

The processes of Cu^{2+} ions adsorption by native and modified pea pods are physical adsorption processes, since the value of the adsorption energy is less than 8 kJ/mol and the Gibbs energy values in the range of −20 to 0 kJ/mol indicate the spontaneous course of physical adsorption. A numerical increase in the negative value of the Gibbs energy indicates that the adsorption process is more preferable.

4 Conclusions

For the first time, six modified sorption materials were produced by treating the surface of pea pods with 1, 2, and 3% solutions of sulfuric acid and sodium hydroxide. It was found that the treatment of pea pod surface with sulfuric acid and sodium hydroxide solutions resulted in a rougher surface. The sorption properties of native and modified pea pods for copper (II) ions were studied. The obtained isotherms of Cu^{2+} ions adsorption by native and modified pea pods indicate that the treatment of pod surface with sulfuric acid and sodium hydroxide solutions leads to an increase in the sorption capacity of the materials for copper (II) ions. Pea pods treated with 2% solution of sulfuric acid (0.97 mmol) and those treated with 3% solution of sodium hydroxide (0.96 mmol/g) have the highest sorption capacity for copper ions. Processing the adsorption isotherms in the framework of the monomolecular adsorption models, i.e. Langmuir, Freundlich, and Dubinin-Radushkevich, showed that almost all the processes studied are best described by the Langmuir model, that is, adsorption occurs on a uniform surface of the sorption material, where all active centers have the same sorption energy and the sorption process occurs uniformly throughout the material surface. In some cases (for pea pods modified with 1% solution of sulfuric acid and sodium hydroxide), there is a better similarity with the Freundlich model. This phenomenon can be accounted for by the fact that the modification of the material in this case did not proceed evenly, which resulted in the formation of active centers with a higher sorption energy and the uneven process of copper (II) ions adsorption. The calculated values of the thermodynamic values of the processes (sorption energy and Gibbs energy) indicate that in all cases spontaneous physical adsorption occurs and in modifying pea pods with sulfuric acid and sodium hydroxide solutions the sorption process is more preferable. The use of waste from the processing of agricultural raw materials, in this case pea pod shells, allows us to solve a double ecological problem: to use production waste as a sorption material to remove heavy metal ions from aqueous media, which reduces the anthropogenic impact on water bodies.

Acknowledgements The work has been carried out within the framework of the Flagship University Development Program on the base of BSTU named after V.G. Shukhov.

References

1. Malik DS, Jain CK, Yadav AK (2017) Removal of heavy metals from emerging cellulosic low-cost adsorbents: a review. Appl Water Sci 7:2113–2136
2. Joseph L, Jun B, Flora JRV, Park CM, Yoon Y (2019) Removal of heavy metals from water sources in the developing world using low-cost materials: a review. Chemosphere 229:142–159
3. Stepanova SV, Shaikhiev IG, Sverguzova SV (2014) Purification of model effluents containing heavy metal ions, wheat husks. Bulletin of BSTU named after V.G. Shukhov, vol 6, pp 183–186
4. Ince M, Ince OK (2017) An overview of adsorption technique for heavy metal removal from water/wastewater. A critical review. Int J Pure Appl Sci 3(2):10–19
5. Mo J, Yang Q, Zhang N, Zhang W, Zheng Y, Zhang Z (2018) A review on agro-industrial waste (AIW) derived adsorbents for water and wastewater treatment. J Environ Manag 227:395–405
6. Acharya J, Kumar U, Rafi PM (2018) Removal of heavy metal ions from wastewater by chemically modified agricultural waste material as potential adsorbent—a review. Int J Curr Eng Technol 8:526–530
7. Ngwenya S, Guyo U, Zinyama NP, Chigondo F, Nyamunda BC, Muchanyereyi N (2019) Response surface methodology for optimization of Cd(II) adsorption from wastewaters by fabricated tartaric acid-maize tassel magnetic hybrid sorbent. Biointerface Res Appl Chem 9:3996–4005
8. Renu MA, Singh K, Upadhyaya S, Dohare RK (2017) Removal of heavy metals from wastewater using modified agricultural adsorbents. Mater Today Proc 4:10534–10538
9. Afroze S, Sen TK (2018) A review on heavy metal ions and dye adsorption from water by agricultural solid waste adsorbents. Water Air Soil Pollut 229:1–50
10. Krstić V, Urošević T, Pešovski B (2018) A review on adsorbents for treatment of water and wastewaters containing copper ions. Chem Eng Sci 192:273–287
11. Bharathi KS, Ramesh ST (2013) Removal of dyes using agricultural waste as low-cost adsorbents: a review. Appl Water Sci 3:773–790
12. Kanthasamy S, Hadibarat T, Hidayat T, Alamri SA, Al-Ghamdi AA (2020) Adsorption of azo and anthraquinone dye by using watermelon peel powder and corn peel powder: equilibrium and kinetic studies. Biointerface Res Appl Chem 10:4706–4713
13. Anastopoulos I, Kyzas GZ (2014) Agricultural peels for dye adsorption: a review of recent literature. J Mol Liq 200:381–389
14. Nasar A, Shakoor S (2017) Remediation of dyes from industrial wastewater using low-cost adsorbents. Mater Res Found 15:1–33
15. Sanghi R, Verma P (2013) Decolorisation of aqueous dye solutions by low-cost adsorbents: a review. Color Technol 129:85–108
16. Zhang L, Cheng Z (2015) Removal of organic pollutants from aqueous solution using agricultural wastes: a review. J Mol Liq 212:739–762
17. Galblaub OA, Shaikhiev IG, Stepanova SV, Timirbaeva GR (2016) Oil spill cleanup of water surface by plant-based sorbents: Russian practices. Process Saf Environ Prot 101:88–92
18. Sobgaida NA, Olshanskaya LN (2010) Sorbents for water purification from oil products. SSTU, Saratov
19. Lin S-H, Juang R-S (2009) Adsorption of phenol and its derivatives from water using synthetic resins and low-cost natural adsorbents: a review. J Environ Manage 90:1336–1349
20. Zolgharnein J, Shahmoradi A, Ghasemi J (2011) Pesticides removal using conventional and low-cost adsorbents: a review. Clean Soil Air Water 39:1105–1119
21. Raulino GSC, da Silva LS, Vidal CB, Almeida ES, Melo DQ, do Nascimento RF (2018) Role of surface chemistry and morphology in the reactive adsorption of metal ions on acid modified dry bean pods (*Phaseolus vulgaris* L.) organic polymers. J Appl Polym Sci 45879:1–11

22. Özcan AS, Tunali S, Akar T, Özcan A (2009) Biosorption of lead(II) ions onto waste biomass of *Phaseolus vulgaris* L.: estimation of the equilibrium, kinetic and thermodynamic parameters. Desalination 244:188–189
23. Saeed A, Iqbal M (2003) Bioremoval of cadmium from aqueous solution by blackgram husk (*Cicer arientinum*). Water Res 37:3472–3480
24. Aravind J, Muthusamy S, Sunderraj SH, Lenin Chandran L, Kanmani Palanisamy K (2013) Pigeon pea (*Cajanus cajan*) pod as a novel eco-friendly biosorbent: a study on equilibrium and kinetics of Ni(II) biosorption. Int J Ind Chem 4:1–9
25. Sigoria EN (2012) Heavy metal pollution in red haricot beans (*Phaseolus vulgaris*) and cowpeas (*Vigna unguiculata*) grown using untreated sewage water in Ruai, Nairobi-Kenya. Kenyatta University, Kenya

Thermal Modification and Combination of Organo-Mineral Wastes for Sorption Purification of Test Solutions from Congo Red and Methylene Blue Dyes

Zh. A. Sapronova, A. V. Svyatchenko, I. V. Bomba, and V. S. Voropaev

Abstract Removing dyestuffs from sewage waters poses a serious problem in environment protection; as such substances are durable, resistant to sunlight, to many oxidizers, thermal action and biodeterioration, and have a complicated and variable structure. The paper presents the research findings in purifying test solutions from methylene blue and congo red dyes with modified organo-mineral food industry waste—carbonaceous saturation sludge, generated in sugar industry, and waste diatomite—the brewing industry waste product, as well as with their mixes. The materials were baked at various temperatures, and then their physical and chemical properties were studied. It has been determined that the highest value of water purification efficiency for congo red dye (87%), is achieved at using a mix of carbonaceous sludge, heat-treated at 600 °C, and waste diatomite, baked at 550 °C at the ratio 5:1; and for methylene blue the best index of purification efficiency (85%) was achieved with a mix of the same sorbents at the ratio 1:5.

Keywords Wastewater purification · Sorbent · Waste recycling · Dyes

1 Introduction

Over 10,000 of coloring agents are widely used in textile, paper, rubber, plastics, leather, cosmetic, pharmaceuticals and food production industries. The overall consumption of dyes in the world makes up over 1.5 mln tons per year. Up to 15% of this amount gets to waste dumps and sewage waters [1, 2].

The molecules of dyestuffs consist of two key components: chromophores, which are mostly responsible for color development, and auxochromes, which not only supplement the chromophore, but also make its molecule water-soluble and

Zh. A. Sapronova · A. V. Svyatchenko (✉) · I. V. Bomba · V. S. Voropaev
Belgorod State Technological University Named After V.G. Shoukhov, Belgorod, Russia

S. V. Klyuev (ed.), *Digital Technologies in Construction Engineering*,
Lecture Notes in Civil Engineering 173, https://doi.org/10.1007/978-3-030-81289-8_29

strengthen its substantivity (attaching) to fibers. The dyes can be classified in different ways, depending on their chemical composition, specification grade and final use [3].

The penetration of dyestuffs even in small amounts into water bodies causes a wide range of negative effects on the ecosystems [2–6].

Removing dyestuffs from wastewaters is an environmental problem, as such substances are durable, resistant to sunlight, many oxidizers, thermal action and biodeterioration and have complicated and variable structure [1–3].

Among many ways of water purification the good results are shown by the adsorption method, as it can be used for removing various types of colored substances [6]. It is also cost-efficient in economic terms [4].

Adsorption is one of the main treatment methods, used for wastewater purification, due to its convenience, ease of operation and high efficiency [7].

At present a lot of adsorbing materials are produced and used all over the world. Sorbents can be divided into several main groups: inorganic, organic and organo-mineral [8].

The most widely spread adsorbent nowadays is activated carbon. It is normally used for removing various pollutants from water, such as dyes or heavy metals. But its widespread use for wastewater treatment is sometimes limited by its high cost and need for regeneration. So, scientists all over the world carry out research in obtaining low-cost adsorbents for wastewater purification [9–11].

This paper presents the research findings in purifying test solutions from methylene blue and congo red dyes with modified organo-mineral food industry wastes—carbonaceous saturation sludge, generated in sugar industry, and waste diatomite, which is the brewing industry waste.

2 Methods and Materials

The carbonaceous sludge of sugar production is generated at the enterprises, which make sugar of sugar-beet. It consists of finely-dispersed $CaCO_3$ particles and organic substances from beet juice, adsorbed on their surface [12].

Waste diatomite is generated in brewing industry at the stage of beer filtering and is one of the main waste products of brewing industry [13].

Natural diatomite is a sedimentary rock, formed as a result of diatomic microalgae life activity, the main chemical component of which is SiO_2 [14]. The organic constituents of diatomite sludge are presented with insoluble components of malt and malt-free materials, brewing yeast cells, proteins, high-molecular glucose polymers and other organic substances [15–17]. This waste product, as well as carbonaceous sludge, is stored at specialized landfills, which results in economic expenditures [13, 18].

For modification of organic and mineral wastes, they were baked at various temperatures, and then their physical–chemical properties were studied.

The heat treatment was performed by heating a proportionated sample in the electric furnace LOIPLF-7/13, keeping it at the maximum temperature for 60 min.

The data of the chemical composition of materials were obtained by means of ARL9900 Intellipower Workstation device, which allows carrying out the X-ray fluorescence analysis of elements.

The organic matter content was determined by comparing the weight of dry residue (obtained at temperature 105 °C) and ignited residue (at 600 °C).

The pH values of water extracts were determined after 3-min boiling of 5 g of material in 50cm^3 of distilled water.

The true density was determined with picnometer method.

For preparing test solutions such dyes as «congo red», its chemical formula is $C_{32}H_{22}N_6Na_2O_6S_2$ [19], and methylene blue, its chemical formula is $C_{16}H_{18}ClN_3S$, were used [20].

The concentration of dyes in the solutions was determined by means of photocolorimetric method (using spectrophotometer «KFK-3», Russia), with plotting calibration curves.

The purification of model solutions was carried out in a static manner by mixing the samples of the sorbent with an aqueous system followed by the filtration of the spent sorption material.

The efficiency (E) of dye solutions purification with the obtained sorbents was determined by Formula (1):

$$E = (C_1 - C_2) \cdot 100/C_1 \tag{1}$$

where C_1—initial concentration of dye in the solution, mg/l; C_2—its final concentration in the solution, mg/l.

3 Results and Discussion

The agricultural products' processing wastes, selected for the research, have a number of similar physical, chemical and mechanical properties, and the related sources of generation in production. Their main feature consists in the presence of a mineral «core», with the organic components, adsorbed on its surface. The comparative characteristics of the technogenic wastes are presented in Tables 1 and 2.

As we can see from the presented data, the carbonaceous sludge contains 25% of organic matter, and diatomite sludge—21%. The oxides, which compose the mineral part of the waste diatomite, are silicon oxide and, to a lesser extent, calcium oxide; the carbonaceous sludge is presented with calcium carbonate, expressed as oxides—49% of calcium oxide and 38.24% of carbon dioxide.

As sorbents in this research the materials of the following thermal modifications were used (Table 3).

Besides, C600, D450 and D550 were mixed at various ratios to study the opportunities and prospects of their combined use.

Table 1 Technological properties of waste sludges

Parameter	Value	
	Carbonaceous sludge	Diatomite sludge
Color	Beige-brown	Light-beige
Organic matter content (%)	25	21
Bulk density (kg/m^3)	1240	670
True density (kg/m^3)	2320	2680
pH of water extract	8.59	6.82

Table 2 Chemical composition of sludge's mineral part

Sludge	SiO_2	CaO	Na_2O_3	Al_2O_3	Fe_2O_3	MgO	CO_2	Other
Diatomite	92.02	0.71	0.86	3.33	1.30	0.51	-	1.27
Carbonaceous	1.71	49.0	0.32	1.70	0.68	3.83	38.24	4.52

Table 3 Variants of sludge modification and conventional notations

Initial substance	Treatment temperature (°C)	Notation
Carbonaceous sludge	600	C600
Diatomite sludge of brewing industry	450	D450
Diatomite sludge of brewing industry	550	D550
Carbonaceous sludge + diatomite sludge at the ratio 1:1	600	CD600

The research in purifying test solutions from congo red dye was carried out; the initial concentration of substance was 10 mg/dm^3, the added sorption material was 10 g/dm^3 (the research was performed in 100 ml volume of solutions, so, the added sorbent made up 1 g). The findings are presented in Fig. 1.

As we can see from the obtained data, the best results in extracting congo red dye from the test solution were demonstrated by sorbents C600 and CD600. Among the sorbents, obtained by mixing materials at various ratios, the best result was received from the combination of C600 and D450 at the ratio 1:1, and the mix of C600 and D550 at the ratio 5:1. The latter shows the best result of all the tested materials, amounting to 87%.

The studies in purification of test solutions from methylene blue dye were carried out similarly to the experiment described above. The initial concentration of the pollutant was 20 mg/dm^3. The findings are presented in Fig. 2.

In the carried-out experiment the best results were demonstrated by the following materials: among the mono-sorbents the best in extracting methylene blue from the solution were C600 and D450. Combining of sorption materials has allowed determining that the best results are shown at using the mix of sorbents: C600 and D550 at the ratio 1:5, C600 and D450 at the ratio 1:1, and C600 and

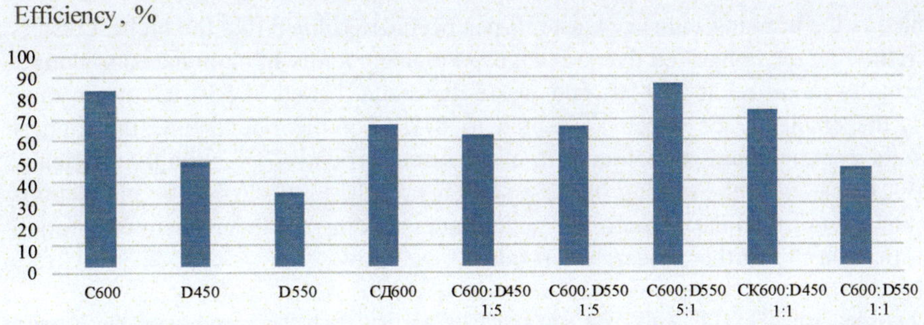

Fig. 1 The efficiency of extracting congo red dye with various sorbents

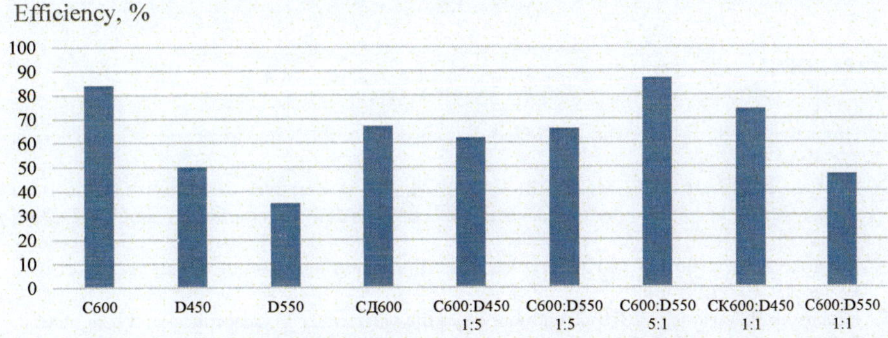

Fig. 2 The efficiency of extracting methylene blue dye with various sorbents

D550 at the ratio 1:1. The best result was achieved at using the mix of C600 and D550 at the ratio 1:5, which amounted to 85%.

As we can see from the carried-out experiments, mixing the sorbents C600 and D550 causes specific interactions, which influences the efficiency of extracting dyes from test solutions. For purifying test solutions from congo red dye the highest efficiency was achieved at using the mix of C600 and D550 at the ratio 5:1, while at purifying test solutions from methylene blue the optimal ratio was 1:5 for the same sorbents. The further research is needed in order to study electrokinetic interactions, which take place in the process of wastewater purification and cause the observed effect.

4 Conclusion

The experiments were carried out in purification of test solutions from congo red dye and methylene blue dye with the use of sorbents, obtained on the basis of carbonaceous saturation sludge, generated at sugar industry, and waste diatomite, which is the brewing industry waste. It has been determined that the highest results in removing the congo red dye are achieved at using a mix of carbonaceous sludge, thermally modified at 600 °C, and waste diatomite, baked at 550 °C, at the ratio 5:1, the efficiency of which amounted to 87%. For methylene blue the highest results were obtained at another ratio of the same sorbents—1:5, with the efficiency 85%. Apparently, the mixing of these sorbents results in electrokinetic interactions in the solution-sorbent system. For better understanding the mechanism of sorption purification the further research is needed.

Acknowledgements This work was supported by a grant from the President of the Russian Federation for state support of young Russian scientists—candidates of sciences and doctors of sciences and leading scientific schools of the Russian Federation, application number MD-1249.2020.5

References

1. Homaeigohar Sh (2020) The nanosized dye adsorbents for water treatment. Nanomaterials 10(295):43
2. Khodaie M, Ghasemi N, Moradi B, Rahimi M (2013) Removal of methylene blue from wastewater by adsorption onto $ZnCl_2$ activated corn husk carbon equilibrium studies. J Chem 6:383985
3. Adegoke KA, Bello OS (2015) Dye sequestration using agricultural wastes as adsorbents. Water Res Ind 12:8–24
4. Himanshu P, Vash RT (2010) Treatment of textile wastewater by adsorption and coagulation. E-J Chem 7(4):1468–1476
5. Sverguzova SV, Stepanova SV, Shajhiev IG (2014) Ochistka model'nyh stokov, soderzhashchih iony tyazhelyh metallov, sheluhoj pshenicy. Bulletin of BSTU named after V.G. Shukhov 6:183–186
6. Šmelcerovic M, Đorđevic D, Novakovic M, Mizdrakovic M (2010) Decolorization of a textile vat dye by adsorption on waste ash. J Serb Chem Soc 75(6):855–872
7. Crini G, Lichtfouse E, Wilson L, Morin-Crini N (2019) Conventional and non-conventional adsorbents for wastewater treatment. Environ Chem Lett 17(1):195–213
8. Rotar OV, Iskrizhitskaya DV, Iskrizhitsky AA, Oreshina AA (2014) Cleanup of water surface from oil spills using natural sorbent materials. Proc Chem 10:145–150
9. Hegazi HR (2013) Removal of heavy metals from wastewater using agricultural and industrial wastes as adsorbents. HBRC J 9:276–282
10. Sabino, De Gisia, Lofrano G, Mariangela GMN (2016) Characteristics and adsorption capacities of low-cost sorbents for wastewater treatment: a review. Sustain Mater Technol 9:10–40
11. Starostina IV, Stolyarov DV, Anichina YN, Porozhnyuk EV (2018) The carbonaceous sorbent based on the secondary silicacontaining material from oil extraction industry. IOP conference series: earth and environmental science, vol 107, p 012075
12. Sapronova ZA, Sverguzova SV, Svyatchenko AV (2018) About a possibility of usage of sugar beet industrial carbonatecontaining byproducts in dry construction mixtures and oil paints manufacturing. Solid State Phenom 284:899–904
13. Hergert GW, Darapuneni MK, Aqeel AM, Wilson RG, Harveson RM, Bradshaw JD, Nielsen GA (2017) Agronomic potential of using precipitated calcium carbonate on early plant growth and soil quality in the intermountain west-greenhouse studies. J Sugar Beet Res 54(1 & 2):35–39

14. Mathias TR, dos S, Mello PPM, Camporese Servulo EF (2014) Solid wastes in brewing process: a review. J Brew Distilling 5(1):1–9
15. Lin KL, Lan JY (2013) Porous ceramic characteristics sintered from waste diatomite and sodium silicate sand. Int J Mater Mech Manuf 1(3):240–244
16. Abass AO (2020) The brewing industry and environmental challenges. J Clean Prod 256(102817):22
17. Yitbarek M, Abdeta K, Beyene A et al (2020) Experimental evaluation of sorptive removal of fluoride from drinking water using natural and brewery waste diatomite. Process Saf Environ Prot 128:95–106
18. Xiaoxi G, Weijun T, Jie B et al (2019) Highly efficient deproteinization with an ammonifying bacteria Lysinibacillus fusiformis isolated from brewery spent diatomite. J Biosci Bioeng 127(3):326–332
19. Khaniabadi YO, Mohammadi MJ, Shegerd M et al (2017) Removal of Congo red dye from aqueous solutions by a low-cost adsorbent: activated carbon prepared from Aloe vera leaves shell. Environ Health Eng Manag J 4(1):29–35
20. Kuang Y, Zhang X, Zhou Sh (2020) Adsorption of methylene blue in water onto activated carbon by surfactant modification. Water 12(587):19

Optimization of Mixing-Crushing Device Design Using CAE-Analysis

S. Y. Lozovaya, N. E. Bogdanov, N. M. Lozovoy, and V. M. Kravchenko

Abstract The design of modern equipment is not limited to its solid-state modeling. Design options should provide static strength and structural rigidity, durability, stability while minimizing weight, which determine the competitiveness of new equipment. The purpose of this paper is to optimize the structural elements of the mixing-grinding device while minimizing the weight, for this purpose it is necessary to determine the strength and stiffness of the main elements. The designed device is used to produce a wide range of fine powders and construction mixtures. For this purpose, the main values of stresses on the elements of the frame structure are calculated. To calculate the frame assembly for strength, a finite element mesh was created for all parts of the frame. The strength characteristics of the most stressed elements of the frame assembly were evaluated. As a result, to reduce the stress on the part in the frame design have been changed, specifically bracket between the working chamber and the bottom plate, which will support the chamber and significantly reduce the load on the other elements of the design.

Keywords Mixing-crushing device · CAE-analysis · Methods of finite elements · 3D model · Optimization

1 Introduction

The automated design system in the process of evolution was divided into separate areas, within which highly specialized tasks were solved. The arsenal of tools for achieving the goal was also expanded. At each stage of production, you can choose the system that is most suitable for your particular case. The technology of creating a 3D model in CAD has significantly accelerated the productions of new products that are designed with the specified characteristics. The solid-state prototype is tested and tested with sufficient accuracy virtually, minimizing the cost of real

---

S. Y. Lozovaya · N. E. Bogdanov · N. M. Lozovoy · V. M. Kravchenko (✉)
Department of Mechanical Equipment, Belgorod State Technological University named after V.G. Shukhov, Kostyukov St., 46, Belgorod 308012, Russia

S. V. Klyuev (ed.), *Digital Technologies in Construction Engineering*,
Lecture Notes in Civil Engineering 173, https://doi.org/10.1007/978-3-030-81289-8_30

testing. The production of modern equipment at the design stage is not limited to its geometric modeling. Without a comprehensive engineering analysis of the projected object, it is impossible to produce competitive products. Developers around the world work hard to ensure that their design solutions provide static strength and rigidity of the structure, its sufficient durability, stability and suitable dynamic characteristics, while minimizing weight, cost, energy consumption at the desired maximum performance, taking into account the quality of the finished product. Optimization of the specified parameters and design is the key to their competitiveness. The use of CAE-analysis tools [1–3] allows modern enterprises to create products that are not inferior to the best world standards and even surpass them. CAE systems—a class of products for computer-aided engineering (computer-aided engineering). The emergence of the ability to create a solid-state model required a detailed description of it, predicting operational loads, including the effects of temperature and environmental resistance.

2 Materials and Methods

When minimizing the weight, it is necessary to determine the strength of the main elements of the devices. Calculation of the assembly of the optimized device is made taking into account the arising stresses from its own weight (3713.04 N), as well as taking into account external influences (Figs. 1, and 2) [3–5].

In the device [3–5], the main loads are applied to the frame 2 (Fig. 1) from its weight (Fig. 2), from the weight of the three vibration exciters 8, from the housing assembly with brackets 3 and 5, as well as the feed hopper 6. Calculation scheme (Fig. 2) with the loads and boundary conditions is made, with the own weight G = 2540 N; force F1 from the impact of three vibration exciters on the frame F1 = 2500 × 3 = 7500 N; force F2 = 133 N from the weight of the hopper with the material. A force acts inside the grinding chamber F3 = 200 N from the weight of the material and grinding bodies. The movement of the frame assembly is limited in the spring mounting area (Fig. 2).

After idealizing the geometry, which includes suppressing small holes, chamfers or roundings, and performing partitioning of the body into small simple volumes for further construction of hexahedral and tetrahedral meshes.

For strength calculation of the frame assembly, a finite element mesh is created for all parts (Fig. 3). When generating the mesh, the elements are applied to the outer surface of the model and then they are created within the geometry of the entire structure or its elements. During mesh generation, the system evaluates its quality according to the following parameters: element disconnection from the ideal form and the value of the Jacobian matrix. In particular, the Jacobian matrix is used when integrating over the area/volume of the element when transitioning from one coordinate system to another. When a mesh element is highly distorted due to partitioning or deformation, the Jacobian may become negative or zero, which is not acceptable [6–8]. If during mesh generation the system detects geometry defects that are smaller than 10% of the mesh element size, they will be ignored by the system.

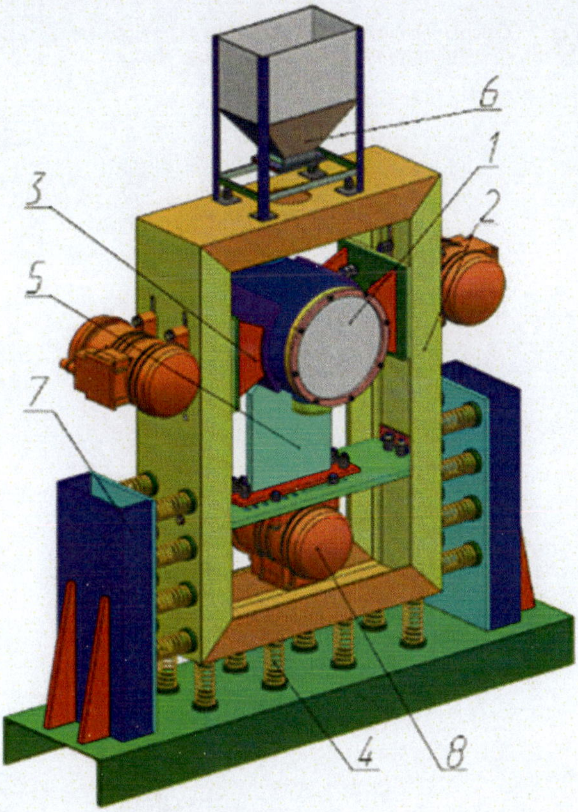

Fig. 1 Digital 3D model of the mixing-crushing device: *1*—housing; *2*—frame; *3*—bracket; *4*—spring support; *5*—bottom bracket with plate; *6*—hopper; *7*—lateral support; *8*—vibrating exciter

3 Results and Discussion

To make a conclusion about the serviceability of the structure under the selected constraints and loads, it is necessary to check the strength and stiffness conditions of all elements (parts) of the structure [9, 10]. According to the results of the calculation, it was found that the most loaded part is the bottom plate (Figs. 1, 3, and 4).

The strength condition is checked by the maximum design stresses, the selected element (part) of the structure. The strength condition involves comparing the maximum design stresses with the allowable stresses for the selected material and determining the safety factor:

$$n = \frac{\sigma_{adm}}{\sigma_{\max}} \geq [n]; \tag{1}$$

where n—safety strength condition;

σ_{adm}—allowable stress for the selected material, $\sigma_{adm} = 245$ MPa;

$\sigma_{\max}$—maximum stress value, $\sigma_{\max} = 68.51$ MPa;

$[n]$—allowable safety strength condition, $[n] = 5$.

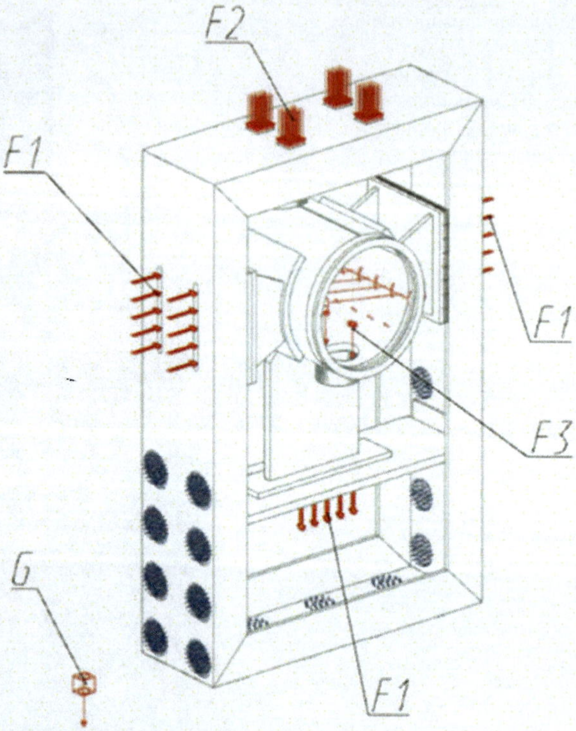

Fig. 2 Calculation scheme of the forces acting in the system

$$n = \frac{245}{68.51} = 3.51 \leq 5;$$

Strength condition $n = \frac{\sigma_{adm}}{\sigma_{\max}} \geq [5]$ is not carried out (Fig. 4).

The stiffness condition is a comparison of the maximum displacements with the allowable displacements:

$$\delta_{\max} \leq [\delta], \tag{2}$$

where $\delta_{\max}$—maximum expected displacement, $\delta_{\max} = 0.005$ mm;

$[\delta]$—permissible movements:

$$[\delta] = \frac{a}{250}, \tag{3}$$

where a—overall construction size, $a = 575$ mm.

$$[\delta] = \frac{575}{250} = 2.296 \text{ mm},$$

Fig. 3 Scheme of the mapped finite element mesh

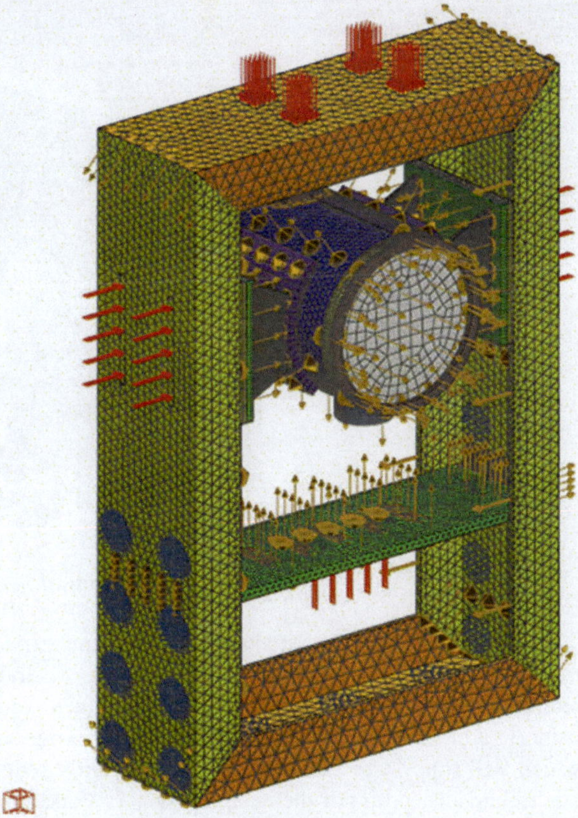

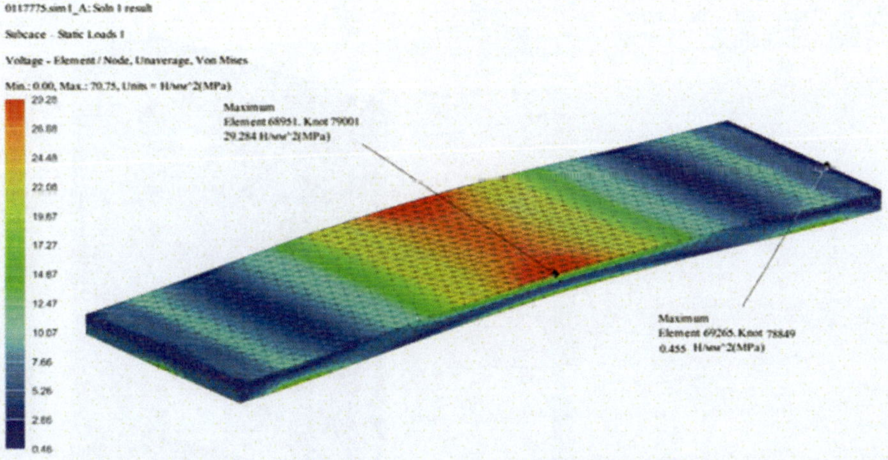

Fig. 4 Analysis of the bottom plate for stresses before making changes in the design

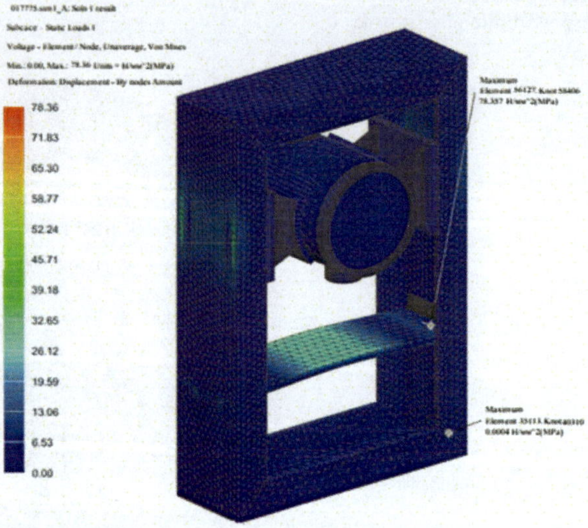

Fig. 5 Analysis of the frame stresses before changes are made

$$0.005 \leq 2.296 \text{ mm},$$

Based on the results of analysis of all elements (parts), calculations of the entire frame structure were made for stresses (Fig. 5) and displacements (Fig. 6). To reduce the stresses on the bottom plate, changes were made in the frame design, namely, a bracket is mounted between the working chamber and the bottom plate, which will support the chamber and significantly reduce the loads on the plate and other elements (parts) of the design (Figs. 7, 8 and 9). To determine the operability

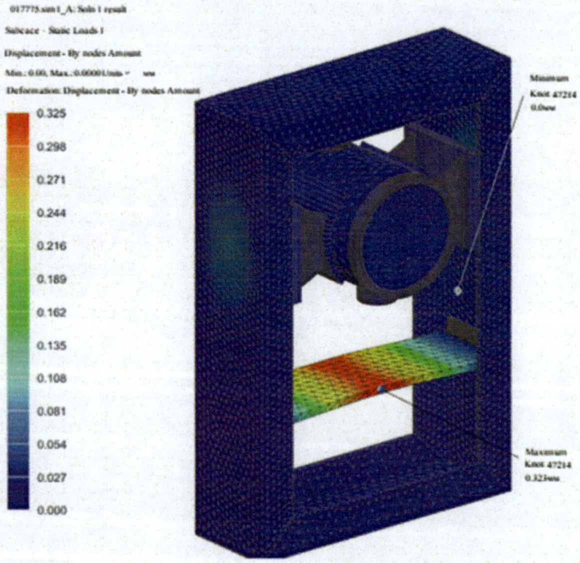

Fig. 6 Analysis of the frame displacement before modifications

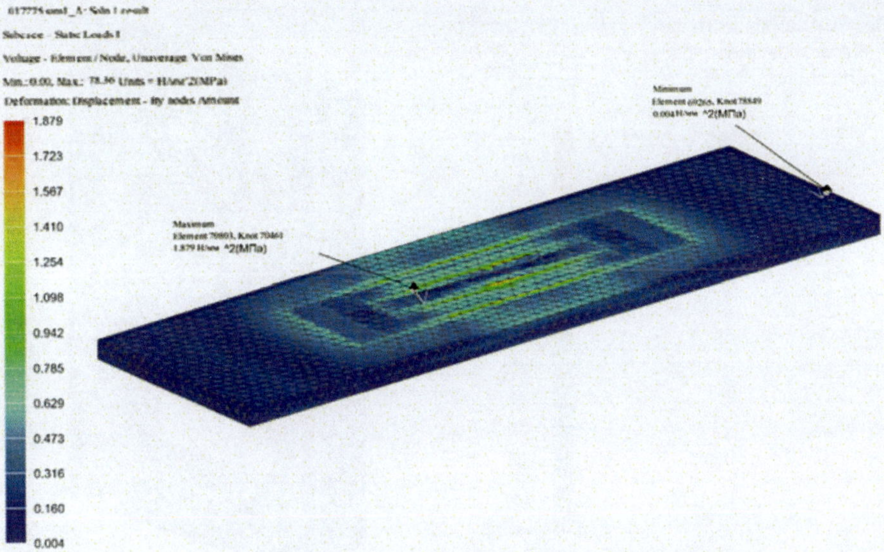

Fig. 7 Stresses occurring in the bottom plate after modifications to the design

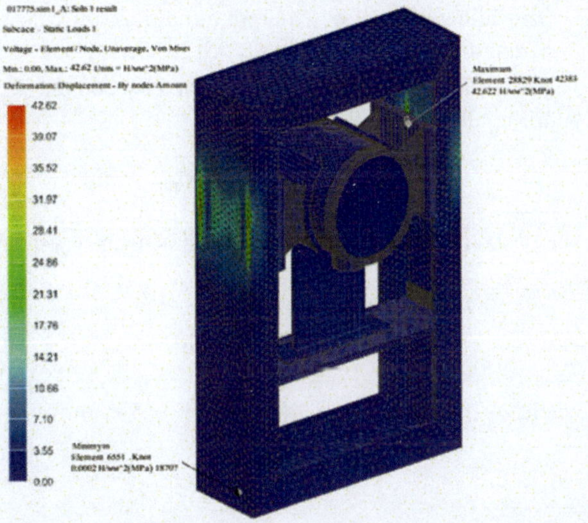

Fig. 8 Analysis of frame stresses after modifications

of the frame structure under the selected constraints and loads, it is necessary to make a second check of the strength and stiffness conditions.

σ_{adm}—the allowable stress for the selected material, $\sigma_{adm} = 245$ MPa;

$\sigma_{\max}$—maximum stress value, $\sigma_{\max} = 1.354$ MPa;

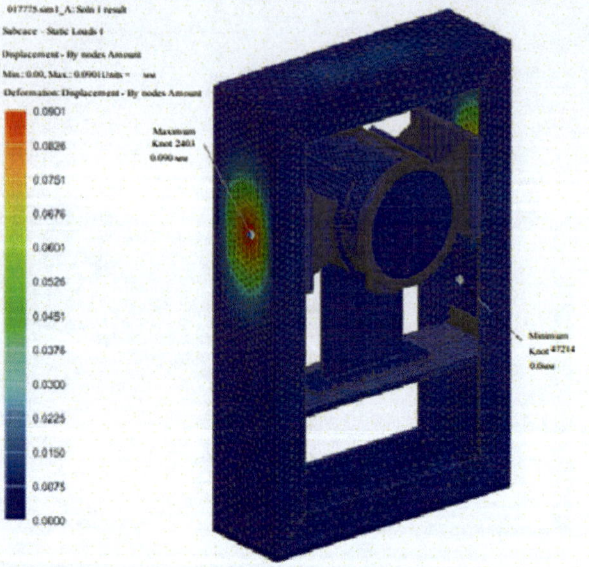

Fig. 9 Analysis of frame displacements after modifications

$[n]$—allowable safety strength condition, $[n] = 5$.

Strength condition $n = \frac{\sigma_{adm}}{\sigma_{\max}} \geq [5]$ is carried out.

The stiffness condition is determined by comparing the maximum displacements with the allowable displacements:

$$\delta_{\max} < [\delta],$$

where $\delta_{\max}$—maximum design displacement, $\delta_{\max} = 0.0901$ mm;

$[\delta]$—permissible movements:

$$[\delta] = \frac{a}{250},$$

where a—overall construction size, $a = 575$ mm.

$$[\delta] = \frac{575}{250},$$

$$[\delta] = \frac{575}{250} = 2.3 \text{ mm},$$

$$0.0901 \leq 2.3 \text{ mm},$$

Stiffness condition $\delta_{\max} \leq [\delta]$ is carried out.

4 Conclusion

During the optimization of structural elements of the mixing-grinding device by means of CAE-analysis, their calculation by methods of finite elements has been conducted, which indicated that in the original design of the frame in some structural elements, in particular in the bottom plate of the frame the conditions of strength are not fulfilled. In order to reduce the stress on the part, changes were made to the frame design, namely, a bracket was mounted between the working chamber and the bottom plate, which would support the chamber and significantly reduce the stress on the other elements of the frame. Analysis of calculation results showed that in the new design of the device the conditions of strength and stiffness are fulfilled, due to redistribution of stresses and displacements on the side struts of the frame.

Acknowledgements This work was realized in the framework of the Program of flagship university development on the base of the Belgorod State Technological University named after V. G. Shukhov. The work was realized using equipment of High Technology Center at BSTU named after V. G. Shukhov.

References

1. Lozovaya S, Lozovoy N, Uvarov V (2019) Computer simulating of strength characteristics for deformable chambers and most loaded mixer elements with a vertical shaft. In: IOP conference series: materials science and engineering, vol 560, pp 1–6
2. Samsonova P, Lozovaya S, Bogdanov N, Lozovoi N (2020) Mixing process simulation of the initial building materials components using the dem solution edem system. In: IOP conference series: materials science and engineering, vol 945, p 012028
3. Bogdanov V (2019) Intensification of the grinding process in vibration mills. J Phys Conf Ser 1353:012041
4. Bogdanov, N. E., Bogdanov, V. S., Lozovaya, S. Yu., Fadin, Yu. M.: Usage of discrete element method in the research of vibrating mills with circular vibrations of the grinding chamber. IOP Conf. Ser: Mater Sci Eng. 786, 012019 (2020).
5. Bukin SL, Bukina AS (2013) Dynamic possibilities of the inertia biharmonic vibratory mill of new type. Adv Technol Syst Mech Eng 1–2:61–71
6. Otsokov KA (2020) Innovative technologies in construction and their use in organizational and technological events. Constr Mater Prod 3(1):7–13
7. Furukawa R, Shiosaka Y, Kadota K, Takagaki K, Noguchi T, Shimosaka A, Shirakawa Y (2016) Size-induced segregation during pharmaceutical particle die filling assessed by response surface methodology using discrete element method. J Drug Deliv Sci Technol 35:284–293
8. Krenzer K, Mechtcherine V, Palzer U (2019) Simulating mixing processes of fresh concrete using the discrete elementmethod (DEM) under consideration of water addition and changes in moisture distribution. Cem Concr Res 115:274–282
9. Golshan S, Zarghami R, Norouzi N, Mostoufi N (2017) Granular mixing in Nauta blenders. Powder Technol 305:279–288
10. Michalczyk J, Cieplok G, Sidor J (2010) Numerical simulation model of the rotary-vibration mill working process. Arch Metall Mater 55(3):321–331

Economic Analysis of Investments in Design Solutions for Ice Palace Water Treatment System

M. A. Razakov, A. V. Gorbenko, G. O. Titova, and E. A. Sergeeva

Abstract There are the investment discusses in the reconstruction of the ice palace water treatment system which has located in the Moscow region. Authors note some features of design solutions for the water treatment process in this particular region. Researchers consider the requirements for water quality before using the reserve osmosis unit in this article. It has been highlighted the features of the technological process in the ice palace—pouring ice. Authors have noted the strong requirements for ice rink water quality than water which has been used in water supply of drinking purpose. There are the laboratory tests results of water source quality which have been used in technological process in the ice palace. The possible sources which increase the content of various water indicators in the water source of building have been identified by authors of this paper. These sources can have the natural or artificial origin. Researchers have shown them in this report. Authors present the analysis of main investments to design solutions for the ice rink water treatment system. It has been considered the potential operating cost for the ice rink technological process by authors of this paper. There are some possibilities of reducing the operating cost for ice rink water treatment system.

Keywords Water treatment · Ice palace · Investments · Water source characteristics · Capital investments

1 Introduction

There is the program "Strategy for the development of physical culture and sports in the Russian Federation until the period to 2020" which was adopted in Russian Federation in 2009 [1, 2]. It was made in accordance with the tasks which were set

M. A. Razakov (✉) · A. V. Gorbenko · G. O. Titova
Moscow Power Engineering Institute, Moscow, Moscow Region, Russia
e-mail: muhammet@nln.ru

M. A. Razakov · E. A. Sergeeva
Moscow State University of Civil Engineering, Moscow, Moscow Region, Russia

S. V. Klyuev (ed.), *Digital Technologies in Construction Engineering*,
Lecture Notes in Civil Engineering 173, https://doi.org/10.1007/978-3-030-81289-8_31

by Russian President V.V. Putin. These tasks included a set of measures which were directed at supporting youth and amateur sports in the Russian Federation. According these tasks there are the constriction of sports and recreation complexes (SRC) for different purposes. One of the most technologically and costly types of facilities in the constriction are ice sports complexes (ISC). In addition to difficulties during constriction, the process of operating this type of buildings is not less difficult. Complex heat and mass transfer processes inside of the building, inconsistent climatic conditions and other physical processes complicate the process of building operation. A special place in the technological process of the ice palace is the selected ice treatment method because of source water composition (SWC). The SWC affect to the type of water treatment [3, 4]. Authors have considered in water treatment methods for ice producing process and compare the investments to possible types of water treatment (WT).

2 Methods and Materials

The required and actual parameters of water source are required to determine the less-costly method of water treatment. There are the information of water characteristics which were used for ice cover. It should be noted that almost all the main characteristics of water for ice cover, must be conformed with the regulatory and rules of drinking water. The hardness indicator is the exception of this statement. It's value must be in a range which is in other regulatory and rules—SP 31-112-2007 "Physical and sports halls. Part 3. Ice palaces". According this document, the hardness indicator should be in range from 0.5 to 1.0 degree of hardness. During 2 years other indicators of the source water has been changed in small ranges and matched to Sanitary regulatory and rules 2.1.4.1074–01 "Drinking water of water supply system", therefore, they were not included in Table 1. There are the installed water treatment system for ice cover in the Fig. 1.

According to laboratory results it could be observed that some indicators excess relatively regulatory documents. These indicators include the following parts: smell; smack; fluoride ions; sulphides and hydrogen sulphide. At first year it could be included an iron too. The hardness indicator of water is in normal drinking quality limits but for ice cover this indicator is out of normal limits. Excessive of smell and smack indicators are consist with and increased value of fluoride ions, sulphides and hydrogen sulphide. This event could happened because of season changes in the composition of water or with discharge of untreated wastewater in near territories. Since water has low hardness indicators, it is possible to use all types of WT. Also it can be used not very popular WT. According to the parameters of the source water, deep WT must be used only in the ice cover. There were also small increase of the iron indicator in 2018. This indicator has stability from in 2020 but it is necessary to provide a method to reduce this indicator in the WT scheme. According to laboratory results in 2018, it is impossible to find out which type of iron was in the water. Iron can be in different forms because of its valence. It

Table 1 The information of water source characteristics

Water indicators	Water sample analysis results in 2020	Water sample analysis results in 2018	Required water quality according to SanPiN 2.1.4.1074-01
General and organoleptic indicators			
Smell at 20 °C, points	3	2	2
Smack, points	3	2	2
Chromaticity, degree of chromaticity	17	16.5 ± 3.3	20
Turbidity, mg/l	<1.0	1.2 ± 0.24	1.5
Hydrogen exponent (pH)	7.76	8.19 ± 0.2	6–9
Total hardness, degrees	5.4	5.2 ± 0.8	7.0 (for ice cover: 0.5–1)
Alkalinity	4.9	-	-
Inorganic and organic substances, mg/l			
Oxidability (permanganate)	2.24	1.1 ± 0.1	5
NH3 and ammonium ion	<0.1	1.5 ± 0.3	2.0 (by NH_3)
Fe (total)	<0.213	0.36 ± 0.09	0.3
Sulphides and hydrogen sulphide	0.020	–	0.003
Nitrates	<0.1	<0.1	45.0
Nitrites	<0.003	0.014 ± 0.007	3.0
Sulphates	84.1	83.3 ± 9.2	500
Fluoride ion	3.11	<0.04	1.5
Chlorides	13.5	16.3 ± 2.0	350
Dissolved oxygen	7.6	–	–

could be in following forms: insoluble in water or elementary metal iron; double valence iron; triple valence iron; organic iron [4]. If this indicator is increase, engineering service must do deep analyses of water source. Also they should design methods for reducing the specific types of iron.

During the examination of installed water treatment system, it was matched that this WTS cannot be used for ice cover. As a result of incorrect WTS selection, we have observe the calcium deposits on ice cover. It has formed because of hardness excess in cold water supply system and after the boiler in the hot water system.

The water treatment system of the ice cover should provide the regulatory indicators. Also this system should be economically useful in these specific conditions of water source which is in this location. According to V.V. Solodyannikov, there are some types of water treatment systems in Table 2 [5] (Figs. 2 and 3).

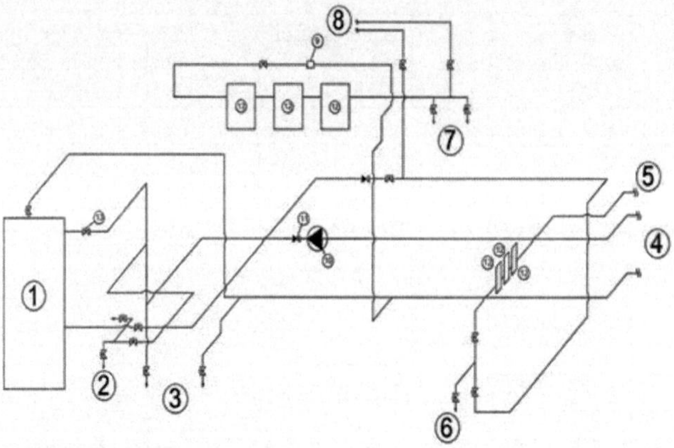

1 – electric boiler; 2 – emergency water drainage; 3 – heating system pipeline; 4 – hot water supply system's pipeline; 5 – cold water supply system's pipeline; 6 – valve in cold water supply system's pipeline; 7 – valves in hot water supply system's pipeline; 8 – hot water supply system's pipeline to sink mixer; 9 – flow register; 10 – pumping station; 11 – check valve; 12 – coarse filtering device; 13 – valve.

Fig. 1 Installed water treatment system (WTS)

Table 2 The main methods of water treatment

Type of water treatment	Method of water treatment	The main indicator	Use limits
Physical	Sedimentation (1 type)	Particle size	Low water consumption
	Bulk filter	Particle size	Coarse filtering
	Membrane filter	Particle size	Fine filtering
	Ultraviolet	Microbial content	For the destruction of microorganisms
Chemical	Acidification	pH, temperature, composition of substances, composition of compounds	Complex action to treatment
	Sedimentation (2 type)	Composition of substances and compounds	High water consumption
Physical and chemical	Sorption	Composition of substances and compounds	Complex action to treatment
	Electrodialysis	Composition of substances and compounds	Complex action to treatment
	Thermal	Composition of substances and compounds	Complex action to treatment

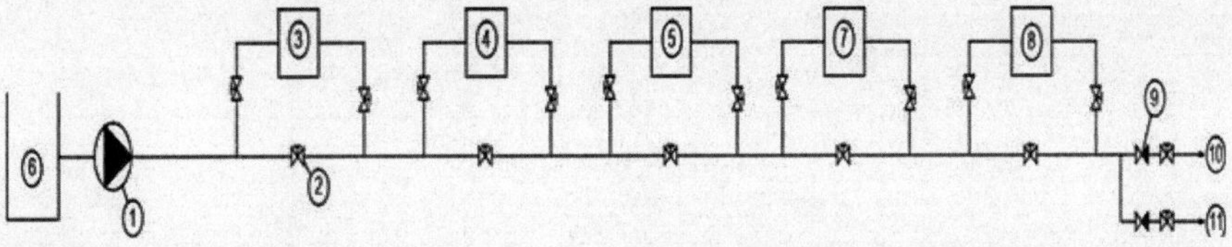

1 – pumping station; 2 – valve; 7 – fine filter; 3 – filter with sorbent filler; 4 – carbon filter; 6 – water tank (1500 liters); 5 – ion exchange filter; 8 – reverse osmotic system; 9 – check valve; 10 – hot water supply system pipeline; 11 – cold water supply system's pipeline.

Fig. 2 Type №1 of proposed WTS

Authors have shown the proposed schemes of water treatment in Figs. 4, 5 and 6. They are used for ice cover. All the methods of water treatment which has been given below, can be classified as complex action methods of source water purification. They are consist of physical and chemical methods of WT.

The reverse osmotic water treatment systems have become widespread at the moment. Some positive features of these water treatment's methods are considered in many scientific papers [6–10]. Despite the high degree of water source purification this WT method also requires high economic and energy costs. The main costs are related to electric energy which is used for pumping equipment and for technology of vacuum creation. Another costs are operational investments for the filter elements. There are the scheme of the combined Physical and Chemical WTS in Fig. 4.

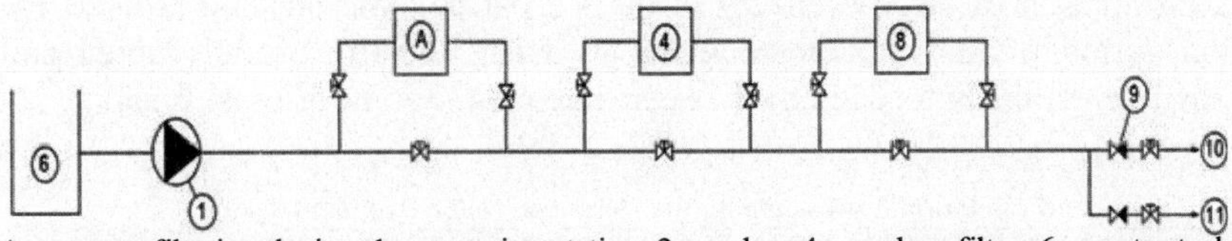

A – coarse filtering device; 1 – pumping station; 2 – valve; 4 – carbon filter; 6 – water tank (500 liters); 8 – reverse osmotic system; 9 – check valve; 10 – hot water supply system's pipeline; 11 – cold water supply system's pipeline.

Fig. 3 Type №2 of proposed WTS

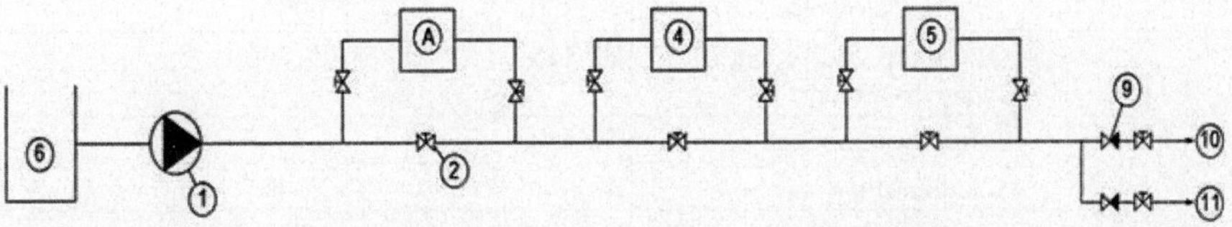

A – coarse filtering device; 1 – pumping station; 2 – valve; 4 – carbon filter; 5 – ion exchange filter; 9 – check valve; 10 – hot water supply system's pipeline; 11 – cold water supply system's pipeline.

Fig. 4 Type №3 of proposed WTS

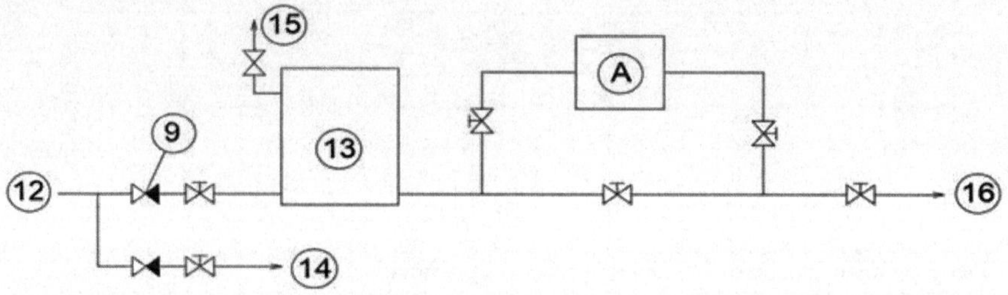

12 – pipeline from the water treatment system ; 13 – boiler in the hot water supply system; 14– valve for filling the ice machine with cold water; 15 – air valve; 16 – valve for filling the ice machine with hot water; A – coarse filtering device.

Fig. 5 Hot and cold water supply system scheme

It should be noted that WTS has been used only for cold water supply system which is consumed for hot and cold water for ice cover in ISC. There are the scheme of technical cold and hot water supply system for ISC in Fig. 5.

3 Results and Discussion

Authors have presented the capital investments distribution of considered water treatment types in Table 3.

The installation of a reservoir tank is the main method of reducing the operational consumption of electric energy in the ice palace. This method reduces the operation period of the pumping equipment. There are the overall capital and operational investments for ice cover water treatment system in Figs. 6 and 7.

Table 3 Capital and operational investments for ice cover water treatment system

Type of water treatment	Type of capital investments	Capital investments, Rubles	Annual operational investments, Rubles
1	The main equipment of water treatment	364,035	27,395
	Pump equipment	74,350	9,125
	Installation work	63,880	5,000
2	The main equipment of water treatment	39,445	13,770
	Pump equipment	50,550	8,300
	Installation work	41,000	5,000
3	The main equipment of water treatment	198,500	9,000
	Pump equipment	60,550	9,125
	Installation work	41,000	5,000

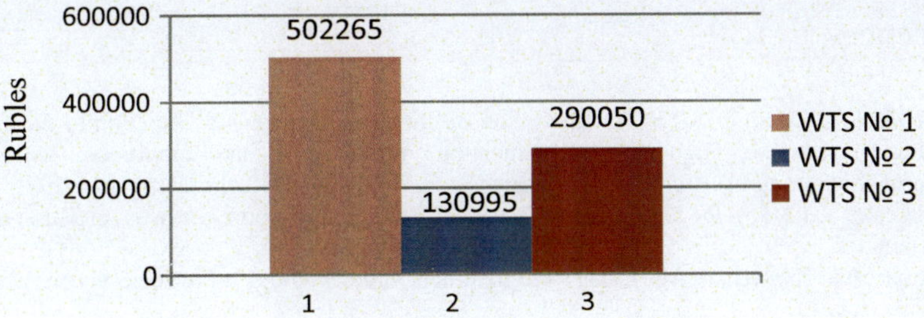

Fig. 6 Contribution of capital investments for considered water treatment systems

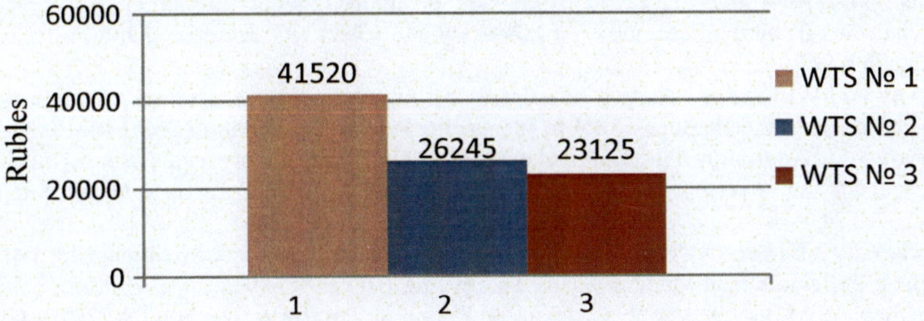

Fig. 7 Contribution of operational investments for considered water treatment systems

4 Conclusion

Reverse osmotic water treatment systems are increasing the capital investments for the industrial ice cover WTS by 3.8 times and the operating investments by 1.8 times. The use of reverse osmotic systems is not economically useful in these specific conditions. The main indicators of water source can reduce only with chemical type of treatment. If the cost of electricity is low it is possible to use ultrasonic or magnetic methods of water treatment for ice cover. During the survey of water sample analysis from water source it was found the significant excess of Fluoride ion by 77.75 times if it compared with previous sample analysis information. This uprise phenomena is not associated with seasonal changes. Therefore, it is necessary to identify and isolate this factor in the future for prevent much more harmful effects.

References

1. Kruglova T, Chudnih A (2007) Integration of "fitness management" direction to the general training system of high physical education specialists in the conditions. Reports of international congress "terms and concepts in field of physical culture" pp 194–197
2. Enchenko I (2015) Development of physical culture and sport services in Russia. PhD dissertation
3. Gurba MG, Kvartenko AN (2007) Pre-treatment biotechnology of surface water. Ecol Ind Russia 4:27–31
4. Zhanna GE, Vadim M, Vasiliy G, Semenovykh A, Volkov A, Pustovgar A (2008) Adamtsevich technology of groundwater biological deferrization. MATEC Web of Conferences, vol 251, pp 03036. https://doi.org/10.1051/matecconf/201825103036
5. Solodyannikov V (2003) Calculation and mathematical modeling of water treatment. Energoatomizdat
6. Spitsov D, Andrianov A (2017) Influence of hydrodynamic properties on the fouling mechanisms of membranes and their efficiency and selectivity decrease. J Ind Pollut Control 33(1):959–967
7. Pervov AG (2015) Precipitation of calcium carbonate in reverse osmosis retentate flow by means of seeded techniques-a tool to increase recovery. Desalination 368:140–151
8. Pervov AG, Andrianov AP, Danilycheva MN (2018) Preliminary evaluation of new green antiscalants for reverse osmosis water desalination. Water Sci Technol Water Supply 18 (1):167–174
9. Jurchevsky EB, Pervov AG (2020) Potentialities of membrane water treatment for removing organic pollutants from natural water. Therm Eng 67(7):484–491
10. Oshchepkov M, Golovesov V, Redchuk A, Popov K, Pervov A, Ryabova A (2020) Gypsum crystallization during reverse osmosis desalination of water with high sulfate content in presence of a novel fluorescent-tagged polyacrylate. Curr Comput-Aided Drug Des 10(4):309

Efficiency of Using a Technogenic Product of Electrolytic Aluminum Production as a Mineralizer in the Burning of Cement Clinker

A. G. Novosyolov, Y. I. Ershova, I. N. Novoselova, and Y. A. Vasina

Abstract The paper considers the possibility of using a technogenic product of electrolytic aluminum production as a mineralizing additive in the burning of cement clinker. A comparison of the efficiency of the mineralizing effect of the technogenic product of electrolytic aluminum production and the traditional mineralizer—calcium fluoride is presented. The effectiveness of mineralizers is achieved due to the earlier binding of calcium oxide into clinker minerals. The temperature of complete binding of calcium oxide for the additive-free raw material mixture is 1400 °C, for the raw material mixture with the addition of calcium fluoride—1300 °C, and for the raw material mixture with the addition of a technogenic product of electrolytic aluminum production—1250 °C. The use of mineralizers, both calcium fluoride and a technogenic product of electrolytic aluminum production, allows reducing the burning temperature of the clinker by 100 °C without changing significantly the phase and mineralogical composition of the cement clinker. Reducing the clinker burning temperature to 1350 °C will lessen the specific heat consumption for clinker burning. The decrease in specific heat consumption is due to a lowering in the temperature of the exhaust gases and heat loss to the environment by the furnace system body. The specific heat consumption, when using mineralizers, is reduced by 175 kJ/kg of clinker.

Keywords Mineralizers · Crushed electrodes · Clinker burning · Specific heat consumption

---

A. G. Novosyolov (✉) · Y. I. Ershova · I. N. Novoselova · Y. A. Vasina
Chemical Technological Institute of Belgorod State Technological University named after V.G. Shukhov, Belgorod, Russian Federation

S. V. Klyuev (ed.), *Digital Technologies in Construction Engineering*,
Lecture Notes in Civil Engineering 173, https://doi.org/10.1007/978-3-030-81289-8_32

1 Introduction

Mineralizers are compounds that, due to their chemical properties, facilitate the flow of reactions and contribute to the acceleration of physical and chemical processes. Mineralizers increase the reactivity of interacting components, breaking the molecular chemical equilibrium and reducing the strength of spatial lattices.

Both natural minerals and products of other industries containing compounds that accelerate the sintering process of clinker and increase the reactivity of the cement raw material mixture can be used as mineralizers [1, 2]. According to scientific research, such compounds include oxides of some metals and various salts. Alkali metal compounds, for example, contribute to the acceleration of calcining during clinker burning and affect the strength characteristics of cement [3, 4], while titanium dioxide reduces the sintering temperature of clinker and changes its phase composition [5]. Heavy metals can also have a mineralizing effect without significantly affecting the quality of the clinker [6]. The greatest minimization of energy costs can be achieved when using fluorinated mineralizers separately or in combination with other compounds [7, 8]. First, the fluorinated mineralizers include calcium fluoride CaF_2, which is a traditional mineralizer. The disadvantage of calcium fluoride is its cost. The other fluorinated compounds can be used as an alternative to the use of expensive CaF_2. The source of such a compound can be a technogenic product of the electrolytic production of aluminum (crushed electrodes) [9]. The crushed electrodes is a technogenic waste of aluminum production, rich in elements such as carbon, fluorine, sodium, and aluminum.

The aim of this work is to study the effectiveness of using a technogenic product of electrolytic aluminum production as a mineralizer in the burning of clinker, as well as to compare the effectiveness of the mineralizing effect of crushed electrodes and calcium fluoride.

2 Methods and Materials

The object of the study was a raw mixture (Table 1), designed to produce an ordinary clinker, characterized by the modules: KN = 0.93; n = 2.18; p = 1.32.

The calcium fluoride reagent and crushed electrodes were used as mineralizing additives, which were inserted into the raw mixture in excess of 100% in an amount of 1% in terms of the content of the fluorine ion (F^-). To evaluate the mineralizing effect of the crushed electrodes, it was used burnt, that is, without the carbon

Table 1 Chemical composition of the raw mixture and clinker, %

Component	SiO_2	Al_2O_3	Fe_2O_3	CaO	MgO	SO_3	R_2O	LOI
Raw mixture	13.92	3.63	2.75	43.21	0.86	0.30	0.54	34.80
Clinker	21.35	5.57	4.22	66.27	1.32	0.46	0.83	–

Table 2 Chemical composition of the burnt crushed electrodes, %

SiO_2	Al_2O_3	Fe_2O_3	CaO	MgO	SO_3	TiO_2	F^-	K_2O	Na_2O	Σ
0.69	21.55	1.38	1.23	0.44	3.20	0.02	38.08	0.30	32.79	99.68

component. The processing temperature is 700 °C, to exclude the sublimation of the compounds that are part of the crushed electrodes. The chemical composition of the burnt crushed electrodes is shown in Table 2.

The main components of the crushed electrodes are fluorine, sodium and aluminum oxides. It is these compounds that will influence the process of clinker formation.

3 Results and Discussion

One of the options for evaluating the effectiveness of mineralizers is to determine the temperature range of the binding of calcium oxide in clinker minerals. To assess the binding of free calcium oxide, cylindrical samples of the prepared mixtures weighing 2 g were pressed, which were burnt in batches in a laboratory furnace in the temperature range of 1100–1450 °C with holding time of 15 min and a sampling step of 50 °C. Further, each sample was examined for the content of free calcium oxide by the ethyl-glycerol method. The kinetic curves of the binding of calcium oxide in the mixtures are shown in Fig. 1.

The complete binding of free CaO for mixture without a mineralizer is fixed at a temperature of 1400 °C.

When CaF_2 is inserted into the raw mix, the content of free calcium oxide is within the tolerable limits at a temperature of 1250 °C, and completely absent at a temperature of 1300 °C. The insertion of the crushed electrodes into the raw mix contributes to better binding of calcium oxide compared to other samples. At 1200 °C, we find 0.3% of free CaO, which indicates an almost complete process of formation of clinker minerals.

Thus, the high efficiency of the action of mineralizers, CaF_2 and the crushed electrodes, on the binding of calcium oxide into clinker minerals, makes it possible to reduce the burning temperature of clinker to 1350 °C.

To test the possibility of reducing the burning temperature of clinkers with the addition of mineralizers, burning was carried out in a laboratory furnace at a temperature of 1350 °C with holding time of 1 h. The samples were cooled to 1150 °C at speed about 20 °C/min and then abruptly in air to ambient temperature. The samples were burnt without mineralizers at a temperature of 1450 °C with holding time of 1 h. The clinker was cooled to 1200 °C at a speed of 20 °C/min and then abruptly in air to ambient temperature. The cooling mode of the synthesized clinkers corresponds to the cooling mode of the clinker in the cooling zone of the rotary kiln and the clinker cooler.

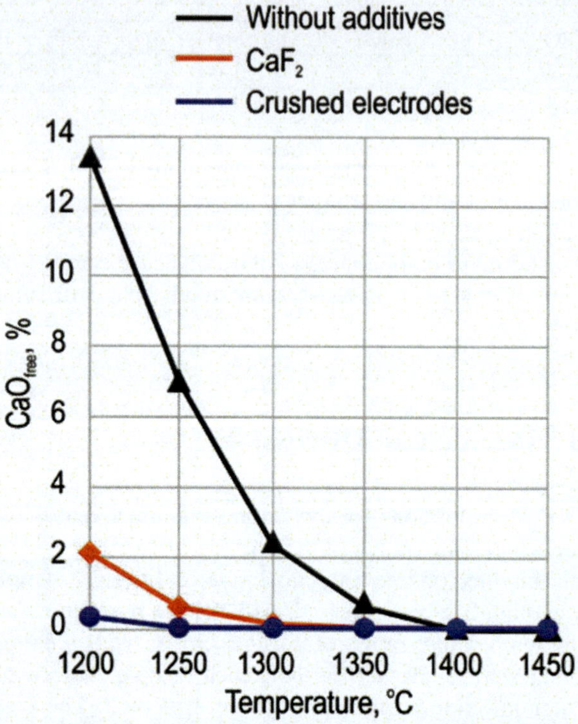

Fig. 1 Kinetics of CaO_{free} binding

All clinkers are characterized by the presence of basic clinker minerals and the absence of free calcium oxide (Fig. 2). Additional phases were not identified. It should be noted that the X-ray image of clinker with the addition of calcium fluoride does not show diffraction reflections corresponding to tricalcium aluminate d, Å = 2.69. This is due to the thermal instability of C_3A in systems with the presence of fluorine ions. Tricalcium aluminate in the presence of CaF_2 dissociates into $5CaO{\cdot}3Al_2O_3$ and free CaO. CaO and $2CaO{\cdot}SiO_2$, dissolving rapidly in large quantities in the liquid phase, cause intensive formation of $3CaO{\cdot}SiO_2$. This is confirmed by the higher intensity of the alite diffraction reflections on the X-ray image d, Å = 3.043; 1.770. At the same time, on X-ray images of clinkers without additives, the diffraction reflections of C_3A are clearly expressed, with the addition of the crushed electrodes, they have a slightly lower intensity relative to the additive-free clinker.

The use of mineralizing additives and the reduction of the burning temperature can have a significant impact on the microstructure of the clinker. To verify this assumption, a petrographic analysis of the synthesized clinkers was performed (Fig. 3).

All clinkers have a clear fine-crystalline homogeneous structure. The alite phase is represented by individual small crystals of regular shape with an average size of 15–20 μm. There are rare individual grains of about 25–30 μm in size, which are

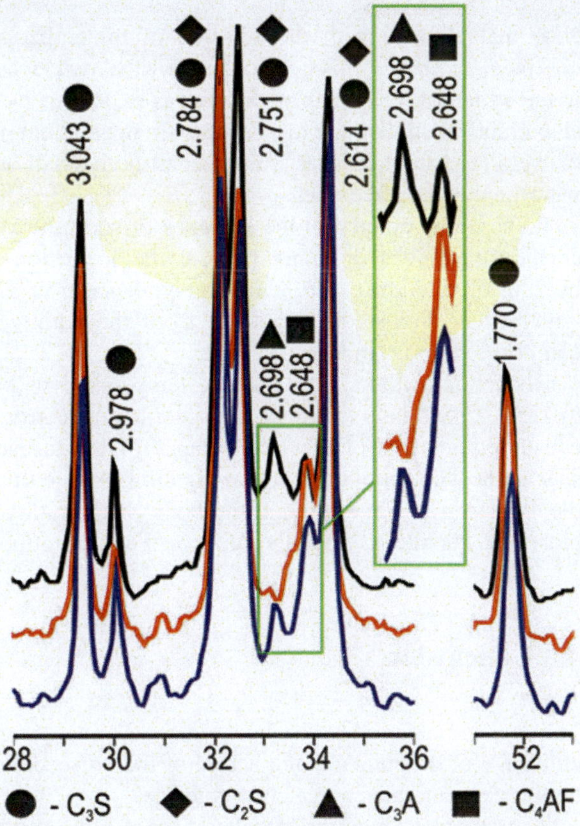

Fig. 2 Phase composition of clinkers

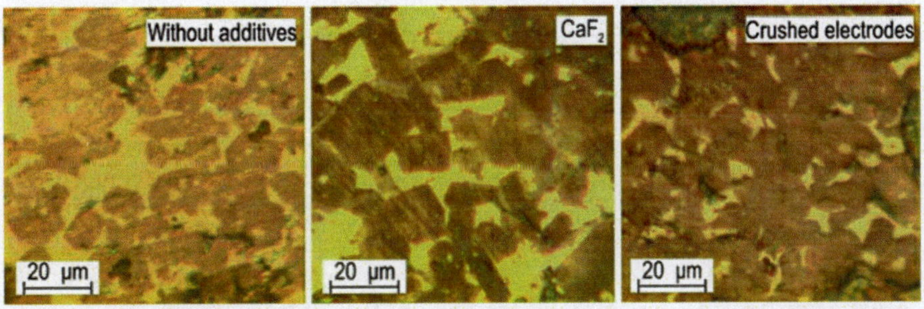

Fig. 3 Microphotographs of synthesized clinkers

mostly elongated quadrangular in shape. Grains of belite are very rare in clinker, they can be individual crystals of about 10–15 μm in size, or small crystals of about 2–3 μm in size, which are included in the structure of alite crystals. The difference between clinkers is the amount of intermediate phase. Visual evaluation makes it clear that clinkers with the addition of mineralizers have a smaller amount of intermediate phase. This is evidenced by the very dense arrangement of alite grains in the structure. Presumably, this fact is explained by an increase in the content of alite grains in clinker due to a decrease in the amount of tricalcium aluminate. These clinkers also differ in a large number of both small and large pores in comparison with the additive-free one.

It should be noted that the presence of sodium oxide in the composition of the crushed electrodes does not prevent the formation of clinker minerals and the binding of calcium oxide. Clinkers synthesized at 1350 °C with the insertion of mineralizers almost do not differ from the additive-free clinker in both phase composition and mineralogical one.

Reduction of the clinker burning temperature with the insertion of mineralizers to 1350 °C will help to reduce the specific heat consumption on burning. The main consumption items of the heat balance of the furnace system, which reduce the specific heat consumption on burning clinker—are the reduction of heat losses with the exhaust gases and through the body of the rotary kiln. According to the computations, the total saving of specific heat consumption will be 175 kJ/kg of clinker.

4 Conclusion

The crushed electrodes can be used as a mineralizer when burning clinker. The efficiency of the mineralizing action of the crushed electrodes is not inferior to the traditional mineralizer—calcium fluoride.

The complete binding of calcium oxide into clinker minerals is completed at 100–150 °C earlier compared to the additive-free mixture when using mineralizers, which allowed reducing the burning temperature of the clinker to 1350 °C. At the same time, the binding of calcium oxide at the insertion of the crushed electrodes is completed at 50 °C lower than at the addition of calcium fluoride CaF_2.

The clinkers obtained at a lower temperature with the addition of the crushed electrodes and calcium fluoride practically do not differ in their phase and mineralogical composition from the additive-free clinker synthesized at a temperature of 1450 °C.

Reducing the clinker burning temperature will reduce the specific heat consumption by 175 kJ/kg of clinker due to the reduction of the temperature of the exhaust gases and heat loss through the body of the furnace system.

Acknowledgements This work was realized in the framework of the Program of flagship university development on the base of the Belgorod State Technological University named after V.G. Shukhov. The work was realized using equipment of High Technology Center at BSTU named after V.G. Shukhov.

References

1. Boughanmi S, Labidi I, Megriche A, El Maaoui M, Nonat A (2018) Natural fluorapatite as a raw material for Portland clinker. Cem Concr Res 105:72–80
2. Mehta PK, James R (1977) Brady: utilization of phosphogypsum in portland cement industry. Cem Concr Res 7(5):537–544
3. Klassen VK, Ermolenko EP, Novosyolov AG, Mishin DA (2013) Problem of impurity of salts of alkali metals in cement raw materials. Middle-East J Sci Res 17(8):1130–1137
4. Novosyolov AG, Vasina YA, Novoselova IN, Goriaynova DN, Ershova YI (2020) Investigation of the possibility of reducing the circulation of volatile compounds in a rotary kiln of a dry production method. Bull BSTU named after V.G. Shukhov 7:84–92
5. Mishin DA, Barbaniagre VD, Cherkasov AV (2017) Clinker formation processes in the presence of mineralizers: monograph. BSTU named after V.G. Shukhov, Belgorod
6. Kolovos KG (2006) Waste ammunition as secondary mineralizing raw material in Portland cement production. Cement Concr Compos 28(2):133–143
7. Kacimi L, Simon-Masseron A, Ghomari A, Derriche Z (2006) Influence of NaF, KF and CaF_2 addition on the clinker burning temperature and its properties. Comptes Rendus Chimie 9 (1):154–163
8. Da Y, He T, Shi C, Wang M, Feng Y (2021) Potential of preparing cement clinker by adding the fluorine-containing sludge into raw meal. J Hazard Mater 403
9. Kulikov BP, Nikolaev MD, Kuznetsov AA, Pigarev MN (2010) Utilization of fluorinated waste from aluminum production in the cement industry. Ecol Ind Russia 5:4–6

Influence of Technological Additives on the Characteristics of Cement Powders

L. D. Shahova, E. S. Chernositova, L. S. Schelokova, and N. G. Uhaneva

Abstract Powders make up 50% of the materials used in industry as raw materials or in the form of finished products. The article deals with the influence of technological additives introduced during cement grinding on the fineness of cement grinding and rheological characteristics. The effects of several individual chemicals on the dry fine grinding of clinker with 5% gypsum natural stone under laboratory conditions were investigated. The influence of certain industrially produced chemicals was studied in dosages in terms of active substances 0.015 and 0.025 wt.%, by weight of the materials loaded for grinding. To assess the mobility of cement powder, such quantitative evaluation indicators as the fluidity of the powder by weight, the material passed through the sieve, and the Hausner index are used. It is established that the effect on the grinding process and the rheological parameters of the mobility of powders depend on the structural features and differences in the functional groups of organic molecules estimated by polarization.

Keywords Technological additives · Glycols · Amines · Fluidity of the powder flow · Hausner index

1 Introduction

In the cement industry, the main technological operations with powders include grinding in ball mills, transportation by pneumatic transport, storage in silos, loading and unloading works. It is known that technological additives, which include grinding aids, during dry grinding of Portland cement clinker change the rheological properties of the powders, which affects the parameters of the grinding process, affect the energy consumption for transportation and the time of loading

L. D. Shahova
LLC «Polyplast Novomoskovsk», Novomoskovsk, Russia

E. S. Chernositova (✉) · L. S. Schelokova · N. G. Uhaneva
Belgorod State Technological University Named After V.G. Shukhov, Belgorod, Russia
e-mail: ES-Helen@yandex.ru

S. V. Klyuev (ed.), *Digital Technologies in Construction Engineering*,
Lecture Notes in Civil Engineering 173, https://doi.org/10.1007/978-3-030-81289-8_33

and unloading works. To optimize the use of grinding intensifiers, it is necessary to know the influence of their structural features and functional groups on the structural and rheological properties of cement powders. The variety of chemicals, materials to be crushed, types of grinding systems and process parameters do not allow systematizing the literature data and achieve a complete understanding of this issue.

The behavior of the flow of fine material during grinding and transportation is multidimensional in nature and depends on many physical characteristics. The fluidity of powdered materials is essentially the result of a combination of the physical properties of the material, the environmental conditions, and the used processing and storage equipment. The particle size and granulometric composition of the material play an important role in fluidity. It is known that even small changes in the particle size can cause significant changes in the resulting mobility [1–3]: the smaller the particles in size, the more the powder tends to decrease fluidity due to an increase in the contact points and the interface per mass unit. The second most important factor affecting the mobility of powders is the adhesion of particles. When grinding mineral powders, the particles are affected by the standard force of the grinding media, which destroys large particles simultaneously and compacts the layer of small particles. These processes are accompanied by autohesion and adhesion to the metal surface of grinding media and hunch plate [4].

It is noted that when identical particles come into contact with the same substrate, ambiguous charges can occur. As the particle size decreases, the electric charges increase, and consequently, the electrical component of the adhesion forces. A particularly significant increase in charges is observed after the separation of relatively small particles from the metal surface. During separation, particles of both signs of charge are detected, and the final sign of the charge does not depend on the sign of the initial charge of the particles. Nevertheless, negatively charged particles prevail, the share of which is approximately 75% of the total number of particles [5]. Additionally, the energy heterogeneity of the surface of the powder from Portland cement clinker is acquired due to the polymineral composition of the clinker and triboelectric phenomena in the grinding process, which increases the adhesive forces. By eliminating or reducing the electrical component, the adhesive strength can be changed. Therefore, in order to exclude the electrical component, it is necessary to inform the surface of the donor or acceptor properties, which can be done by modifying the surface, which consists in replacing some surface molecular groups with others. This is possible by using organic substances that are prone to polarization, with different molecular structures and the presence of different functional groups, the so-called grinding aids [6–8].

The aim of this work is to study the effects of chemicals of different classes of organic compounds according to the classification [9] structural and rheological characteristics of powders.

2 Methods and Materials

As part of this work, the effects of several specific chemicals on the dry fine grinding of clinker with 5% gypsum natural stone in a laboratory ball mill with a volume of 25 L were investigated. In the experiment, all the parameters during grinding were kept constant: the loading of the mill with material and grinding media, the time of grinding and unloading. For the experiments, an industrial clinker with the mineralogical compositions indicated in Table 1 was selected.

The influence of specific industrially produced chemicals was studied (Table 2) in dosages in terms of active substances 0.015 and 0.025 wt.%, of the weight of the uploaded materials.

The fineness of cement grinding was determined by the residues on sieves with cell sizes in the light of 80 and 45 μm, and the specific surface area was evaluated using a Blaine device (Blaine, Tonidustrie, Germany). To increase the statistical reliability of the results on the fineness of grinding, three samples were analyzed in parallel.

The fluidity of the powder was determined according to the adapted method of ASTM C1565-09, which consists in sifting the mass of the powder at the specified parameters of mechanical action. The fluidity measurements were carried out under ambient conditions and repeated twice for each sample. The fluidity index was expressed in % of the residue on the sieve with a cell of 500 μm.

The internal friction index for cohesive powders was estimated by the Hausner index, which is determined by the ratio between the density of the powder in the compacted state and the bulk density.

Table 1 Mineralogical composition of clinker, wt.%

C_3S	C_2S	C_3A	C_4AF
65.89	9.9	8.28	12.23

Table 2 Characteristics of individual chemicals

Class of compounds	Name	Short description	Chemical formula	Molar mass, g/mol	*Polarization, $\cdot 10^{-24}$, cm^3
Glycols	Ethylene glycol	EG	$C_2H_6O_2$	62.07	5.07 ± 0.5
	Glycerin	Gl	$C_3H_8O_3$	92.09	8.13 ± 0.5
	Diethylene glycol	DEG	$C_4H_{10}O_3$	106.12	10.06 ± 0.5
	Diethanolamine	DEA	$C_4H_{11}NO_2$	105.14	10.80 ± 0.5
	Methyldiethanolamine	MDEA	$C_5H_{13}NO_2$	119.16	12.68 ± 0.5
Amino alcohols	Triethanolamine	TEA	$C_6H_{15}NO_3$	149.19	15.13 ± 0.5
	Diethanol-Isopropanolamine	DEIPA	$C_7H_{17}NO_3$	163.21	16.95 ± 0.5
	Triisopropanol-amine	TIPA	$C_9H_{21}NO_3$	191.27	20.59 ± 0.5

*calculated in the ACDLab program

3 Results and Discussions

The results of the experiment are presented in Table 3 and in Figs. 1 and 2.

As it is shown in Fig. 1, all the studied substances lead to a change in the particle size in comparison with the powder ground without any addition. Thus, the degree of particle size reduction depends on both the type and dosage of the additive.

Unlike amines, glycols increase the specific surface area by 30–40 m^2/kg or by 10–12%. The binding strength between the residues on the sieves and the specific surface area on the Cheddock scale varies from 0.2 to 0.5 for amines and from 0.75 to 0.86 for glycols. The absence of a close relationship between particle size and specific surface area is consistent with the practice of obtaining fine-ground mineral particles [10]. There is also a low correlation between the content of particles smaller than 45 μm and the polarizing properties of the additives. But there is a strong inverse dependence of the specific surface on the polarization (Table 2).

The effect on the structural and rheological properties of the powder of specific chemicals is shown in Fig. 2. According to Fig. 2, grinding cement without any additive results in a cement fluidity of 66–73% and, consequently, a powder with high cohesive bonds.

The cohesive nature of the powder flow is changed by the most part of amines, while glycols do not show a significant dependence of fluidity on the dosage of additives within the studied dosage range. On the contrary, a strong dependence on the dosage of the additive is visible in the case of the input during TIPA and DEIPA grinding. Here, low dosages of 0.015 wt.% lead to the formation of powders with fluidity values of 89% and 87%, respectively. On the contrary, easily flowing powders (T = 97–100%) are obtained at higher dosages of 0.025% additives. There is a strong direct correlation of fluidity and an inverse dependence for IHR on the polarization of organic substances.

The results of this study demonstrate a significant influence of functional groups, structural features and the tendency to polarization of organic molecules on the grinding process and the structural and rheological properties of crushed powders. According to the effect on the fineness of the product, the studied chemicals are arranged in a row: amines > glycols.

According to [11], the asymmetric distribution of polar groups within amine molecules favors the success of grinding, and vice versa, a coarser product is obtained in the presence of symmetric glycol molecules.

Insufficient dosing is more likely when using organic compounds with a low ratio of polar and non-polar atoms, such as glycols. This indicates that the surface of the particle must be covered with functional groups to a certain extent in order to achieve the best fineness result. Therefore, molecules with several evenly distributed polar atoms affect a more complete and protective adsorption layer. In fact, it is difficult to determine how effectively the molecules of a substance are adsorbed, in what area they cover the surface of the particle, and how steric effects affect the distribution and adsorption of other molecules.

Table 3 Actual mineralogical composition of clinkers

Indicators	Dosage (%)	Control sample	EG	Gl	DEG	DEA	MDEA	TEA	DEIPA	TIPA	Correlation coefficient with polarization
Particle content less than 45 microns (%)	0.015	81.9*	81	81.1	81.4	82.6	85.7	83.2	83.4	81.6	0.34
	0.025	82.7**	82.6	78.9	84.2	84	86.7	83.7	84	82.7	0.27
Surface area (m^2/kg)	0.015	287*	318	322	314	310	299	303	291	289	-0.93
	0.025	295**	320	308	307	304	296	298	273	269	-0.95
Flowability T (%)	0.015	66*	77	69	73	85	86	82	87	89	0.75
	0.025	73**	80	72	77	88	93	89	97	100	0.86
IHR	0.015	1.284*	1.252	1.268	1.244	1.223	1.221	1.239	1.212	1.204	-0.82
	0.025	1.251**	1.234	1.335	1.236	1.239	1.194	1.208	1.177	1.17	-0.72

Note To obtain different initial fineness, the control additive-free composition was milled for 20 (*) and 25 (**) minutes

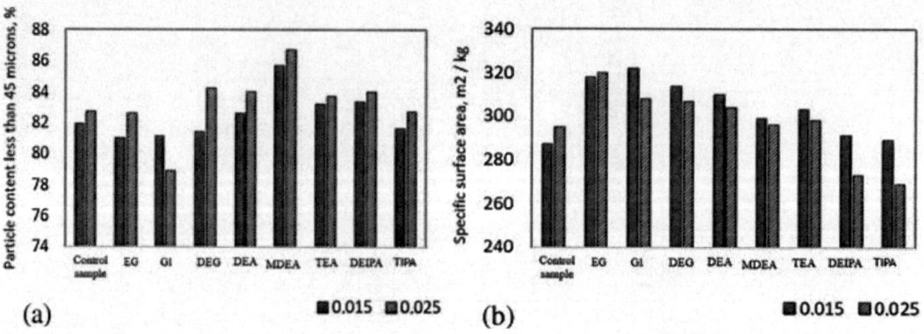

Fig. 1 Influence of the type and dosage of the additive during grinding on the fineness of cement grinding: **a** particle content of more than 45 μm, wt.%; **b** Blaine specific surface area, m^2/kg

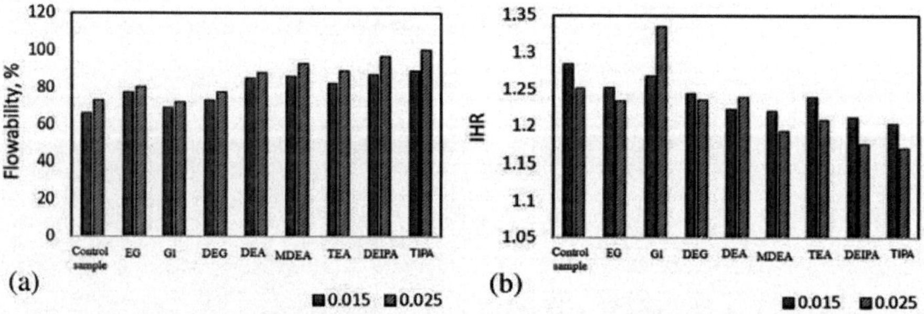

Fig. 2 Influence of the type and dosage of the additive during grinding on the fluidity of cement powder: **a** fluidity, %; **b** Hausner index IHR

The polarization of the molecules of the studied substances was calculated in the ACDLab program (Table 2). The presence of a nitrogen ion in the structure has a significant effect on the polarization of amines, which, under the influence of an external electric field, shifts the electrons in the functional groups [12]. For example, when the molecular weight of DEG and DEA is equal, the polarization values are 10.06 and 10.8, respectively. This can explain the difference in all indicators for the fineness of grinding and fluidity of powders. The higher the value of the polarization of the molecule and the more structural units, the more expressed the effect on the specific surface and structural and rheological parameters of the powders (Table 3) is due to the change in the surface energy of mineral particles by adsorbed organic matter molecules and the steric effect [13]. The value of the Hausner index correlates with the data on the fluidity and specific surface area of crushed cements well. The values for cements in the presence of glycerol fall out of this series. This may be due to the high viscosity of glycerol, which is almost two times higher than that of other glycols, which affects the rate of distribution of molecules (diffusion of molecules) over the surface of cement particles. The cement powder obtained in the presence of glycerol is well compacted, perhaps this is due to the low content of particles in the powder less than 45 microns.

4 Conclusion

The results obtained indicate the influence of structural features, functional groups, and polarization of organic molecules of various classes of chemical compounds on the grinding process and rheological properties of powders.

Glycols have a positive effect on the specific surface area of powders and the Hausner index and a negative effect on their fluidity.

In the presence of amines, the particle content of less than 45 μm in the powders increases, but the specific surface area hardly change in comparison with the control additive-free composition.

The fluidity of powders with amines is higher than on glycols, while a decrease in the Hausner index is noted.

Acknowledgements This work was realized in the framework of the Program of flagship university development on the base of the Belgorod State Technological University named after V.G. Shukhov, using equipment of High Technology Center at BSTU named after V.G. Shukhov.

References

1. Fitzpatrick JJ, Ahrne L (2005) Food powder handling and processing industry problems, knowledge barriers and research opportunities. Chem Eng Proc 44:209–214. https://doi.org/10.1016/j.cep.2004.03.014
2. Erica E, Oliver J, Pugsley T, Sharma J, Zhou J (2009) Flowability of moist pharmaceutical powders. Powder Technol 189:409–415. https://doi.org/10.1016/j.powtec.2008.06.017
3. Geldart D, Harnby N, Wong AC (1984) Fluidization of cohesive powder. Powder technology 37:25–37. https://doi.org/10.1016/0032-5910(84)80003-0
4. Shahova LD, Schelokova LS, Chernositova ES (2021) Influence of clinker microstructure on grinding efficiency in the presence of grinding intensifiers. Lecture Notes Civ Eng 95:23–29. https://doi.org/10.1007/978-3-030-54652-6_4
5. Zimon AD (1976) Adhesion of dust and powders. Chemistry, Moscow
6. Shahova LD, Chernositova ES, Denisova JV (2017) Investigation of the effect of technological additives on the rheological properties of cement powder. Bull BSTU named after V.G. Shukhov 10:123–128
7. Shahova LD, Chernositova ES, Denisova JV (2017) Flowability and durability of cement containing technological additives during grinding process. AIME. AER-Adv Eng Res 133:162–167
8. Elistratkin MYu, Minakov SV, Shatalova SV (2019) Composite binding mineral additive influence on the plastisizer efficiency. Constr Mater Prod 2(2):10–16
9. Shakhova LD, Cherkasov RA (2014) Classification of technological additives in cement grinding. Fundam Res 12:295–299
10. Macaulay S, Krutikov DM (2011) Transition from Blaine measurements to granulometric analysis by laser diffraction. Cement Appl 2:115–119
11. Dombrowe H, Hoffmann B, Scheibe W (1982) Über Wirkungsweise und Einsatzmöglichkeiten von Mahlhilfsmitteln. Zement-Kalk-Gips 11:571–580
12. Schultz J, Lavielle L (1989) Interfacial properties of carbon fiber-epoxy matrix composites. ACS Symp Ser 391:185–202
13. Prziwara P, Breitung-Faes S, Kwade A (2017) Impact of grinding aids on dry grinding performance, bulk properties and surface energy. Adv Powder Technol 29(2):416–425

New Methodology for Studying the Drying Behavior of Sludge

D. A. Mishin, D. V. Smal, and S. V. Kovalev

Abstract The proposed methodology, as well as the traditional one, is based on the study of the behavior (drying) of sludge on the contact surfaces. The study was carried out using cement raw sludge from OJSC "MMC "Norilsk Nickel". The traditional facility for studying the sludge drying process is a metal drum rotating on rollers, in which fasteners are placed, with chains attached to them. The proposed method for studying the drying behavior of sludge involves the use of plates with smooth, rough and ribbed surfaces. The surface of the plates has a significant impact on the sludge drying process and the critical moisture value. The most active sludge is collected on a plate with a rough surface; the critical moisture is 36%. The smooth surface of the plate allows the sludge to gain slowly on its surface, but at the same time go away quickly, at a critical moisture of 24%. In the ribbed plate, the material collection is relatively slow. The material sliding is accompanied by rapid surface cleaning; the value of critical moisture was 32%. The values of the obtained ribbed plate drying curve are comparable with the curve obtained for drying sludge in the experimental facility; the value of critical moisture in both cases was 32%.

Keywords Cement raw sludge · Moisture · Drying · Wet method · Heat exchange

1 Introduction

Currently, the main method of cement production in the world is the dry method. However, a significant number of cement plants in Russia and the CIS countries use the wet method [1]. Therefore, work is underway in the field of research of chain systems [2–4] and sludge [5–7].

One of the main directions of optimizing the costs of wet cement production is the introduction of energy-saving technologies. The study of the features of sludge drying in a chain system allows choosing the optimal type of heat exchangers that can help reduce the consumption of burner oil, thereby reducing costs. Optimization

D. A. Mishin (✉) · D. V. Smal · S. V. Kovalev
Belgorod State Technological University named after V.G. Shukhov, Belgorod, Russia

S. V. Klyuev (ed.), *Digital Technologies in Construction Engineering*,
Lecture Notes in Civil Engineering 173, https://doi.org/10.1007/978-3-030-81289-8_34

of the complex of heat exchange devices ultimately leads to a reduction in the heat consumption for firing clinker, an increase in the productivity of the furnace and the resistance of the refractory lining, a reduction in dust entrainment from the furnace and the degree of environmental pollution [8, 9].

Due to the complexity of using the traditional method of determining the properties of sludge, it is very relevant to develop an alternative, simplified method for studying the features of sludge drying.

The proposed methodology, as well as the traditional one, is based on the study of the features of the behavior (drying) of sludge on the contact surfaces, which in turn can be considered as a model of heat and mass transfer processes observed in the cold part of the furnace of the wet method of cement production.

2 Methods and Materials

The study was carried out using cement raw sludge of OJSC MMC "Norilsk Nickel", the characteristics of which are presented in Table 1.

In addition to the standard methodology and experimental facility (Figs. 1 and 2), a new approach was used to determine the adhesive-cohesive properties and the value of the critical moisture content of the sludge. Namely, samples were prepared, which are plates simulating various surfaces of chains (Fig. 3), with which the drying behavior of the sludge was studied.

The traditional facility for the study of the sludge drying process is a metal drum rotating on rollers, in which fasteners are placed, with chains attached to them. Through the hole in the drum, a constant amount of fixed-temperature heat transfer agent is supplied, which ensures the drying of the sludge.

3 Results and Discussion

The proposed method for studying the drying behavior of sludge involves the use of plates with smooth, rough and ribbed surfaces (Fig. 3).

So, in a container with a total volume of 500 ml of sludge, plates with different surfaces were immersed at certain intervals, and so on several times. After that, they were weighed and sent to the drying cabinet. Then the dried plates with the material on them were re-weighed, thereby determining the change in the moisture content

Table 1 Characteristics of the sludge

Sludge content (%)				Moisture (%)	Spread rate (mm)
Lime	Clay	Cinders	Ash		
78	18	1.5	2.5	W = 45	108

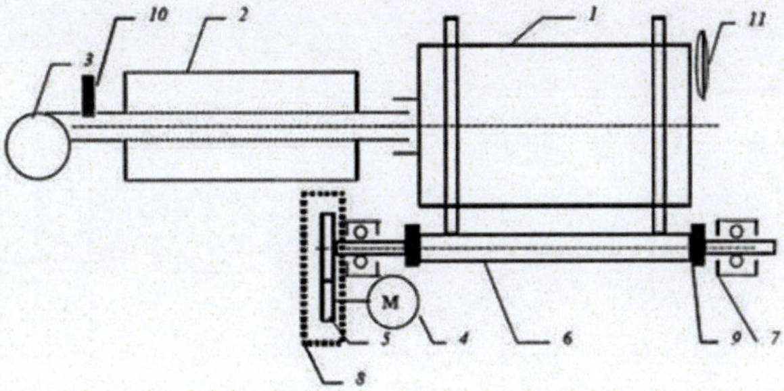

Fig. 1 Experimental facility for the study of the sludge drying process: *1*—drum; *2*—muffle cabinet; *3*—fan; *4*—electric motor; *5*—gear drive; *6*—track roller; *7*—rolling bearings; *8*—gear drive body; *9*—limiter; *10*—gate; *11*—thermometer

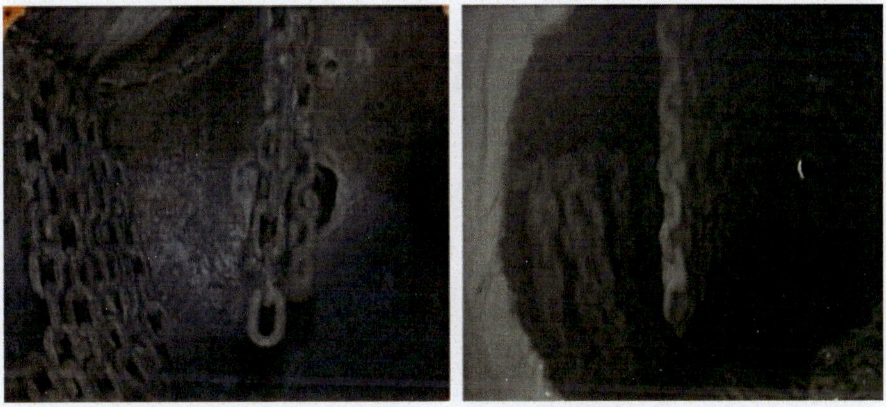

Fig. 2 View of the internal structure of the drum of the experimental facility, on the left—empty one, on the right—with sludge

Fig. 3 Plate samples simulating different surfaces: *1*—ribbed; *2*—rough; *3*—smooth

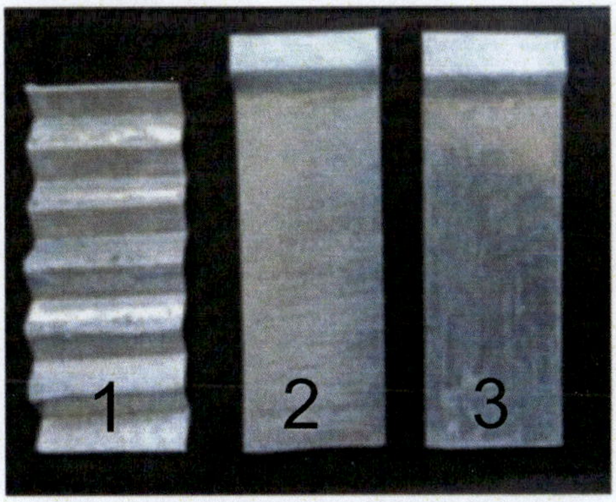

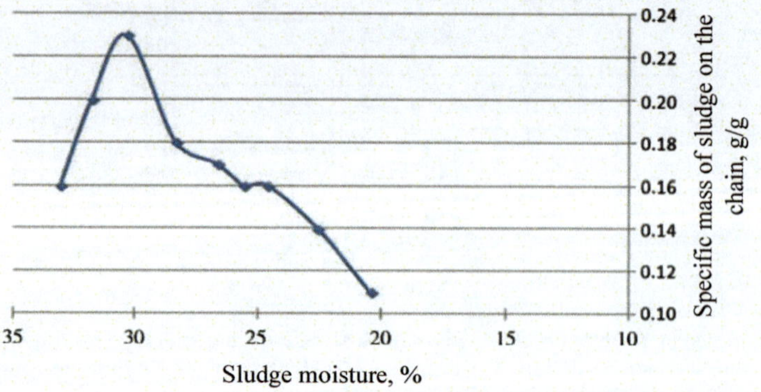

Fig. 4 Sludge drying in the experimental facility

of the sludge over time and the specific mass of the material on the plates. The results of sludge drying on are shown in Fig. 5

The study is carried out in the traditional way [10], using a laboratory facility (Figs. 1 and 2). In the course of the experiment, the specific mass of the material on the chains and plates was estimated at different moisture of the sludge. The selection of chains and the evaluation of the mass and moisture content of the material was carried out by analogy with the plates.

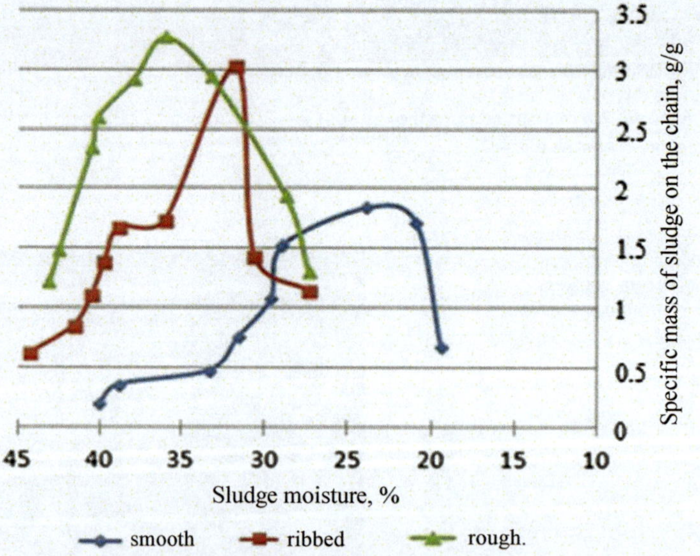

Fig. 5 Drying of sludge on plates with different surfaces: smooth, ribbed, rough

The results of the study of sludge drying in a traditional facility (in a drum) are presented in Fig. 4. According to the behavior of the curve, the following is observed: the largest mass of sludge on the chains is 0.23 g, achieved at a value of the critical moisture content of the sludge of about 32%. At the same time, there is an intensive collection of material on the chain and as it dries, it gradually disappears. This behavior of the sludge indicates its possible tendency to dust formation.

The data obtained when drying the sludge on plates with different surfaces are shown in Fig. 5.

The sludge is collected on a plate with a rough surface most actively among all the plates; the critical moisture is 36%. In turn, the smooth surface of the plate allows the sludge to gain on its surface slowly, but at the same time go away quickly, at a critical moisture of 24%.

In the ribbed plate, due to the complex geometric surface, the material collection is relatively slow. The material slide is accompanied by a rapid cleaning of the surface; the critical moisture value was 32%. The surface of the plates has a significant impact on the sludge drying process and the critical moisture value. The values of the obtained drying curve using a ribbed plate are comparable to the curve obtained when drying the sludge in the experimental facility, the critical moisture value in both cases was 32%.

4 Conclusion

The use of the proposed methodology allows studying the drying behavior of the sludge on the chains, it is characterized by less labor costs and the duration of the experiment, allows determining the value of the critical moisture and justifying theoretically thc choice of the chain system scheme of an industrial furnace.

Acknowledgements The study is implemented in the framework of the Flagship University Development Program at Belgorod State Technological University named after V.G. Shukhov, using the equipment of High Technology Center at BSTU named after V.G. Shukhov.

References

1. Klassen VK (2012) Technology and optimization of cement production. BSTU, Belgorod
2. Fedorenko BZ, Gorlov AS, Petrashev VI (2018) Models of aerodynamic processes in the chain veils of cement kilns. Bull BSTU named after. V. G. Shukhov 7:103–115. https://doi.org/10.12737/article_5b4f02c143dc21.96604314
3. Fedorenko BZ (2001) Modeling, analysis and optimization of heat-technological processes in chain systems of cement kilns. In: Proceedings of the seventh academic readings of the RAACS "Modern problems of construction materials science". 2:347–359, BelSTACM, Belgorod

4. Moskvichev DS (2008) Intensification of heat transfer processes in the hot part of the chain heat exchanger of a rotary kiln for firing cement clinker. Chem Chem Technol Success 22(7)(87):36–39
5. Ломаченко, Д.В., Шаповалов, Н.А., Яшуркаева, Л.И.: Модифицирование свойств цементных шламов с использованием различных добавок (2013) Bull BSTU named after. V. G. Shukhov 3:137–139
6. Shahova LD, Chernositova ES, Berezina NM, Danilin AA (2016) Composite thinners raw sludge in cement production. Bull BSTU named after V. G. Shukhov10:191–196. https://doi.org/10.12737/22103
7. Besedin PV, Andruschak SV, Kozlov VK (2015) Research and simulation of the motion cement slurry under laboratory conditions. Bull BSTU named after V.G. Shukhov 4:113–119
8. Borisov IN (2003) Management of the processes of agglomeration of materials and the formation of coating in rotating furnaces of the cement industry. Belaudit, Belgorod
9. Barbanyagre VD, Smal DV (2008) Increasing the reactivity of a cement raw material mixture with coarse-grained quartz. Cem Appl 1:162–166
10. Borisov IN, Moskvichev DS, Smal DV (2010) Investigation of heat transfer in chain systems of rotating furnaces on the basis of a mathematical model. In: Proceedings of the international scientific and practical conference "Innovative materials and technologies (XX Scientific readings)". Belgorod State Technological University named after V.G. Shukhov, Belgorod, pp 7–11

Activity of High Porous Fillers for Heat Insulating Dry Mixtures

V. I. Loganina, S. N. Kislitsyna, and M. V. Frolov

Abstract The efficiency of using four materials as highly porous fillers in the composition of heat-insulating lime dry building mixture is estimated: glass hollow microspheres, ash microspheres of aluminosilicate, expanded vermiculite sand, expanded perlite sand. The oxide composition of the highly porous fillers used in the work is given. The pozzolanic activity of the investigated fillers was determined by the absorption of $Ca(OH)_2$ from a lime solution. It was revealed that the activity of glass hollow microspheres is 480.5 mg/g. It was revealed that new formations—calcium hydrosilicates—are formed on the surface of hollow glass microspheres during the reaction with lime. The high efficiency of the use of hollow glass microspheres as a filler in heat-insulating plaster solutions has been confirmed. A lower specific strength of composites with a filler of expanded perlite and vermiculite sand and a higher specific strength with a filler of hollow glass microspheres was found.

Keywords Dry building mixtures · Pozzolanic activity · Calcium hydrosilicates · Glass hollow microspheres · Specific strength

1 Introduction

To improve the operational properties of lime dry building mixtures (DBM), various modifying additives are introduced into their composition [1–3]. Pozzolanic activity is one of the characteristics that make it possible to assess the effectiveness of their use in the composition of the DBM [4–6]. The action of the pozzolanic effect of modifying additives in lime-based dry mixes is manifested in the chemical interaction of amorphous silica contained in the additives with lime.

It should be noted that amorphous silica can be contained not only in modifying additives, but also in other components of lime-based dry mixes, for example, in

V. I. Loganina (✉) · S. N. Kislitsyna · M. V. Frolov
Penza State University of Architecture and Construction, 28, Titov Street,
Penza 440028, Russia

S. V. Klyuev (ed.), *Digital Technologies in Construction Engineering*,
Lecture Notes in Civil Engineering 173, https://doi.org/10.1007/978-3-030-81289-8_35

fillers. When used as fillers with high pozzolanic activity in the composition of lime DBM, low-basic calcium hydrosilicates are formed in the composites, which strengthen the lime composites.

The purpose of the work is to evaluate the effectiveness and select of porous fillers in the DBM formulation.

2 Materials and Methods

The work evaluated the pozzolanic activity of 4 different highly porous fillers:

- glass hollow microspheres MS-V with a bulk density of 130 kg/m^3;
- ash microspheres of aluminosilicate bulk density of 400 kg/m^3;
- expanded perlite sand of the M-150 grade with a bulk density of 150 kg/m^3;
- expanded vermiculite sand with a bulk density of 150 kg/m^3.

The oxide composition of the highly porous fillers used in the work is given in Table 1.

These fillers were used in the development of heat-insulating dry mixes of the following composition: fluff lime, modifying additive, white cement, ground waste from the production of aerated concrete, redispersible powder, plasticizing additive, water repellant, light highly porous filler [7–10].

The method for determining the pozzolanic activity of fillers is based on the ability of amorphous silica to absorb lime from a lime solution, binding it into low-basic calcium hydrosilicates. The experimental technique is as follows. The investigated filler was dried to constant weight at a temperature of 100–105° C. Then the material was ground in a mortar until it passed through a sieve No. 008. After that, 2 g of filler was placed in a container, into which was then added 100 ml of a saturated solution of lime with a CaO concentration of 0.80–0.85 g/l. During the experiment, the container was periodically shaken. Every 2 days, 50 ml of the titration solution was taken from the container. For titration, a 0.05 N solution of hydrochloric acid HCl with methyl orange indicator was used. After titration, 50 ml of a lime solution with a CaO concentration of 0.80–0.85 g/l was added to the

Table 1 Oxide composition of highly porous fillers

Filler	Content of oxides (%)						
	SiO_2	Na_2O	Al_2O_3	MgO	Fe_xO_y	K_2O	Pr
Glass hollow microspheres	73.0	26.5	0.2	0.0	0.2	0.0	0.1
Ash microspheres, aluminosilicate	55.0	1.5	35.0	1.5	2.5	2.9	1.6
Expanded perlite sand	73.9	0.0	13.2	0.0	0.8	4.9	7.2
Expanded vermiculite sand	40.6	3.8	14.9	24.5	5.9	10.1	0.2

cylinder. The activity of the fillers was determined by the amount of $Ca(OH)_2$, which was absorbed by 1 g of the filler in 30 days.

To determine the specific strength, samples of coatings with a size of 20 × 20 × 20 mm were made. Highly porous fillers were taken in 40% of the mass of lime. After 28 days of hardening in air-dry conditions, the compressive strength and the average density of the mortar composites were determined, and then the specific strength was determined using the formula (1):

$$R_{sp} = \frac{R_{com}}{\rho_{av}} \tag{1}$$

where

R_{sp} specific strength, m^2/s^2;
R_{com} compressive strength, Pa;
P average density, kg/m^3.

3 Research Results

The high pozzolanic activity of glass hollow microspheres, amounting to 313.0 mg/g (Fig. 1, curve 1), is caused by a high content of SiO_2 in this filler—73.0%, while a significant part of SiO_2 is contained in the walls of microspheres in amorphous silica. The lower pozzolanic activity of ash microspheres (Fig. 1, curve 2) than in glass hollow microspheres is explained by the lower SiO_2 content in this filler—55.0%.

Expanded perlite sand is characterized by a rather high pozzolanic activity—147.7 mg/g (Fig. 1, curve 3), which is caused by a high SiO_2 content of 73.9% in it. The low pozzolanic activity of the introduced vermiculite sand (Fig. 1, curve 4) is explained by the lower SiO_2 content of 40.6% than that of the other investigated fillers.

The experiment to determine the pozzolanic activity of glass hollow microspheres was continued until the amount of calcium hydroxide absorbed by the filler ceased to change. Hollow glass microspheres continued to actively absorb $Ca(OH)_2$ up to 70 days, then the absorption rate slowed down and by 90 days the activity reached 480.5 mg/g.

Electron microscopic images of glass hollow microspheres were made using an electron microscope VEGA3 TESCAN before and after the experiments (Figs. 2 and 3).

Hollow glass microspheres are closed gas-filled balls of different diameters, almost ideal spherical shape (Fig. 2). These balls have clearly defined borders. In the photograph of the filler obtained after the experiment to determine the pozzolanic activity, there is no clear border in the glass microspheres (Fig. 3). On the surface of the balls, neoplasms are visible, presumably represented by calcium

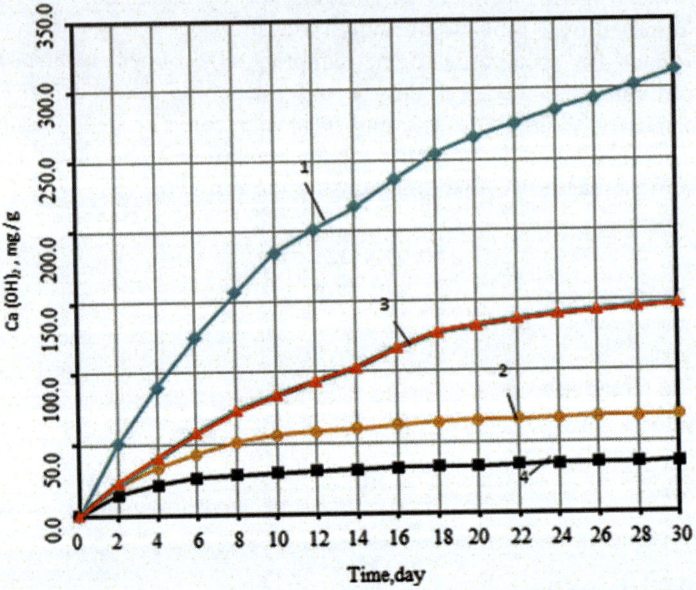

Fig. 1 Activity of highly porous fillers for the absorption of $Ca(OH)_2$ from a lime solution: *1*—glass hollow microspheres; *2*—aluminosilicate ash microspheres; *3*—expanded perlite sand; *4*—expanded vermiculite sand

Fig. 2 Electron microscopic image of glass hollow microspheres before the experiment to determine the activity of fillers

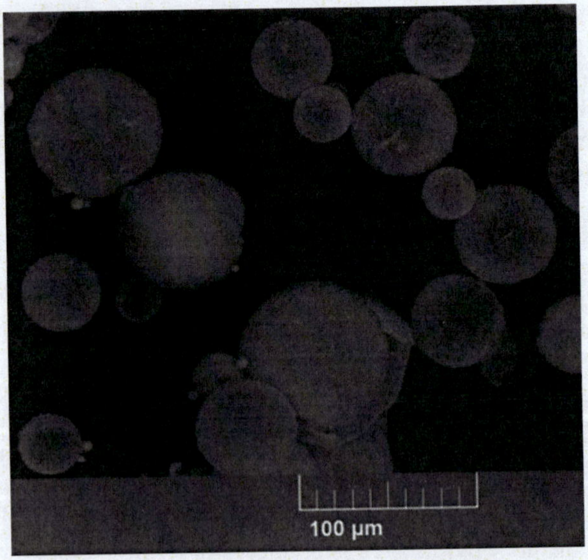

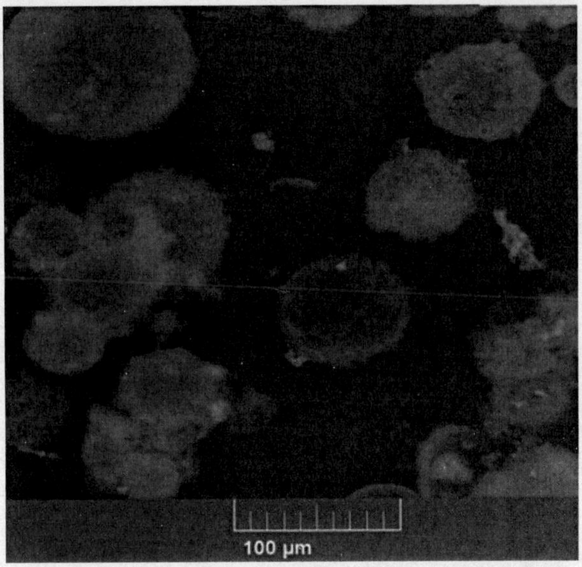

Fig. 3 Electron microscopic image of glass hollow microspheres after the experiment to determine the activity of fillers

hydrosilicates formed as a result of the reaction of calcium hydroxide and amorphous silica contained in the walls of the microspheres.

Composites filled with glass hollow microspheres are characterized by the highest specific strength R_{sp} = 5500 m^2/s^2 in comparison with composites filled with other highly porous fillers. This can be explained by the high pozzolanic activity of this filler and the absence of an uneven distribution of stress concentrations around the spherical particles of microspheres (Table 2).

The lower strength of composites filled with expanded perlite and vermiculite sand can be explained by the high moisture capacity of these fillers and the uneven distribution of stress concentrations around the filler particles due to their complex non-isometric shape. In this case, the specific strength of composites filled with expanded pearlite sand is higher due to the high pozzolanic activity of this filler in comparison with vermiculite sand.

Table 2 Properties of mortar composites

Filler	Average density (kg/m^3)	Compressive strength (MPa)	Specific strength (m^2/s^2)
Glass hollow microspheres	600	3.30	5500
Ash microspheres, aluminosilicate	860	4.25	4942
Expanded perlite sand	800	2.97	3713
Expanded vermiculite sand	660	2.10	3182

4 Conclusion

The data presented in the article indicate the high efficiency of the use of hollow glass microspheres as a highly porous filler in heat-insulating drywalls. it was found that composites filled with hollow glass microspheres are characterized by the maximum specific strength. This is due to the high pozzolanic activity of the fillers.

References

1. Loganina VI, Frolov MV (2020) Research of the dependence of the moisture regime in the walls of gas concrete on the characteristics of the external finishing. In: IOP conference series: materials science and engineering, vol 939, pp 012043
2. Loganina VI, Frolov MV (2019) Research of cracking resistance of thermal insulation coatings for aerated concrete. Mater Sci Forum 974:458–463
3. Loganina VI, Frolov MV, Skachkov YuP (2018) Substantion of selection of componets at creation of thermal insulating dry building mixtures. Int J Eng Technol 7(4):4341–4344
4. Semenov V (2017) Properties of dry masonry mixtures based on hollow aluminosilicate microspheres. MATEC Web Conf 117(24):00149
5. Demirboğa R, Örüng I, Gül R (2001) Effects of expanded perlite aggregate and mineral admixtures on the compressive strength of low-density concretes. Cem Concr Res 31 (11):1627–1632
6. Ferrándiz-Mas V, Bond T, García-Alcocel E, Cheeseman CR (2014) Lightweight mortars containing expanded polystyrene and paper sludge ash. Constr Build Mater 61:285–292
7. Suryavanshi AK, Swamy RN (2002) Development of lightweight mixes using ceramic microspheres as fillers. Cem Concr Res 32(11):1783–1789
8. McBride SP, Shukla A, Bose A (2002) Processing and characterization of a lightweight concrete using cenospheres. J Mater Sci 37(19):4217–4225
9. Wu Y, Wang J-Y, Monteiro PJM, Zhang M-H (2015) Development of ultra-lightweight cement composites with low thermal conductivity and high specific strength for energy efficient buildings. Constr Build Mater 87:100–112
10. Bentz DP, Peltz MA, Durán-Herrera A, Valdez P, Juárez CA (2011) Thermal properties of high-volume fly ash mortars and concretes. J Building Phys 34(3):263–275

Determination of the Boundary Parameters of Equipment Operation in the Process of Destruction of Carbonate Sediments by Inertial-Cutting Tools

I. R. Bondarenko, D. Yu. Volkov, and L. A. Kovalev

Abstract Layers of carbonate sediments that occur on the inner surface of pipes during the operation of heat exchange units, as well as on pipeline systems operating in various industries, affect negatively their operational characteristics. Mechanical destruction of these defective layers is a simple, effective and cost-effective method of solving this problem. In this regard, the tools and equipment used for this purpose are widely used, but their performance is limited by a number of factors, the most important of which are the strength and physical–mechanical properties of the material of the defective layers. Thus, the used equipment requires the need of effective operating modes that allow achieving maximum performance, which depends on the properties of the defective layers. In this regard, this paper considers the issue of determining the boundary parameters of the equipment used for the destruction of carbonate defective layers on the inner surface of heat exchange pipes. The theoretical study is based on the assumption that the tool speed limit value is determined by the strength characteristics of the machined surface as well as by the tool wear value. As a result of the study, a condition for determining the boundary minimum value of the speed of rotation is obtained. In order to confirm the obtained theoretical dependence, a series of verification experiments was conducted. The limit value of the tool speed was determined by estimating the amount of material destroyed in the defective layer.

Keywords Inner surface of pipes · Carbonate sediments · Purification · Destruction · Tool · Minimum frequency

I. R. Bondarenko (✉) · L. A. Kovalev
Belgorod State Technological University named after V.G. Shukhov, Belgorod, Russia

D. Yu. Volkov
Teploenergoservice OAO (Open Joint-Stock Company), Belgorod 308024, Russia

S. V. Klyuev (ed.), *Digital Technologies in Construction Engineering*,
Lecture Notes in Civil Engineering 173, https://doi.org/10.1007/978-3-030-81289-8_36

1 Introduction

Carbonate sediments on the inner surface of the pipes of various heat exchange units, as well as pipeline systems involved in various industries, affect negatively their performance figures [1–8].

Mechanical operation of the inner surface of pipes is one of the simplest and most popular ways to restore the operability of heat exchangers, as well as various pipeline services [7–9].

Specialized units and machining tools are widely used in this process (see Fig. 1) [10].

The use of rational, calculated modes of the equipment operation, allowing providing maximum performance at minimum cost, while ensuring the required quality of purification of heat-exchange tubes is an urgent task.

2 Materials and Methods

Undoubtedly, the technological performance of tools and equipment are within certain limits due to their design parameters and physical and mechanical characteristics of the machined material.

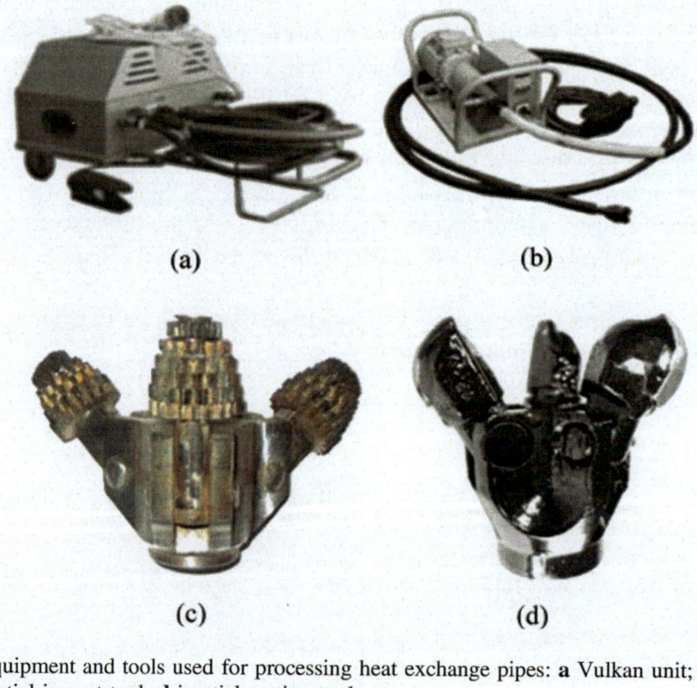

Fig. 1 Equipment and tools used for processing heat exchange pipes: **a** Vulkan unit; **b** typhoon unit; **c** inertial impact tool; **d** inertial cutting tool

During tool operation, its working element initially penetrates the sediment layer and subsequently removes it during rotary movement. It is known that the determining physical and mechanical properties affecting the material fracture process are its strength and hardness. The hardness determines the resistance to the penetration of the tool into the carbonate sediment layer and, therefore, to ensure a continuous destruction process, it is necessary that the mechanical stresses p in the contact zone of the tool with the surface of the removed layer are greater than the hardness of the sediments [11], which can be expressed by the condition:

$$p > p_{sh,} \tag{1}$$

there p—contact pressure of the tool cutting plate on the material,
p_{sh}—the value of the hardness of the material of carbonate sediments.

The pressure capacity of the working elements to the machined surface is determined by the inertial force F, which depends on the square of the rotational speed of the tool F $\sim$ ω^2 [12].

In accordance with the Dalembert principle [12], it is possible to create an equation of moments relative to the point O through which the tool attachment axis passes (see Fig. 2):

$$M_N + M_{tr} + M_F = N_P l_N - F_{tr}\left(\frac{D_{vn}}{2} - h\right) - m\omega^2 R_F l_F = 0 \tag{2}$$

there M_N—the moment of force of the normal reaction of the surface of the material; M_{tr}—the moment of the friction force acting along the pipe axis; M_{F}—the moment of normal inertia force, N_{p}—the strength of the normal reaction of the material surface; l_N—the arm of the force of the normal reaction of the sediment surface relative to the attachment axis; F_{tr}—friction force acting along the axis of the pipe, which can be defined as $F_{\text{tr}} = fN_{\text{p}}$; f—coefficient of friction of the cutting plate of the working element against the material to be broken; D_{v}—inner diameter of the tube; h—thickness of the layer to be destroyed, m—weight of the working element; ω—tool rotation speed; R_{F}—distance from the pipe axis to the point of application of the inertia force F; l_{F}—the arm of the inertia force F relative to the attachment axis.

Fig. 2 Diagram for determining the pressure force

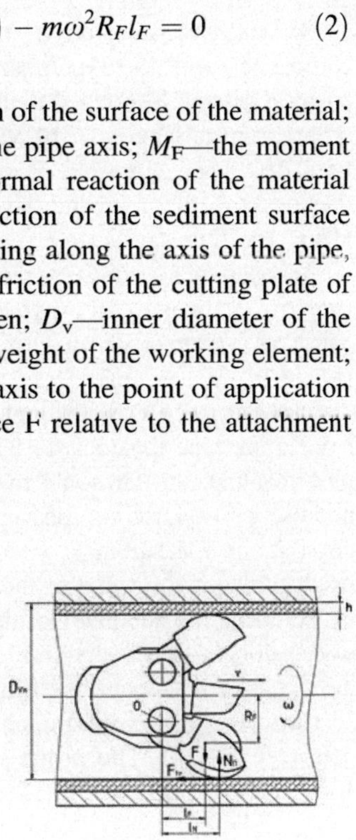

The parameters l_N, R_F, l_F depend on the geometric and structural parameters of the tool, the inner diameter of the pipe and the layer thickness, and can be measured in the CAD system if there are drawing files.

Then, taking into account that from the law of equality of action and reaction $N_p = P$, there P is the pressure force of the cutting plate on the surface of the material, from Eq. (2) we determine the contact pressure of the working element on the surface of the deposit p as:

$$p = \frac{P}{S} = \frac{m\omega^2 R_F l_F}{S\left(l_N - f\left(\frac{D_{vn}}{2} - h\right)\right)}, \tag{3}$$

there S—the area of contact of the working element with the surface, which at the moment preceding the introduction of the tool into the material can be assumed to be equal to the area of wear of the blade edge.

Taking into account the condition (1) and the dependence (3), we obtain the condition for the minimum value of the tool speed value

$$\omega \geq \sqrt{\frac{p_{sh} S\left(l_N - f\left(\frac{D_{vn}}{2} - h\right)\right)}{m R_F l_F}}. \tag{4}$$

For further comparative analysis of the results of theoretical studies and their comparison with the experiment, the tool rotation frequency was calculated. The calculation data corresponded to the initial experimental data: compressive strength of the sediment material σ_{szh} = 14 MPa, die hardness p_{sh} = 270 MPa, coefficient of friction f = 0.45, inner diameter of the pipe Dvn = 54 mm, sample length 1 = 200 mm, thickness of the removed layer h = 5 mm, pipe material steel 20 with hardness HB = 156.Для дальнейшего сравнительного анализа результатов теоретических исследований и сопоставления их с экспериментом был произведён расчёт частоты вращения инструмента. Данные расчёта соответствовали начальным данным эксперимента: прочность материала отложения при сжатии σ_{szh} = 14 mPa, die hardness p_{sh} = 270 MPa, coefficient of friction f = 0.45, pipe inner diameter D_{vn} = 54 mm, sample length l = 200 mm, thickness of the removed layer h = 5 mm, pipe material steel 20 hardness HB = 156. To control the integrity of the pipe material, an ultrasonic thickness gauge A1209 was used (see Fig. 3a). MPB-2 microscope was used to measure the size of the wear site (see Fig. 3b), with the basic error at any measurement interval up to 2 mm equal to 0.01 mm, the measurement results showed the value of the wear area equal to S = 0.5 mm^2.

To check the adequacy of the obtained calculated dependence (4), a series of trial exploratory experiments on the removal of sediment layers was carried out. During the experiments, sections of pipes with artificial sediments deposited on their inner surface were processed, the characteristics of which corresponded to the calculated data (see Fig. 4). The processing was carried out at a speed equal to: n = 1200, 1350, 2000, 2500 rpm.

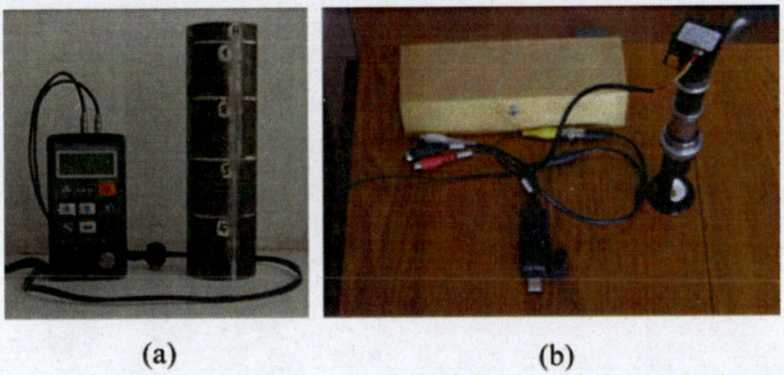

Fig. 3 Devices used for visual and measuring control: **a** ultrasonic thickness gauge, **b** PB-2 microscope

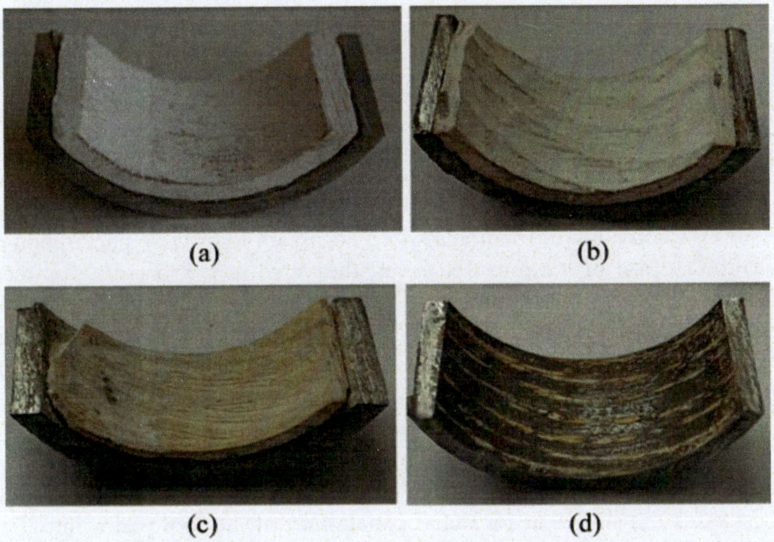

Fig. 4 Cuttings from the experimental samples: **a** machined at n = 1200 rpm; **b** machined at n = 1350 rpm; **c** machined at n = 2000 rpm; machined at n = 2500 rpm

The purification quality of the inner surface of the pipe was evaluated by the value of the residual contamination σ (Fig. 5.) defined as

$$\sigma = \frac{m_{ost}}{S_{vn}} = \frac{m_{ost}}{\pi D_{vn} l}, \quad (5)$$

there m_{ost}—residual mass of the material on the inner surface, g;
S_{vn}—the area of the inner surface of the pipe, m^2.

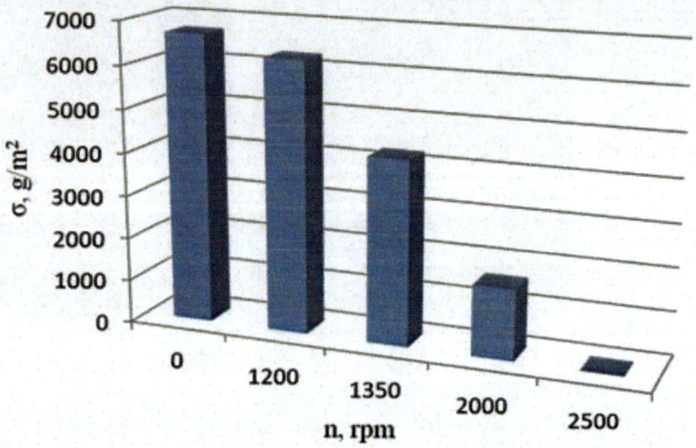

Fig. 5 Dependence of the residual contamination σ on the number of revolutions of the tool *n*

The amount of removed material was determined by weighing the sample before and after purification. The number of parallel experiments at a fixed rotational speed was taken to be three.

3 Results and Discussions

As a result of calculation of rotational speed necessary to provide penetration of the tool into the surface of the removed layer, the value $\omega = 135$ rad/s was received that corresponds to $n = 1289$ rpm.

The results of the experiment (see Table 1) show that at $n = 1200$ rpm the process runs with low intensity, the mass of the removed material is insignificant and amounted to 5.8–6.7% (see Fig. 4a).

At $n = 1350$ rpm (see Fig. 4b), the weight of the removed material was 35.7–37%.

The increase in the mass of broken material can be explained by the fact that the contact stresses are sufficient for radial penetration of the tool plates into the sediment. The number of revolutions of the tool $n = 2000$ rpm ensured the removal of material at the level of 74–78.2% (see Fig. 4c).

At $n = 2500$ rpm, intensive destruction of sediments occurs, the removal of the defective layer is 100% (see Fig. 4d), which allows saying that the developed

Table 1 Results of an experiment on purification of the inner surface of pipes

Number of revolutions of the tool n (rpm)	1200	1289 (calculated value)	1350	2000	2500
Degree of purification in %	6.7	–	36.5	74	100
	5.8	–	37	78.2	100
	6.3	–	35.7	75.3	100

pressure forces cause stresses in the contact zone, ensuring the introduction of cutting plates, at which high process productivity is achieved.

The amount of residual contamination (see Fig. 5) decreases slightly in the range of n = 0–1200 rpm, and at the same time changes significantly with an increase in the number of revolutions from 1200 to 1350 rpm.

Visual measurement control, as well as thickness measurement of the pipe wall, showed that the main metal of the pipe remains intact.

4 Conclusion

Summing up, it can be noted that the results of the calculation are sufficiently consistent with the experimental data. The processing results correspond to the boundary value of n, justified by theoretical studies. This is confirmed by a multiple increase in processing performance with an increase in the tool speed above the theoretical value, which corresponds to the minimum value of the frequency required to ensure the destruction of the surface of carbonate sediments.

All of the above makes it possible to conclude that the obtained theoretical approach can be applied to the issue of assigning modes of effective destruction during mechanical purification of the internal surface of heat-exchange tubes.

Acknowledgements This work was realized in the framework of the Program of flagship university development on the base of the Belgorod State Technological University named after V.G. Shukhov. The work was realized using equipment of High Technology Center at BSTU named after V.G. Shukhov.

References

1. Minko VA, Seminenko AS, Gunko IV, Elistratova YuV, Koltsa LN, Tkach LV (2014) The influence of scale on the operation of heating systems. Bull BSTU named after V.G. Shukhov 6:21–23
2. Minko VA, Seminenko AS, Gunko IV, Elistratova YuV (2014) Influence of sediments on the working surfaces of the heating system on the performance of the system elements. Bull BSTU named after V.G. Shukhov 5:32–35
3. Minko VA, Feoktistov Ayu, Gunko IV, Elistratova YuV, Tarasenko NV, Tkach LV (2015) Methods of conducting and effectiveness of measures to combat scale formation in heat consumption systems Bull BSTU named after V.G. Shukhov 2:16–19
4. Babintseva TV, Mayzel IG (2010) A method for restoring the throughput capacity of pipelines. Mod Technol Syst Anal Model 2(26):17–24
5. Neredko AV, Karnitsky NB, Chizh VA (2007) Thermophysical properties and structure of sediments on heating surfaces of power equipment. Proc High Edu Inst Energy Assoc CIS Power Eng 1:55–60
6. Cüneyt E, Necdet Ö (2012) Optimization of heat exchanger cleaning cycle on a ship. J Naval Sci Eng 8(1):33–46

7. Hou TK, Kazi S, Mahat AB, Teng CB, Al-Shamma'a A, Shaw A (2017) Industrial heat exchanger operation and maintenance to minimize fouling and corrosion. Heat Exch Adv Features Appl 193–207. https://doi.org/10.5772/66274
8. Putman RE Optimizing the cleaning of heat exchangers [electronic resource]. Available at www.conco.net/sites/default/files/userfiles/files/techical-papers/optimizing-cleaning-heat-exchangers.pdf
9. Bondarenko IR, Volkov Dyu, Kovalev LA (2019) Calculation of power and energy indicators of the process of materials rapture during their processing with the help of an inertial tool with a curved cutting rim. In: MEACS 2018 IOP conference series materials science and engineering, vol 560, p 012141. https://doi.org/10.1088/1757-899X/560/1/012141
10. Saturn-Service Tools [electronic resource]. Available at: http://saturntool.ru/catalog/tool
11. Balitsky PV (1975) Interaction of the drill string with the bottom of the well. Nedra 12:21–25
12. Yablonsky AA, Nikiforova VM (1999) Course of theoretical mechanics statics. Kinematics dynamics. Saint-Petersburg

Finite Element Analysis of Concrete Beams with Carbon Fibers and Fiberglass

D. Z. Aljaboobi, O. V. Umnova, and V. P. Yartsev

Abstract In structural analysis, most problems are complex and accurate solutions are obtained for their governing equations only and for limited types of structures of load characteristics and simple engineering. Thus, numerical procedures such as FEM are used to obtain approximate solutions for realistic types of problems. The computer program ANSYS is a sturdy and interesting engineering finite element package that may be utilized to solve a lot of problems. FEM has become a most commonly method used to study the stress, deformation, and other engineering parameters. FEM uses complicated mathematical equations to accurately approximate how the complex structure reacts to a certain load or condition. FE bundles like ANSYS solve thousands or millions of these equations to find a solution for a model. Handling all these equations as a whole be difficult and mostly impossible to solve manually. ANSYS is an inclusive general-purpose FE computer program that contains different elements implemented in the program. This article describes the application of nonlinear finite element models to the analysis of rectangular concrete beams. In this work, the material parameters that must taking into consideration to investigate the CFRP beams behavior that are young modulus (Ec), compressive strength f′c, and the tensile strength (ft). Also, young modulus (Es) and the yielding strength (fy) of steel are considered. Furthermore, the geometry factors considered are the beam width, effective depth, longitudinal reinforcement area and shear reinforcement area.

Keywords Beam · Concrete · Bending · Strength · Reinforcement · Glass

D. Z. Aljaboobi (✉) · O. V. Umnova · V. P. Yartsev
Tambov State Technical University, Tambov, Russia
e-mail: Aldjabubi.dzm@mail.tstu.ru

S. V. Klyuev (ed.), *Digital Technologies in Construction Engineering*,
Lecture Notes in Civil Engineering 173, https://doi.org/10.1007/978-3-030-81289-8_37

1 Introduction

In structural analysis, most problems are complex and accurate solutions are obtained for their governing equations only and for limited types of structures of load characteristics and simple engineering. Thus, numerical procedures such as FEM are used to obtain approximate solutions for realistic types of problems. FEM is an important technique provided solutions to a varied problem in all engineering fields. This work applies a nonlinear FEA for CFRP beams exposed to static load to study the behavior of these beams. A lot of phenomena in solid mechanics are nonlinear. However, in several applications, using linear formulation will be proper and practical solution to obtain an engineering solution. Otherwise, other problems may require analysis with nonlinear property if obtained realistic results such as high deflection and post-yielding. Depending on nonlinear sources, the nonlinear problems involve three types [1], these types are:

a. Problems including material nonlinearity.
b. Problems including geometric nonlinearity.
c. Problems including both materials and geometric nonlinearity.

In finite element method, the complex structure is first divided (discretized) into a limited number of individual non-overlapping components known as ‘elements’ over which the variables are interpolated. ANSYS provides several options to characterize different kinds of material behavior, such as bilinear isotropic (with work hardening) and multi-linear isotropic hardening. For concrete, the concrete crushing in compression algorithm is similar to a plasticity law [2]. This algorithm is similar to a multi-linear work hardening uniaxial stress–strain relationship based on rate independent Von-Mises yielding criterion. Rate independent plasticity constitutes an irreversible straining that occurs in a material once the yield surface is reached. The yield term, which means that value of stress that causing yield to the material, while, the rule of flow after the yield point determines the plastic straining direction. The hardening rule defines the changes in yield surface with gradually changing in yield.

2 Concrete

Solid 65 was used to model the concrete. This element has eight nodes with three degrees of freedom. Each node has translations in the x, y, and z directions. This element is capable of plastic deformation, cracking in three orthogonal directions and crushing. A schematic representation of the element is shown in Fig. 1. Cracking is modeled by the smeared crack approximation.

The accepted dependence of uniaxial deformation under compression of concrete and mortar is a multi-linear isotropic stress–strain curve shown in Fig. 2, and calculated from the ratios. and calculated by the relations [4]

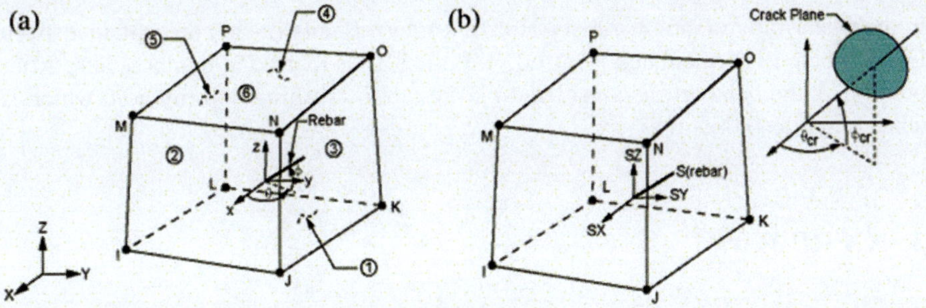

Fig. 1 SOLID 65 element for concrete representation [3]: **a** Geometry SOLID 65. **b** Voltage output SOLID 65

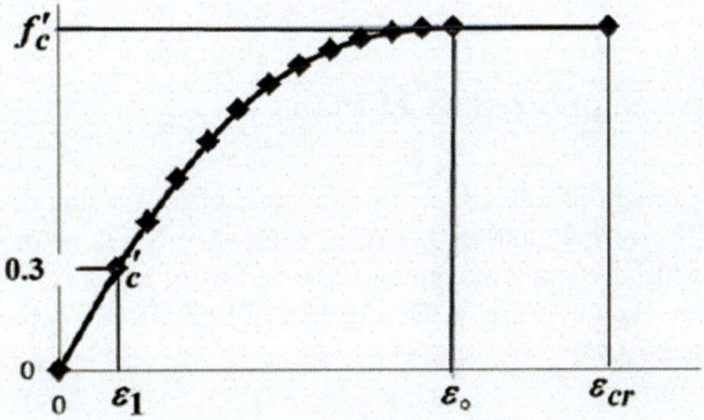

Fig. 2 Multi-line uniaxial compression curve for concrete

$$f_c = \varepsilon \cdot E_c \quad \text{For } 0 \leq \varepsilon \leq \varepsilon_1 \tag{1}$$

$$f_c = \frac{\varepsilon \cdot E_c}{1 + \left(\frac{\varepsilon}{\varepsilon_0}\right)^2} \quad \text{For } \varepsilon_1 \leq \varepsilon \leq \varepsilon_0 \tag{2}$$

$$f_c = f_c' \quad \text{For } \varepsilon_0 \leq \varepsilon \leq \varepsilon_{cu} \tag{3}$$

$$\varepsilon_1 = \frac{0,3f_c'}{E_c} \tag{4}$$

$$\varepsilon_0 = \frac{2f_c'}{E_c} \tag{5}$$

where; f_c' = voltage at any voltage ε, MPa; ε = stress at strain f; ε_0 = deformation at ultimate compressive stress f_c'; E_c = modulus of elasticity of concrete, MPa.

A simplified stress–strain curve for each beam model is constructed from (13) points connected by straight lines. The curve starts at zero stress and strain. Point (1) at $0.3 f_c'$ is calculated for the dependence of stress on concrete deformation in the linear range (should satisfy Hooke's Law). Points (2–11) are obtained from Eq. 2, where ε_0 is calculated from Eq. 4. Point (12) is located at points ε_0 and. After point (12), the behavior is considered plastic, up to crushing deformation, which is taken equal to 0.003.

3 CFRP Bars

The element shown in Fig. 3, can be used in many engineering tasks. LINK 180 is a uniaxial tension–compression element. In this element, bending is not taken into account, plasticity and stiffness under stress are incorporated. The geometry of the location of nodes and the coordinate system for LINK180 are shown in Fig. 3 [3].

4 Supported Steel Plates

Element SOLID185 is used for steel plates on supports and in load areas of a beam. This element has eight nodes, each node has three degrees of freedom. SOLID 185 with x, y and z movement. The geometry and location of the nodes of this element are shown in Fig. 4. Description of SOLID 185 is taken from the ANSYS element library [3].

5 Geometry and Material Properties

The prototypes of the beam (Fig. 5) were made of B 30 class heavy concrete on M500 cement. As a coarse aggregate, we used crushed granite of 15–25 mm fraction, fine—quartz sand. Tests of control concrete cubes were tested in accordance with the methodology of GOST 10180 [5], the class of concrete was B 30.

The deflections of concrete beams during the test were measured with a PSK-MG 4.01 digital deflection meter. According to the readings of the sensors of the testing machine, the displacement of the traverse was determined. The data were

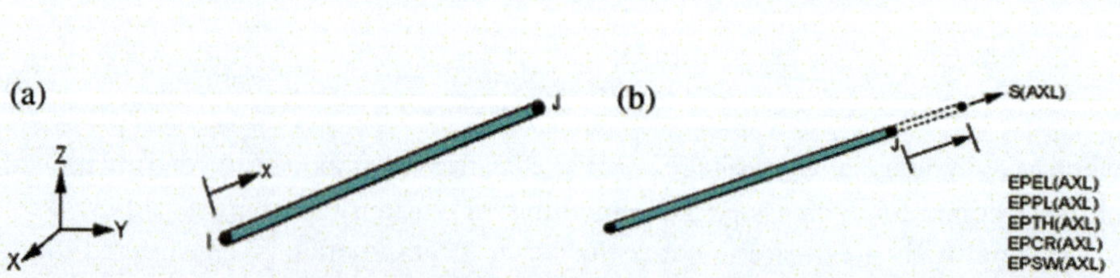

Fig. 3 LINK180 to represent CFRP reinforcement [3]: **a** LINK 180 geometry. **b** LINK 180 voltage output

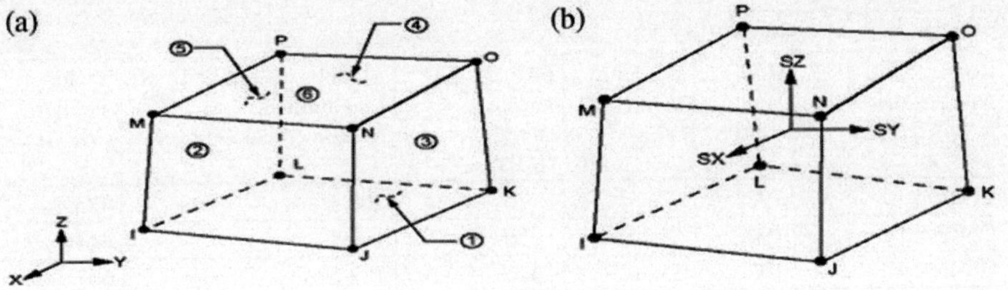

Fig. 4 Representation of SOLID 185 for modeling steel plates and supports [3]: **a** Structural solid geometry. **b** Withdrawal of stresses in a solid structure

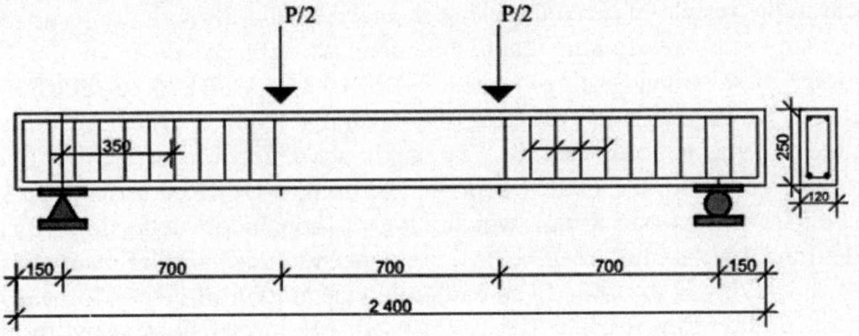

Fig. 5 Load and geometry of the analyzed beam

recorded after applying the load of each stage. Concrete deformations were recorded using a TDS-530 multichannel measuring complex manufactured by Tokyo Sokki Kenkyujo Co., Ltd., Japan. The measuring instruments used were conductive strain gauges with a pol-yester-based substrate of the PL-60-11 type with a measuring base length of 60 mm. The sensors were glued ac-cording to a special technique on a previously prepared concrete surface [6]. Four concrete beams with different types of longitudinal reinforcement were tested before failure on special stands at four-point bending [7]. All tested beams were made with CFRP, GFRP fibers and the characteristics of reinforced concrete with controlled destruction due to crushing of concrete in the compression zone, and not longitudinal rupture of the bar. These information parameters as shown in Table 1 are used as input data in ANSYS application to analysis of bars by using Finite element.

6 Results and Discussion

Comparison of practical results in the laboratory and finite element analysis mentioned in Table 2 [6].

Table 1 Basic parameters of bars

Rod type	Bar diameter (mm)	Rod length (mm)	Breaking load (kN)	FRP design tensile strength (MPa)	Modulus of elasticity of longitudinal reinforcement (MPa)	Yield strength (MPa)
CFRP	10	192	47.3	730	200.4	425
CFRP	14	192	54.2	826	109.7	444
GFRP	10	192	55.43	944	132.6	507
GFRP	14	192	62.5	1050	136.0	549

Through Table 2 found the practical results (experiments) that are largely identical to the results of the finite element analysis, and through the results found that the larger the reinforcing bars, the more strength. Analysis of the "load–deflection" plots when testing beams 1-CFRP10, 2-CFRP14, 3-GFRP10 and 4-GFRP10 shows that after reaching the breaking load, it gradually decreases, brittle fracture was not observed [8]. The maximum deflections of 1-CFRP10 series beams were 13.7 mm, 2-CFRP14 series—12.5 mm, 3-GFRP10 series—12.8 mm, and 4-GFRP14 series—15.8 mm, which allows to conclude—deflections of carbon fiber reinforced glass reinforced beams, it has been noticed similar results.[6]. The Fig. 10 shows this is the zone of maximum concentration of cracks located in the lower part of the beam [6]. The CFRP-D10 series beam has been tested to failure. The load versus deflection in the middle of the span was plotted according to the test results, see Fig. 6. The resulting ultimate load was 93.01 kN with a corresponding deflection of 15.7 mm. The deflection in the middle of the span at break was 15 mm. The percentage increase in bending capacity was 66%. The percentage of the decrease in the values of ultimate ductility at fracture was 62%. The beam

Table 2 Test results maximum load

Beam	*Vu, Expert.* (kN)	*Vu, ANS.* (kN)	*Vu,Experiment* / *Vu,ANS.*
1-CFRP10	93.01	94.43	0.98
2-CFRP14	101.55	104.55	0.97
3-GFRP10	95.83	98.72	0.97
4-GFRP14	100.32	108.13	0.92

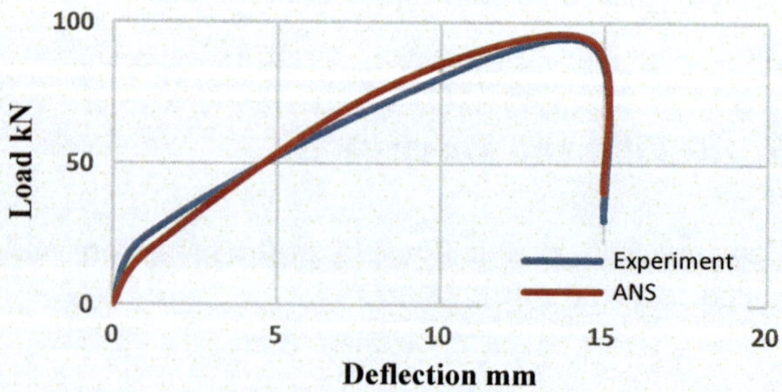

Fig. 6 Dependence of deflection on load for beam-CFRP10

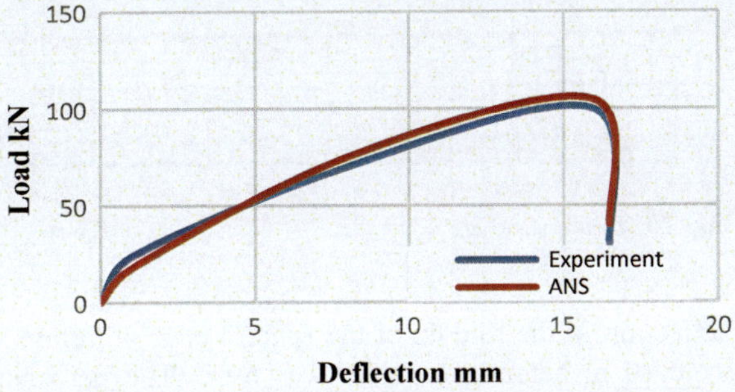

Fig. 7 Dependence of the deflection on the load for the beam-GFRP10

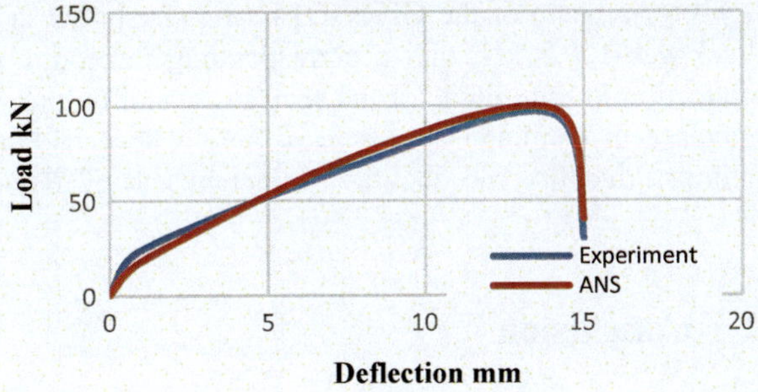

Fig. 8 Dependence of deflection on load for beam-GFRP10

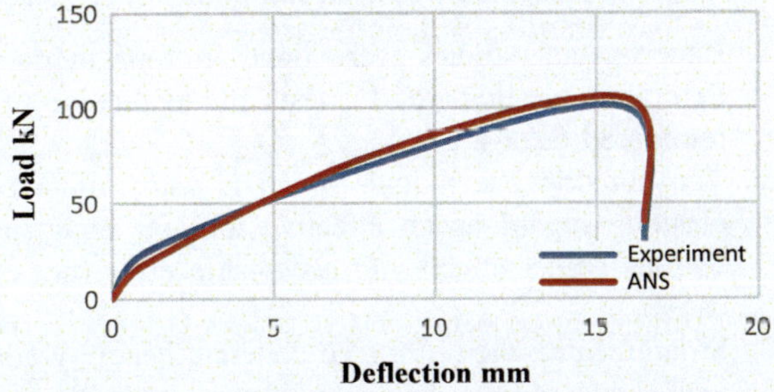

Fig. 9 Dependence of deflection on load for beam-GFRP14

collapsed due to deformation of the CFRP with large flexural cracks as shown in Fig. 6. The Fig. 7 shows CFRP-D14 series beam was also tested to failure. As a result of the test, the dependences of the load on the deflection in the middle of the span were obtained, see Fig. 7. The resulting ultimate load was 101.55 kN with a corresponding deflection of 17.2 mm. The deflection in the middle of the span in case of failure was 17.6 mm. The percentage increase in bending capacity was 75%. The percentage of the decrease in the values of ultimate ductility at fracture was 71%. Test results of the GFRP-D10 series beam are shown in Fig. 8. The resulting ultimate load was 95.8 kN with a corresponding deflection of 15 mm. The

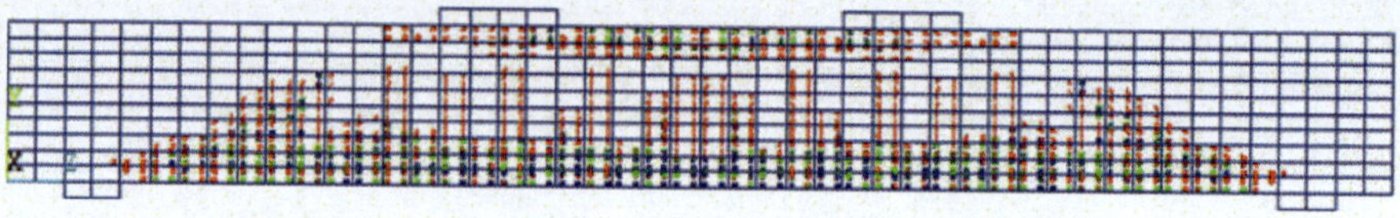

Fig. 10 Crack structure

deflection in the middle of the span in case of failure was 15.3 mm. The percentage increase in bending capacity was 69%. The percentage decrease in the values of ultimate ductility and ductility at fracture was 71% and 70%, respectively as shown in Fig. 8. The Fig. 9 shows dependence of the load on the deflection in the middle of the span for a beam of the GFRP-D14 series is shown in Fig. 9. The resulting ultimate load was 100.3 kN with a corresponding deflection of 17.4 mm. The amount of deflection in the middle of the span in case of failure was 17.6 mm. The percentage increase in bending capacity was 75%. The percentage decrease in the values of ultimate ductility and ductility at fracture was 62% and 63%, respectively.

7 Conclusion

The presented research includes an analysis of concrete beams reinforced with bars composite of carbon and glass.

From the presented work, several points of neutrality can be deduced as follows:

1. Experimental studies were done in four models with different armaments. Samples were prepared for tests and evaluation of the bending behavior of the reinforced beams.
2. The numerical technique which is using finite element method via ANSYS program regard as an effective tool for analyzing composite structure with various defect effect with acceptable error (not exceeded 2.7% for the worst cases).
3. Strengthening the reinforced concrete beam by adding carbon fibers and fiber-glass increased the deflection at ultimate load.

References

1. Madenci E, Guven I (2015) The finite element method and application in engineering using ANSYS. Springer, New York Heidelberg Dordrecht, London
2. Fanning P (2001) Nonlinear model of reinforced and post-tensioned concrete beams. Electron J Struct Eng 2:111–119
3. ANSYS (2007) Analysis guide. Version 11, Swanson Analysis System Inc.

4. Desayi P, Krishnan S (1964) Equation for the stress-strain curve of concrete. J Am Concr Inst 61:345–350
5. GOST 10180–2012 (2013) Concrete. Methods for determining the strength of control samples. Standartin form Publishing House, Moscow
6. Aldjaboobi DZ, Kryukova AA, Yartsev VP (2020) Influence of organic and steel fiber on the strength properties of cement concrete. In: Modern science: theory, methodology, practice. materials of the 2nd All-Russian (National) scientific and practical conference, 28–29 May, FGBOU VO "Tambov State Technical University. Publishing house of IP Chesnokova A.V., Tambov, pp 103–105
7. SP 63.13330.2012 (2015) Concrete and reinforced concrete structures. Basic provisions. Updated edition of Snip 52–01–2003. Ministry of Construction of Russia, Moscow
8. Klyuev SV, Klyuev AV, Vatin NI (2018) Fiber concrete for the construction industry. Mag Civ Eng 84(8):41–47

Development and Research of Technological Equipment for Production of Polymer Composite Materials with Basalt Fiber Fillers from Man-Made Raw Materials

V. S. Sevostyanov, P. Yu. Goryagin, and V. A. Babukov

Abstract One of the most important state tasks in the world is the creation of resource-and-energy-saving technologies for the complex processing and utilization of man-made materials, including the production of new materials and products from secondary raw materials. This article discusses the directions of using technogenic raw materials in the production of new composite materials and products with fiber fillers. Rational methods of processing polymer waste and directions of constructive and technological improvement of technological equipment have been determined. The results of research and development in the field of creating patent-protected units for the complex processing of man-made polymer and fibrous materials are presented. Analytical expressions are given for the high-speed action of the working units of technological equipment on the material being ground. As a result of multifactorial experimental studies, it was established the influence of the percentage of basalt fiber fillers on the physical and mechanical characteristics of composite mixtures and products made from them.

Keywords Composite materials · Fiber fillers · Complex processing · Man-made materials · Resource saving · Innovative equipment

---

V. S. Sevostyanov · P. Yu. Goryagin (✉) · V. A. Babukov
Belgorod State Technological University of V.G. Shukhov, Belgorod 308012, Russia
e-mail: inventor.konf@yandex.ru

V. S. Sevostyanov
e-mail: svs-3149@mail.ru

V. A. Babukov
e-mail: babukov-v@mail.ru

S. V. Klyuev (ed.), *Digital Technologies in Construction Engineering*,
Lecture Notes in Civil Engineering 173, https://doi.org/10.1007/978-3-030-81289-8_38

1 Introduction

At present, in Russia, as well as throughout the world, the volumes of production and consumption of polymer composite materials (PCM) are steadily growing [1, 2]. The main advantages of PCM products are: low density of products, high physical and mechanical characteristics, resistance to corrosion and aggressive environments, the possibility to create materials with specified characteristics and properties, etc. [3, 4]. One of the promising areas is the development of resource-saving production technologies from man-made PCM raw materials with basalt fiber fillers. In this regard, it is necessary to develop energy-efficient specialized equipment for the production of powdered polymer materials from man-made raw materials.

2 Materials and Methods

The scientific team of the department "Technological complexes, machines and mechanisms" of BSTU named after V.G. Shukhov developed and manufactured a rotary-centrifugal combined-action unit (RCA CD) for the processing of polymer waste (Fig. 1). The technical characteristics of it are presented in Table 1. The unit is equipped with needle-milling working bodies that provide a combined effect on the processed materials, a high degree of grinding—i > 100. At the same time, the energy-efficient impact of the rod elements is carried out. The elements with their needle-like surface, provide a mechanism for the destruction of polymers. RCA CD contains a housing consisting of two grinding chambers 1 and 2, horizontally placed one after another (Fig. 1a) [5, 6]. Inside the first grinding chamber 1 there is a bandage installed eccentrically relative to the axis of the cylindrical body. It is lined with removable elements (profile plates). On the shaft 3 there is a rotor 4 with cutting elements in the form of disc cutters.

In the second grinding chamber 2, needle-milling working bodies 5 are installed, which are made of a set of rod elements assembled in packages and rigidly fixed to the holder.

The RCA CD implements the principle of stage-by-stage grinding with the implementation of a combined effect on the processed material. This ensures a decrease in energy consumption for the grinding process, and also expands the technological capabilities of the unit.

When developing energy-efficient equipment for shredding plastic waste, its elastoplastic characteristics should be taken into account.

The most effective methods of plastic destruction are cutting, abrasion, impact after cryogenic treatment, high-speed grinding [7].

With a high-speed action on a polymer material, a relaxation mechanism of wear arises. It consists in the fact that with an increase in the frequency of action on the surface, the highly elastic deformation of the polymers does not have time to

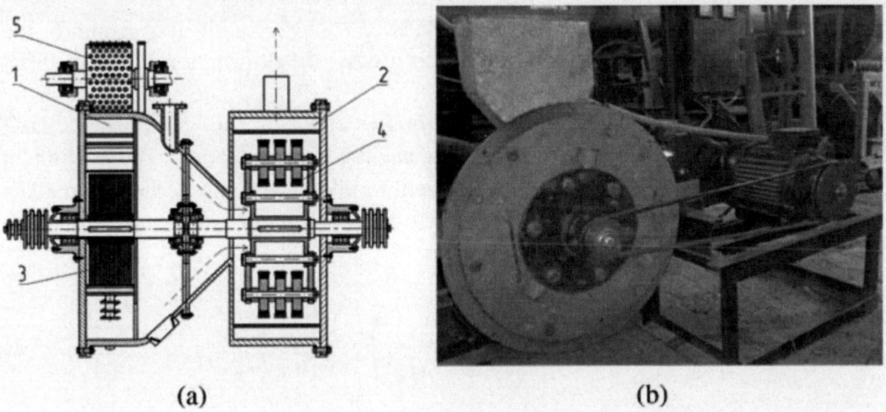

Fig. 1 Scheme (**a**) and experimental setup (**b**) RCA CD: 1, 2—grinding chambers, 3—disc cutters; 4—needle milling working bodies; 5—studded rolls

Table 1 Technical characteristics of the RCA CD experimental setup

Characteristic	Dimension	Designation	Value
Circular frequency			
Disc cutters	min^{-1}	$n_{d.c}$	$(1–1.5) \cdot 10^3$
Needle-milling working bodies		$n_{n.c}$	$(2.5–3) \cdot 10^3$
Grinding chamber E-motor power	kW	N_1	5.5
		N_2	7.5
Overall dimensions			
Length	m	L	1.5
Width		B	0.93
Height		H	1.3

develop and the material becomes brittle. The relaxation mechanism of fracture is possible if the exposure time is shorter than the stress relaxation time in polymer particles. If we take into account the periodic nature of the impact of needle-milling working bodies, then the destructive stress [8]:

$$\sigma = \sigma_0 \sin(\omega t) = \sigma_0 \sin\left(2\pi \frac{t}{T}\right), \tag{1}$$

where σ_0 is the peak voltage, Pa;

ω—exposure frequency s^{-1};

t—time, s;

T—exposure period, s.

Equation (1) shows that at constant values of t and σ_0 with decreasing T, the stress in the polymers will increase and, accordingly, the destruction of the material particles will be more likely.

To determine the speed of the working bodies effect on the value of the maximum force and amplitude stress σ_0, one can use the Hertz equation. The coefficient of elasticity $\delta = 4(1-\mu_1^2)/E_1$ for metal needle-milling working bodies is small in comparison with the coefficient of elasticity for polymers, so its value can be neglected. The expression σ_0 has the form:

$$\sigma_0 = 1.65E_2\left(\frac{\gamma_1}{\gamma_2\left(1-\mu_2^2\right)^4}\right)^{1/5}\left(\frac{\vartheta_0}{\vartheta_{\text{l.d.}}}\right)^{2/5} \tag{2}$$

where E_2 is the dynamic modulus of elasticity of polymeric materials, Pa;

γ_1—material density of needle-milling working bodies, kg/m^3;

γ_2—density of polymeric materials, kg/m^3;

μ_2—poisson's ratio for polymeric materials;

ϑ_0—speed of needle milling working bodies, m/s;

$\vartheta_{\text{l.d.}}$ is the propagation velocity of longitudinal disturbances, numerically equal to the speed of sound propagation in polymeric materials, m/s, $\vartheta_{\text{l.d.}} = \sqrt{\frac{E_2}{\gamma_2}}$.

From Eq. (2) it follows that at the speed $\vartheta_0 \geq \vartheta_{\text{l.d.}}$ of the needle-milling working bodies, the stresses in the material particles increase greatly.

From Eqs. (1) and (2) it follows:

$$\sigma = \sigma_0 \sin(\omega t) = 1.65E_2\left(\frac{\gamma_1}{\gamma_2\left(1-\mu_2^2\right)^4}\right)^{1/5}\left(\frac{\vartheta_0}{\vartheta_{\text{l.d.}}}\right)^{2/5}\sin(\omega t) \tag{3}$$

For the destruction of man-made polymer materials, it is necessary to overcome the critical speeds at which particle size reduction occurs according to the relaxation mechanism. It is known that for the most widespread polymeric materials $\vartheta_{\text{пр.в}} = (2.0\text{–}2.35)\cdot 10^3$ m/s.

As a result of physical modeling, the optimal parameters of the process of grinding polymeric materials have been determined, the analytical dependences of high-speed grinding and the efficiency of using the working bodies that have a bursting and abrasive effect on the material (disk cutters, needle-milling working bodies) have been confirmed.

An important area of the technology for the production of composite mixtures is the development of special equipment for the production of fiber fillers from basalt fibrous waste (BVO) [9].

Due to the low bulk density (200–250 kg/m^3), low flow ability, increased moisture capacity, caking, high adhesive capacity, etc., the process of obtaining basalt fiber from secondary raw materials (production waste) causes certain difficulties [10]. To obtain basalt fibers and composite mixtures, we have developed a vibro-centrifugal unit of combined action (Fig. 2) [11, 12]. The BWO processing procedure in this unit takes place in several technological stages. In the upper prismatic chamber 1, which makes a reciprocating motion, the material is deagglomerated by means of chain curtains. In the middle chamber 2 (elliptical movement)—fine grinding of materials or a combination of grinding and mixing. In the lower chamber 3 (circular trajectory of movement)—obtaining highly concentrated microfiber fillers.

The indicated design features of the developed unit make it possible to obtain highly concentrated granular fiber fillers using small size BVO (1–5 mm) in the third chamber.

To study the efficiency of using basalt fibers in composite mixtures, we carried out special experimental studies. The following materials were used: sand of the Olshansk deposit, perlite sand, basalt fiberglass, bentonite clay, Portland cement

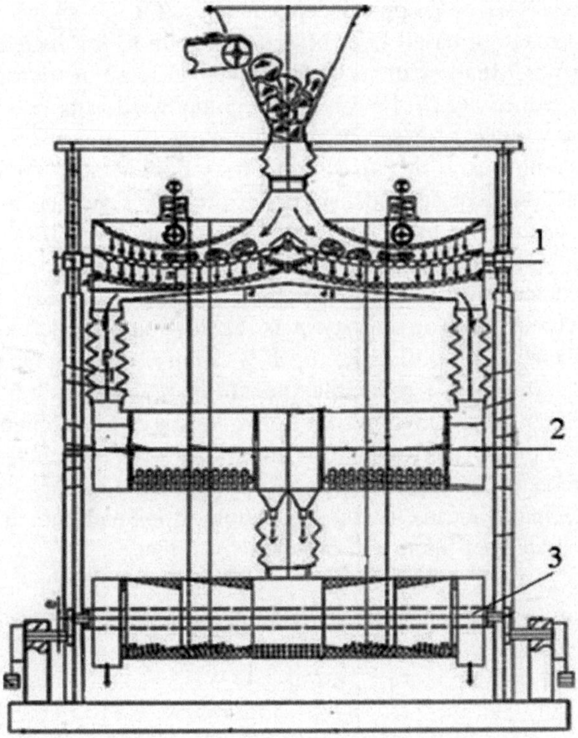

Fig. 2 Vibro-centrifugal combined action unit: 1—deagglomeration chamber; 2—grinding and homogenizing chamber; 3—fiber agglomeration chamber

M500. To impart strength to the samples of heat-insulating mixtures, quartz sand and basalt fiber were used. In order to improve the thermal insulation properties, expanded perlite sand and bentonite were used as a hydrophobizing additive and plasticizer. As a binder—cement PC-500 was used. To determine the specified physical and mechanical characteristics of heat-insulating mixtures, experimental studies were carried out on molded samples (moistened heat-insulating materials) at 28 days of age. To reduce the number of experiments and obtain reliable information data, the central compositional rotational plan of the second order TsKRP-2^4 was adopted as the experimental plan, which makes it possible to implement various combinations of parameters at various levels of their variation. As variable factors we used: the percentage of perlite (from 6 to 34%); percentage of bentonite (from 0.5 to 2.5%); the percentage of basalt fibrous fibers (up to 8%), as well as the water–solid ratio (0.45–0.85). For a qualitative and quantitative assessment of the experimental results, the following output parameters of the molded samples were determined in each experiment: density (ρ) [kg/m^3], compressive strength (σ_c) [MPa], thermal conductivity (λ) [W/(m K)].

3 Results and Discussions

Analysis of graphical dependencies (Fig. 3) shows that an increase in the percentage of basalt fiber filler samples up to 8% increases the compressive strength (σ_c) [MPa], with a simultaneous decrease in density (ρ) [kg/m^3] and thermal conductivity (λ) [W/(m K)] of compressed samples. Thus, with an increase in the percentage of basalt fibers in the range from 4 to 6%, the increase in compressive strength (σ_c) [MPa] is: from $\sigma_c = 0.45$ MPa to $\sigma_c = 0.50$ MPa, with a simultaneous decrease in density from $\rho = 420$ kg/m^3 up to $\rho = 390$ kg/m^3 and thermal conductivity from $\lambda = 0.022$ W/(m K) to $\lambda = 0.019$ W/(m K).

With a further increase in the fiber content in the range of 6–8%, an increase in the compressive strength is observed from $\sigma_c = 0.48$ MPa to $\sigma_c = 0.55$ MPa (by 5.6%). The decrease in density ρ is 4.2% (from $\rho = 400$ kg/m^3 to $\rho = 380$ kg/m^3) and thermal conductivity λ – by 17% (from $\lambda = 0.018$ W/(m K) to $\lambda = 0.013$ W/(m K)).

Thus, it was found that the introduction of basalt fiber filler into the composition of composite mixtures up to 8% increases the strength characteristics of the tested samples. The results of carried out scientific and technical developments and research confirm the possibility of creating resource-saving technologies and technical means for the production of polymer-containing composite mixtures with basalt fiber fillers and products from them.

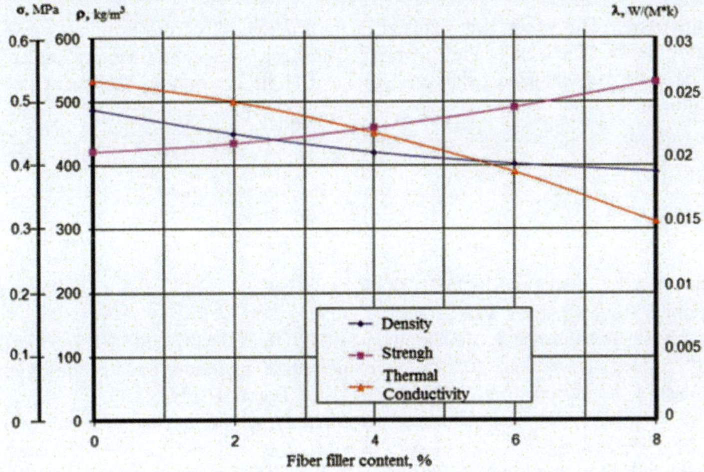

Fig. 3 Influence of the content of basalt fiber filler [%] on the density (ρ), strength (σ) and thermal conductivity (λ) of the samples

4 Conclusion

1. On the basis of the carried out R&D, patent-protected designs of resource-saving equipment for the processing of man-made polymer and fibrous materials have been developed, taking into account their specific physical and mechanical characteristics and physicochemical properties.
2. To reduce the specific energy consumption for the grinding process in the RCA CD, a multistage combined effect of the working bodies on the processed material is implemented.
3. Theoretical studies of the processes of high-speed grinding of TPM have been carried out. Equations of stresses are obtained for relaxation fracture of polymeric materials by needle-milling working bodies. It has been established that an effective way of dispersing TPMs, taking into account their specific physical and mechanical characteristics, is a high-speed action on the material of needle-milling working bodies.
4. As a result of experimental studies of the processes of multistage grinding of polymer waste, the features of their processing were studied, the effectiveness of high-speed action on the material of working bodies with a developed working surface was confirmed.
5. The choice of the optimal reinforcement parameters that meet the technological requirements is a fundamental issue in the manufacture of composite mixtures with basalt fiber fillers. As a result of experimental studies, the optimal value of the volumetric content of fibers was established—8% in composites based on pearlite, bentonite and cement. With this content of basalt fiber fillers, the strength of composite materials increases from $\sigma_c = 0.48$ MPa to $\sigma_c = 0.55$ MPa—by 5.6%.

Acknowledgements This work was realized in the framework of the Program of flagship university development on the base of the Belgorod State Technological University named after V.G. Shukhov. The work was realized using equipment of High Technology Center at BSTU named after V.G. Shukhov.

References

1. Malakhovsky SS, Mishkin SI (2019) Main trends in the production and use of secondary carbon fibers (review). Proc VIAM 9(81):73–79
2. Kablov EN (2015) Innovative developments of FSUE "VIAM" SSC RF for the implementation of Strategic directions for the development of materials and technologies for their processing for the period up to 2030. Aviat Mater Technol 1(34):3–33
3. Raskutin AE (2017) Development strategy for polymer composite materials. Aviat Mater Technol S 344–348
4. Daskovsky MI, Doriomedov MS, Skripachev SY (2016) Systematization of the basic factors that prevent the introduction of polymer composite materials in Russia (review). Proc VIAM 5(41)
5. Sevostyanov VS, Sivachenko TL, Sevostyanov MV, Goryagin PY, Babukov VA (2018) Scientific and practical basis for the creation of multi-purpose needle-milling grinders. Bulletin of BSTU named after V.G. Shukhov 3:107–116
6. Sevostyanov VS, Shein NT, Sevostyanov MV, Goryagin PY, Ermilov RA (2020) Rotary-centrifugal combined action unit for processing organic and mineral materials. Invention patent 2724667
7. Gogol EV, Mingazetdinov IKh, Gumerova GI, Egorova OS, Maltseva SA, Grigorieva IG, Tunakova YuA (2013) Analysis of existing methods of disposal and recycling of polymer waste. Bull Kazan Technol Univ 10:163–168
8. Chapalda DI (2007) Development of a technology for recycling used car tires. Abstract of the dissertation for the degree of candidate of technical sciences. USPTU, Ufa
9. Petropavlovskaya VB, Novichenkova TB, Bur'yanov AF, Soloviev VN, Petropavlovskiy KS (2017) Utilization of waste mineral fibers in the production of gypsum products. MGSU Bull 12(111):1392–1398
10. Shein NT, Sevostyanov VS, Obolonsky VV, Sevostyanov MV, Goryagin PY, Babukov VA (2019) Resource and energy-saving technologies of complex processing and utilization of technogenic materials. IOP Conf Ser: Mater Sci Eng 552(1):012042
11. Sevostyanov MV, Poluektova VA, Sevostyanov VS, Sirota VV, Uralsky VI, Martakov IG, Babukov VA (2019) Device and method for processing fibrous man-made materials to obtain fibrous fillers (options). Patent for invention 2692624
12. Sevostyanov MV, Sinitsa EV, Uralsky AV, Shkarpetkin EA, Babukov VA (2020) Vibro-centrifugal units for complex processing of man-made materials. Bulletin of BSTU named after V.G. Shukhov 1:115–124

Impregnation Composition to Increase the Ice-Phobic Properties of Concrete Roads

R. E. Lukpanov, D. S. Dyussembinov, S. B. Yenkebayev, and D. V. Tsygulyov

Abstract The article presents the results of research of ice-phobic impregnating composition designed for road pavements. The impregnating composition was developed based on animal farming waste with the addition of water-soluble polymer: 60% keratin-containing raw material, 40% acrylic latex. To determine the optimal water concentrate of the composition, laboratory tests of water absorption capacity of samples coated with the composition of different concentrations were performed: 5, 10, 15 и 20%. Frost resistance tests were performed to assess the effect of the composition on the durability of concrete. Analysis of the suitability of the composition was carried out by comparing the results of covered and uncovered specimens. The optimum solution of the composition in water corresponds to the 5% concentration, which corresponds to the minimum coefficient of water absorption, as well as the minimum value of the concrete strength loss after cyclic freezing. The results obtained are of practical value, aimed at improving the operating conditions of vehicles in winter.

Keywords Impregnation composition · Ice-Phobic coating · Pavement · Water absorption · Strength · Frost resistance

1 Introduction

Road construction is an integral part of the economic development of any state. Modern road construction should provide safety and comfort of movement, as well as reliability and durability of roads [1]. To ensure the above requirements apply different kinds of technologies and methods of production, which depend on the type of pavement. The most common road surfaces include asphalt and concrete, both in the first case and in the second have their advantages and disadvantages [2].

The advantages of concrete roads are much higher than asphalt roads, this is primarily durability and strength and as a consequence of less need for repair and

R. E. Lukpanov · D. S. Dyussembinov · S. B. Yenkebayev · D. V. Tsygulyov (✉)
Eurasian National University of L. N. Gumilyov, Nur-Sultan, Kazakhstan

S. V. Klyuev (ed.), *Digital Technologies in Construction Engineering*,
Lecture Notes in Civil Engineering 173, https://doi.org/10.1007/978-3-030-81289-8_39

restoration work [3]. Concrete is an environmentally friendly material as it is produced from mineral binders. Resistance to high and low temperatures, the formation of local deformations make concrete roads the most preferable in terms of intensive operating loads [4]. The disadvantages of concrete roads can be attributed primarily to its cost relative to asphalt roads. High water absorption of concrete as a material can lead to icing of the surface when exposed to negative temperatures [5].

The icing of the road surface in the wintertime is a problem for most cities around the world [6]. Ice crust on the road increases the risk of accidents and makes driving unsafe. In this regard, issues to improve the conditions of operation of motor vehicles in winter are urgent [7].

Today there are many technological and technical proposals to solve this problem, ranging from the modernization of the road surface as a material to the application of abrasive material on the road surface [8]. The first can be implemented in several ways, from the inclusion of special ice-phobic additives to the production of composite material with ice-phobic permanent pavement [9]. Note at once that this solution applies to new construction, does not solve the problem of existing roads, the economic efficiency of which is noticeable after long-term operation of the road. The use of abrasive material on the surface can be implemented by different methods from the installation of roll materials to the application of bulk materials [10]. The latter has a particularly wide application because it is less expensive and technically easy to implement, despite the short-term effect and consequences associated with environmental pollution.

This problem determined the purpose of the research work, which aims to develop an ice-impregnating composition for impregnation of existing roads, using animal waste.

2 Materials and Methods of Research

The ice-phobic composition includes keratin-containing components of animal waste and water-soluble polymer (acrylic latex).

Figure 1 shows the technological scheme for the production of impregnation composition, which includes a hydrolysis procedure in an alkaline medium (to dissolve keratin-containing raw materials), followed by neutralization of the alkaline medium. The production technology is based on the industrial production of keratin.

The reactor is loaded with 1/3 of water and dosed amount of keratinous raw material. Then caustic soda or its aqueous solution of the desired concentration is loaded into the reactor. Ratio (keratin) raw material: liquid phase—1: 3, then required amount of urea (urea) is loaded into the reactor and the rest of the water is added, i.e. 2/3 of it. After that, the reactor is hermetically closed, and the loaded raw material is exposed to heat. The temperature during hydrolysis should not exceed 135 °C. The duration of hydrolysis depends on the type of keratin and can be 4–8 h. After the hydrolysis process is completed, the hydrolysate should cool down

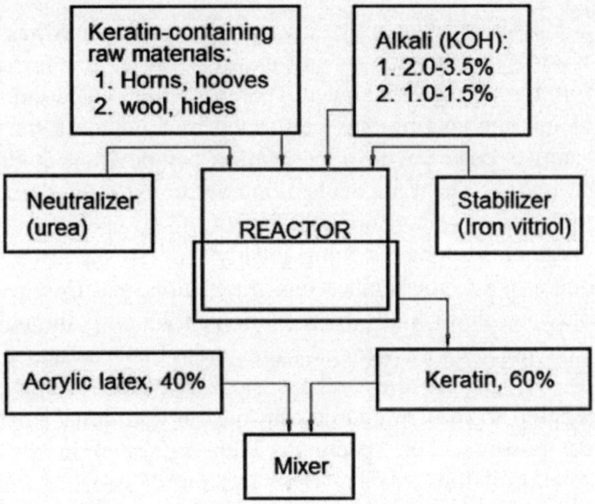

Fig. 1 The technology of impregnation composition production

to ambient temperature. A 1-L sample is taken from the cooled hydrolysate solution, the initial pH is measured, and it is neutralized with sulfuric iron—($Fe_2(SO_4)_3$). After neutralization of the resulting hydrolysate, it is filtered by passing the product through a filter press. The keratin-containing composition is then mixed in a mixer with acrylic latex, in the ratio of 60:40, respectively.

Research methods include laboratory testing of reinforced concrete specimens with and without impregnation:

- measurement of water absorption of specimens;
- measurement of frost resistance of specimens.

The nature of the adhesion of ice to concrete depends on the contact area of the two components (ice and concrete). The adhesion strength (adhesion resistance) of ice and concrete is the result of the conditional additive effect of the following factors: the first factor is the adhesion of ice to concrete as a material; the second factor is the ice's entrapment (anchorage) in the pore (micro and macro) structure along the concrete surface. There is a synergy effect (a mutual influence of factors), which leads to increasing the overall adhesion resistance of the ice crust. Consequently, by excluding the first factor—ice adhesion, we significantly reduce the effectiveness of the second factor—ice crust anchorage. Therefore, the adhesion resistance of the ice crust to the concrete directly depends on the hydrophobic properties of the concrete, for which water absorption tests were performed. The effectiveness of the impregnation compound will depend on the degree of its penetration into the concrete road surface. Since concrete is a hydrophilic material that absorbs water, a greater effect will be achieved when using a concentrate of impregnation compound in water, rather than using the impregnation compound in its pure form. That is, water that tends to penetrate into the concrete structure will entrain the suspended mixture of the polymeric component of the impregnation composition. The polymeric component, which has less penetrating ability than water, will plug macro and micro pores on the surface of the concrete road, thus forming a protective layer. Therefore, the evaluation of the optimal solution of the ice-phobic coating was performed by evaluating the water absorption of concrete samples coated with impregnation composition of different concentrations: 5, 10, 15 и 20%. The main evaluation criterion was the ratio of ice-phobation efficiency to the optimal water concentration of the composition.

Since seasonal freezing greatly reduces the strength characteristics of concrete due to the cyclic temperature fluctuation, and frost resistance is directly dependent on water absorption (the reduction of which is the purpose of the work), then tests were made on the strength of concrete frost resistance. In other words, the duration of the concrete life cycle is the exhaustion of strength, but the strength has no relation to the ice-phobic, but has an additional positive effect on the use of the composition. The specimens were saturated in a 5% aqueous sodium chloride solution before freezing. The duration of freezing was 2.5 h, and thawing 2.0 h.

3 Results and Discussion

Figure 2 shows the results of water absorption tests on the samples. Figure 2a shows the results of individual and average values of densities of dry and wet (after full water saturation) uncoated specimens, as well as the values of water saturation coefficients. Figure 2b shows the same parameters of the samples coated with the ice-phobic composition.

In general, the results obtained have a relatively close relationship with each other, with a medium to a high degree of convergence and a consistent relationship. For the uncoated specimens, the average water absorption was 6.25%; with a standard deviation of 0.39, the coefficient of variation was 6.20%. At a confidence level of 95%, the coefficient of reliability of the test results is 0.93 (93%). For specimens with ice-phobic coating, the following pattern was obtained: the lower the concentration of the impregnation composition in water, the greater the hydrophobic effect of the coating. That is, the lowest water absorption is achieved at a lower concentration of the composition in the water. The effectiveness of impregnation depends not on the amount of concentrate, but on the amount of water that entrains the mixture of the polymer component of the impregnation composition, clogging the micro and macropores of the concrete. The obtained correlation coefficient of change in water absorption by the change in water concentration of the composition confirms the presence of a proportional regularity, which is 0.96. The obtained value is close to 1, which indicates the presence of a strict regularity, and the positive value of the direct proportionality of the variables. The obtained data points also have high convergence and close correlation: the standard deviations are from 0.08 (5%) to 0.21 (10%), the coefficients of variation are from 1.92% (15%) to 4.47% (10%). At a confidence level of 95%, the reliability coefficients of the test results range from 0.95 to 0.98 (95–98%).

(a)

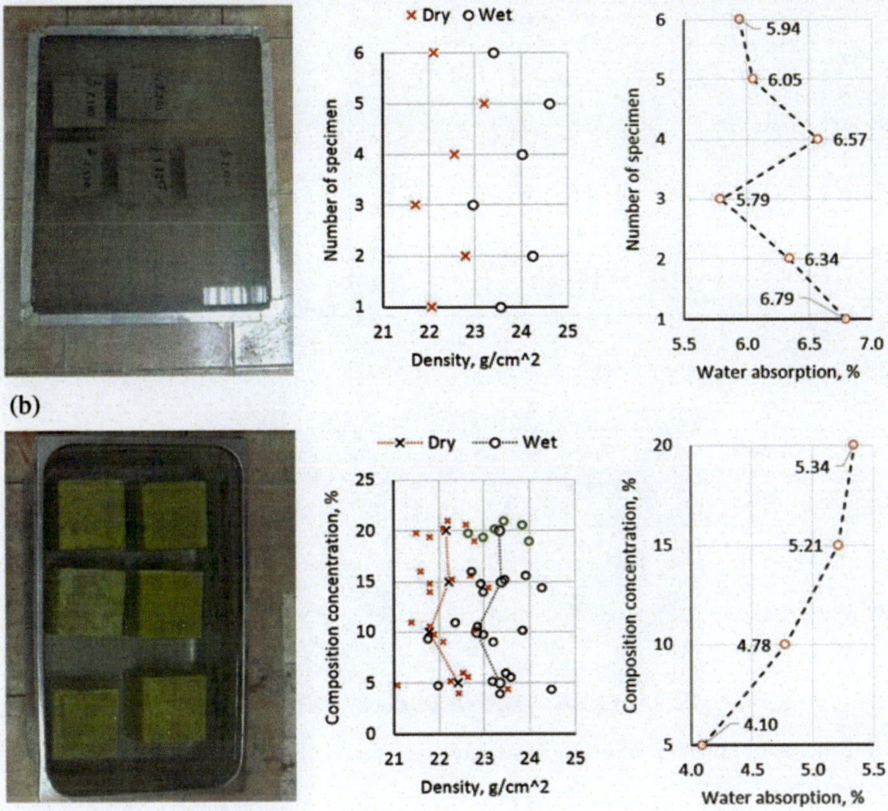

Fig. 2 Results of water absorption measurements

Figure 3 shows the results of frost resistance measurements. Figure 3a shows the variation of strength loss of specimens under different numbers of cyclic freezing of specimens, where the curves characterize the variation of strength loss as the concentration of impregnation composition changes. Figure 3b shows the increment of strength loss as a function of the number of freezing cycles, where the curves characterize the increment of strength loss as a function of the number of cycles. Table 1 shows the statistical data of the test results.

In general, all data points have a close relationship, the coefficients of variation are: for samples without impregnation from 2.6 to 5.8%, for samples with 5% concentration from 3.4 to 5.2%, for samples with 10% concentration from 2.7 to 4.9%, for samples with 15% concentration from 3.3 to 6.4%, for samples with 20% concentration from 4.1 to 6.7%. There is an obvious reduction of strength loss of samples with the lowest values of water absorption. For specimens of optimum (best) concentration (5%) strength loss, on average, decreases 4 times to specimens without impregnation.

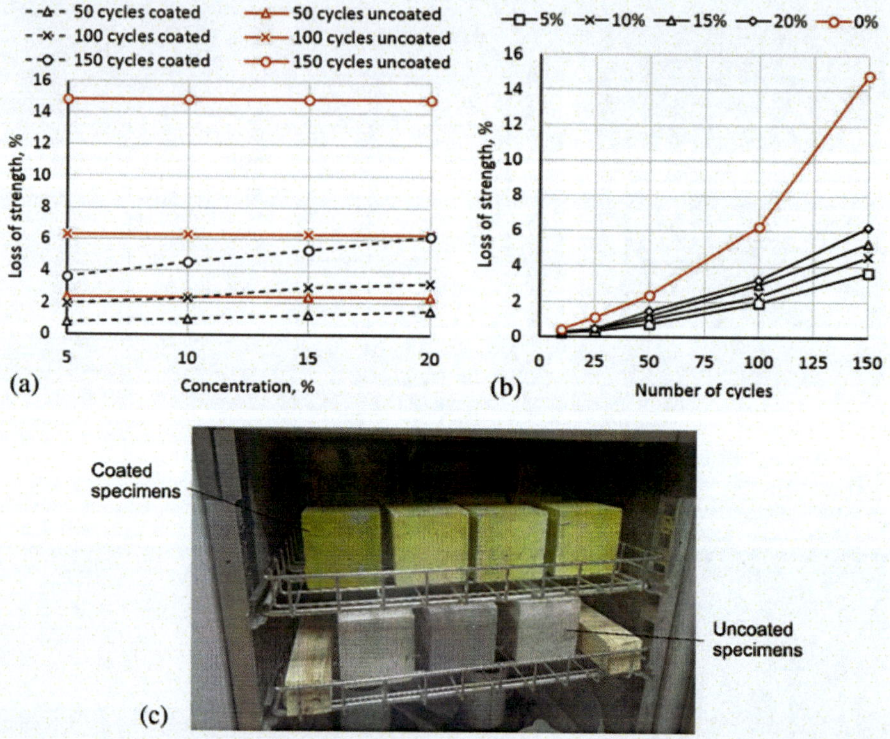

Fig. 3 Results of the frost resistance measurements

Table 1 Statistic data of the test results

Parameters	The concentration of composition (%)				
	0	5	10	15	20
50 cycles of freezing					
Average lost strength δ, %	2.36	0.78	1.02	1.23	1.48
Standard deviation σ	0.13	0.02	0.02	0.03	0.02
Variation coefficient	5.32	2.57	2.26	2.32	1.33
100 cycles of freezing					
Average lost strength δ (%)	6.28	1.98	2.34	2.98	3.24
Standard deviation σ	0.30	0.06	0.07	0.08	0.11
Variation coefficient	4.72	2.87	3.04	2.76	3.26
150 cycles of freezing					
Average lost strength δ (%)	14.87	3.67	4.58	5.32	6.21
Standard deviation σ	1.11	0.24	0.38	0.40	0.42
Variation coefficient	7.48	6.41	8.32	7.54	6.78

4 Conclusion

Proposed impregnating composition for the coating of concrete and cement-concrete roads, made on the basis of keratin-containing components with the addition of water-soluble polymer (acrylic latex), which has ice-phobic properties.

The obtained results of the study confirm the advantage of using the composition to reduce the water absorption capacity of concrete and, as a consequence, the potential reduction of icing adhesion resistance on the surface of the concrete road. The results of water absorption measurements showed that the optimal concentration of the composition in water is 5%. For specimens impregnated with the composition of optimum concentration, the decrease of water absorbency in relation to specimens without impregnation, is 34.4%, and the decrease of loss of strength at cyclic freezing is—75.4%.

References

1. Awwad T, Shakhmov ZA, Lukpanov RE, Yenkebayev SB (2019) Experimental study on the behavior of pile and soil at the frost condition. https://doi.org/10.1007/978-3-030-01902
2. Ramachandran R, Kozhukhova M, Sobolev K, Nosonovsky M (2016) Anti-icing superhydrophobic surfaces: controlling entropic molecular interactions to design novel icephobic concrete. Entropy 18(4). https://doi.org/10.3390/e18040132.
3. Medeiros M, Helene P (2008) Efficacy of surface hydrophobic agents in reducing water and chloride ion penetration in concrete. Mater Struct/Materiaux et Constr 41(1):59–71. https://doi.org/10.1617/s11527-006-9218-5
4. Shen Y, Wu X, Tao J, Zhu C, Lai Y, Chen Z (2019) Icephobic materials: fundamentals, performance evaluation, and applications, vol 103. Elsevier Ltd, 01–06–2019
5. Furmidge CGL (1962) Studies at phase interfaces. I. The sliding of liquid drops on solid surfaces and a theory for spray retention. J Colloid Sci 17(4):309–324. https://doi.org/10.1016/0095-8522(62)90011-9.
6. Shakhmov Z, Lukpanov R, Tleulenova G, Mineev N, Tulebekova A (2018) Comparison of experimental data of model piles in normal and seasonally freezing soil. In: 11th international conference on geosynthetics, vol 1, pp 399–402
7. Lukpanov RE, Awwad T, Orazova DK, Tsigulyov DV (2019) Geotechnical research and design of wind power plant. https://doi.org/10.1007/978-3-030-01920-4_19
8. Wang Z, Xie J, Gao L, Liu M, Liu Y (2020) Improvement of acoustic model and structural optimization design of porous asphalt concrete based on meso-structure research. Constr Build Mater 265. https://doi.org/10.1016/j.conbuildmat.2020.120327
9. Wang Q (2020) Design optimization of ultra-thin asphalt pavement on cement concrete bridge deck based on artificial intelligence. Paper presented at the 2020 IEEE international conference on industrial application of artificial intelligence, IAAI 2020, pp 184–189. https://doi.org/10.1109/IAAI51705.2020.9332883
10. Yao J, Wang Z, Tang D (2019) Development and application of water-saving and moisture-retaining membrane made from controllable high polymer materials for concrete curing. J Perform Constr Facil 33(1). https://doi.org/10.1061/(ASCE)CF.1943-5509.0001261

EDT Drill-Injection Piles as a Design for Strengthening Weak Foundations of High-Speed Highways (HSH)

N. S. Sokolov

Abstract Ensuring reliable operation of high-speed highways is an important geotechnical task. To ensure the normalized values of roadbed sediment in the presence of weak engineering and geological elements in the bases, the issues of increasing their load-bearing capacity are particularly relevant. Existing technologies allow avoiding their negative impact. For this purpose, modern construction has a number of geotechnical technologies, as a rule, these include buried reinforced concrete structures that are submerged in the ground in finished form or manufactured in the ground. The electric-discharge geotechnical technology of strengthening weak bases with the use of EDT drill-injection piles considered in the paper is one of the approaches to solving this problem.

Keywords High-speed highway · Buried reinforced concrete structures · Drill-injection piles · Electric discharge technology EDT

1 Introduction

The construction of any objects on weak foundations requires special approaches associated with increasing the strength and reducing their deformability. Using modern national geotechnical technologies, it is possible to solve the problems of increasing the bearing capacity of weak bases.

High-speed highways are among the most responsible objects, which are subject to the minimum maximum permissible precipitation $S_u = 15.0$ mm. At the same time, their bases very often contain engineering-geological elements with increased values of deformation characteristics and weak strength properties.

N. S. Sokolov (✉)
Chuvash State University Named After I. N. Ulyanov, Moskovskiy prosp., 15, Cheboksary 428015, Russian Federation

N. S. Sokolov
NPF (LLC SPC) "FORST", ul. Kalinina, 109a, Cheboksary 428000, Russian Federation

S. V. Klyuev (ed.), *Digital Technologies in Construction Engineering*,
Lecture Notes in Civil Engineering 173, https://doi.org/10.1007/978-3-030-81289-8_40

In this paper, we consider the case of weak base projection amplification (FSM) on the Moscow-Nizhny Novgorod section.

The construction site under study belongs to territories with complex engineering and geological conditions.

The planned route of the high-speed highway runs within the Volga-Ural Arch [1–12].

2 Methods and Materials

The construction of objects in complex engineering and geological conditions in the presence of layers with reduced values of strength and increased values of deformation characteristics requires a special approach. Besides, in addition to the construction of buried reinforced concrete structures, it is necessary to ensure their high load-bearing capacity on the ground and the strength of their cross-section. It is often necessary to deal with cases when it is necessary to use buried geotechnical building structures to ensure the load-bearing capacity. Modern geotechnical construction in its arsenal has methods and technologies for solving these complex problems.

The use of advanced computer programs allows developing geotechnical objects of any complexity. To identify the most appropriate geotechnical technology, it should be mandatory to use the interactive design method. This is "a developed project—a pilot site—a real project". At the same time, this type of design should be multivariate with the use of various geotechnical technologies and buried geotechnical reinforced concrete structures. The feasibility study of all the elements in the interactive design is also important. Any stage in the design process must be economically sound and technically feasible. It is often necessary to deal with the geological conditions of construction sites, in the engineering-geological sections of which there are alternating layers with weak physical and mechanical indicators. The presence of such EGE reduces significantly the load-bearing capacity on the ground of any pile, and sometimes leads to negative friction due to the overhang of the soil on the piles. Of course, there are many modern methods of structural transformation of soil properties in order to improve their physical and mechanical properties. But these technologies are very busy, requiring significant material costs. In addition, the implementation of such measures also requires the use of modern drilling outfits and mechanisms.

It is known that foundations using drill-injection piles using discharge-pulse technology have a number of competitive technical advantages in comparison with other foundations using other types of drill piles. One of the distinctive parameters of some drilling piles from others is the technological possibility of including them in joint work with the surrounding soil. At the same time, the soil pressure of the walls of the well is carried out with the help of camouflage broadening arranged using discharge-pulse technology.

3 Results and Discussion

During the Quaternary period, different parts of the territory under consideration were in different climatic conditions and were exposed to various physical and geographical processes. The northern part of it was repeatedly subjected to glaciations, and the southern part—to the impact of transgressions of the Caspian Sea. These different conditions left their mark on the nature of the Quaternary cover of the territory. In the north-western part of it, glacial and water-glacial formations are widely developed, in the middle—alluvial and eluvial-deluvial, and in the southernmost territory—marine estuarine sediments. In the thickness of the Quaternary sediments there are peat, brick clays, sand, sand with layers of gravel, loam and clay (in the above-floodplain terraces of the Volga, Sura and Tsivil).

In the areas of the Nizhny Novgorod and Cheboksary Volga regions, the cover deposits are represented by light loams and heavy sandy loams of a characteristic yellowish or brownish-fawn color with clearly expressed signs of loess (macroporosity, columnar separateness, etc.) and subsidence properties. The water content of the rocks of the complex is insignificant and has a sporadic character.

Modern marsh sediments (bIV) are distributed unevenly throughout the study area. In the vast depressions of the relief, confined to river valleys or lowlands in the forest zone, especially large peat lands-peat basins are formed. On the territory under consideration, lowland and upland peat lands are found in approximately equal numbers. The deposits are represented by peat and frozen soils. The average thickness of marsh sediments is from 2 to 3 m and can reach up to 10 m. In some areas where marsh deposits are developed, marsh waters lie directly from the surface.

In order to select the option for strengthening weak foundations, both sections of the highway are considered for which embankments of types No. 3 and 4 are designed. Engineering-geological foundations of the embankments are shown in Tables 1 and 2, and in Figs. 1 and 2—vertical linking of the embankments in the engineering—geological section.

The algorithm for determining the sediment of high-speed highway embankments in the presence of weak bases is given below.

1. Determination of loads from the rolling stock of supporting structures, from the inter-track, the body of the embankment at the level of its sole;
2. Determination of the stabilized deformation by one of the methods: (1) the method of layer-by-layer summation, (2) the method of a linearly deformable layer of finite thickness, (3) the method of an equivalent layer;
3. If the calculated deformation of the bases is higher than the maximum permissible value, the depth of the reinforced part of the base is assigned;
4. The type of buried reinforced concrete structure is assigned;
5. The average P_{IImt} pressure on the sole of the conditional foundation and the value of the stabilized deformation are determined by one of the methods (see pos. 2);

Table 1 Physical and mechanical characteristics of the foundation soils under the type 3 embankment

№	Symbols	Name of EGE	ρ_n (g/cm^3)	φ_n (degree)	C_n (kPa)	E_0 (MPa)	K_f, (m/day)	Thickness of EGE (m)
1	$11_б2$	Fine sand of medium density, moist, water-saturated	2.0	35.0	–	25.7	3.4	3.0
2	11_e2	Plastic sandy loam	1.05	24.0	25.0	25.4	$3 * 10^{-3}$	2.0
3	$21_з3$	Refractory clay	1.76	12.0	38.0	6.0	$1 * 10^{-5}$	1.0
4	$21_з2H$	Medium hard clay	1.74	13	46	18	$1 * 10^{-5}$	2.5
5	$24_ч7$	Very low-strength dolomite	2.12	–	–	–	0.005	

Table 2 Physical and mechanical characteristics of the foundation soils under the type 4 embankment

№	Symbols	Name of EGE	ρ_n (g/cm^3)	φ_n (degree)	C_n (kPa)	E_0 (MPa)	K_f, (m/day)	Thickness of EGE (m)
1	$3_ж4$	Soft-plastic loam with admixtures of organic inclusions	1.94	17.0	18.0	11.2	0.05	2.0
2	$20_ж4$	Soft-plastic loam	1.94	4.0	18.0	5.18	$3*10^{-4}$	1.0
3	$20_б2$	Fine sand of medium density, water-saturated	2.00	36.0	4.0	37.0	4.0	2.5
4	$20_з1H$	Solid clay	1.95	16	66	39.0	$1*10^{-5}$	

6. The time of stabilization of the deformations of the base is determined for the case of loading the surface of the base with a continuous (intensity of a uniformly distributed load P_{IImt};

Tables 3 and 4 below shows the calculations of the stabilized deformations of the bases of types No. 3 and 4, and Table 5 shows the algorithm for determining the stabilized deformation of the same bases by the equivalent layer method (ELM).

The need to determine the ELM deformation is that when the normalized deformation exceeds the maximum permissible value, the time is calculated.

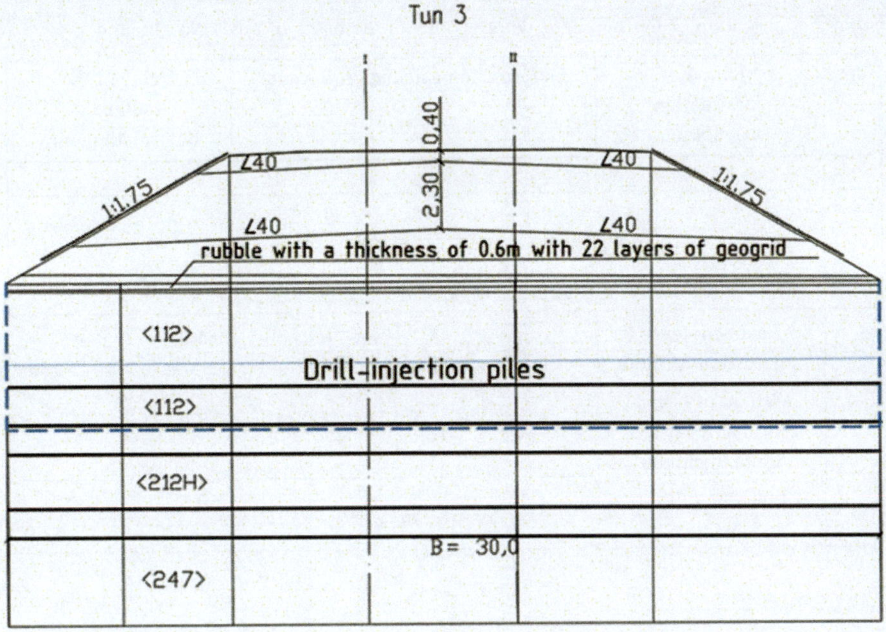

Fig. 1 Section of the roadbed with a vertical reference to the engineering-geological section

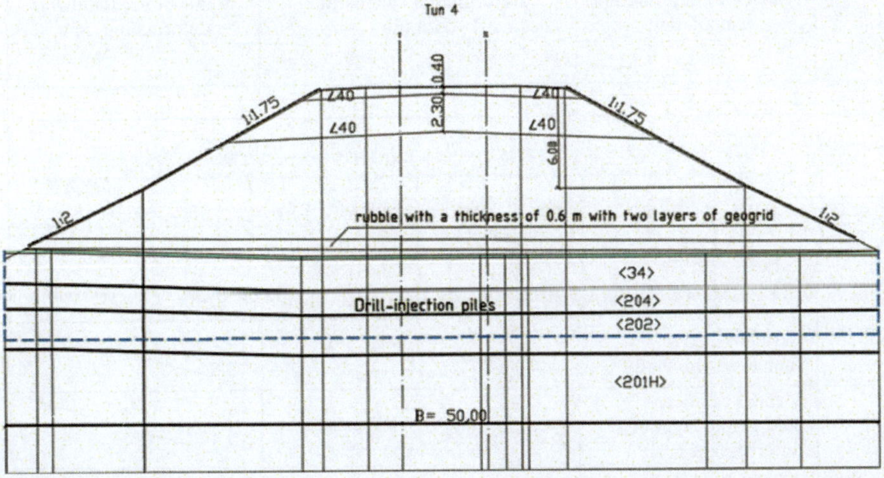

Fig. 2 A cross-section of a type 4 roadbed with a vertical reference to an engineering-geological section

Table 3 Calculation of the stabilized deformations of the bases of types No. 3 and 4

№	Method of deformation calculation	Deformation calculation formula	Value of the stabilized deformation, mm
1	2	3	4
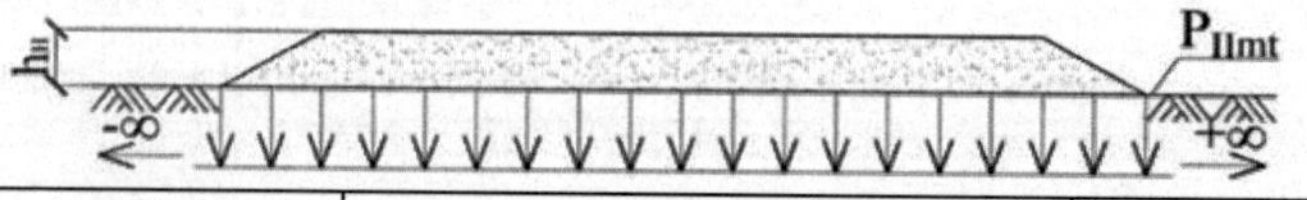			
1	Layer-by-layer summation method	$S = 0.8 \cdot \sum[\frac{(\sigma_{zpi} \cdot h_i)}{E_{0i}}]$	85.0 / 73.0
2	Method of linear-deformable layer of finite thickness	$S = (\frac{P \cdot b_{kc}}{k_m}) \cdot \sum[\frac{(k_i - k_{i-1})}{E_{0i}}]$	78.0 / 69.5
3	Equivalent layer method	$S = h_э \cdot m_{vm} \cdot P_0$	69.5 / 58.0

Table 4 Calculation of the stabilized deformations of the bases of types No. 3 and 4

№	Method of deformation calculation	Deformation calculation formula	Value of the stabilized deformation, mm
1	2	3	4
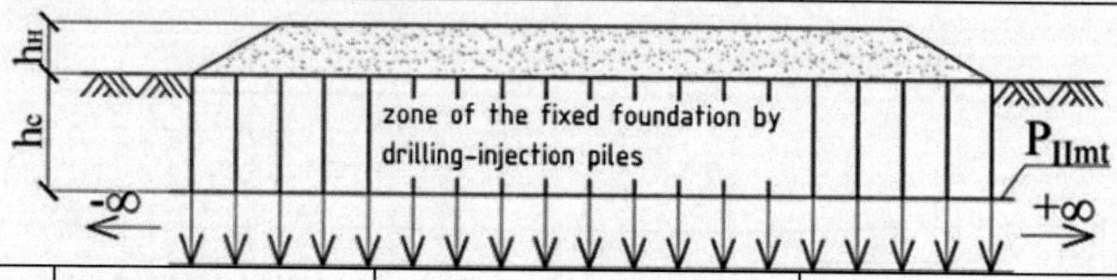			
1	Layer-by-layer summation method	$S = 0.8 \cdot \sum[\frac{(\sigma_{zpi} \cdot h_i)}{E_{0i}}]$	14.8 / 13.8
2	Method of linear-deformable layer of finite thickness	$S = (\frac{P \cdot b_{kc}}{k_m}) \cdot \sum[\frac{(k_i - k_{i-1})}{E_{0i}}]$	13.0 / 14.9
3	Equivalent layer method	$S = h_э \cdot m_{vm} \cdot P_0$	14.5 / 12.8

Table 5 The algorithm for determining the deformation by the equivalent layer method

№	Type of enbankment	Type of base	Name of soil	h_i (m)	E_{0ed} (MPa)	m_v (MPa)	k_ϕ (m/day)	P_{IImt} (kPa)	$h_э$ (m)	S (mm)
1	2	3	4	5	6	7	8	9	10	11
1	№3	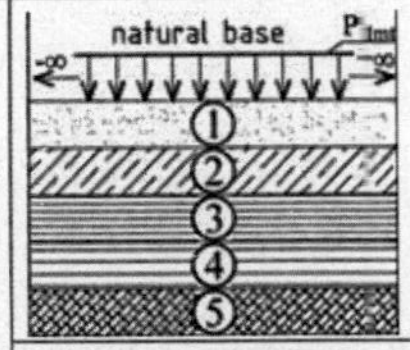	Sand	3.5	7.3	0.101	3.4	1200	8.0	69.5
			Loam	1.5	6.4	0.116	$3 \cdot 10^{-3}$			
			Clay	1.0	1.3	0.300	$1 \cdot 10^{-5}$			
			Clay	2.0	4.0	0.100	$1 \cdot 10^{-5}$			
			Dolomite							
		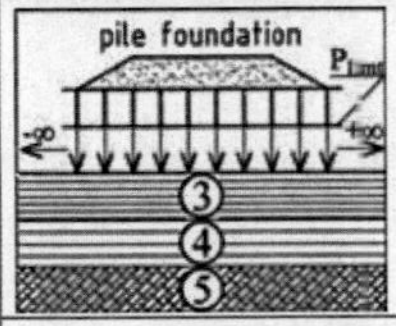	Clay	1.0	1.3	0.300	$1 \cdot 10^{-5}$	300	3.0	14.5
			Clay	2.0	4.0	0.100	$1 \cdot 10^{-5}$			
			Dolomite							
2	№4	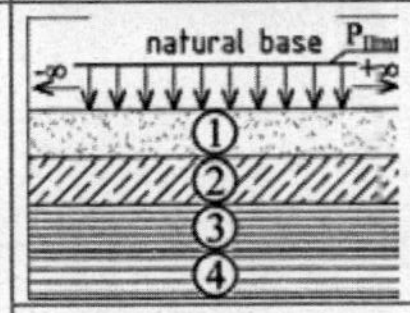	Loamy soil	2.0	3.2	0.194	0.05	220	15.0	58.0
			Loamy soil	2.0	1.5	0.413	$3 \cdot 10^{-4}$			
			Sand	3.0	10.6	0.070	4.0			
			Clay		6.5	0.062	$1 \cdot 10^{-5}$			
		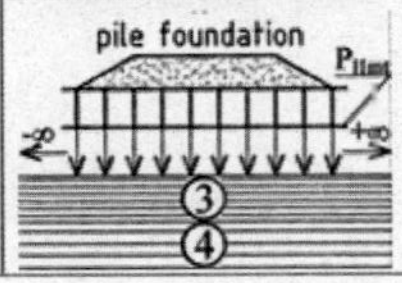	Sand	1.0	10.6	0.070	4.0	400	15.0	12.8
			Clay		6.5	0.062	$1 \cdot 10^{-5}$			

4 Conclusion

The developed algorithm for determining the deformation of weak foundations of high-speed highway embankments in the presence of weak bases in the foundations allows developing technical and technological measures for their safe operation.

References

1. Il'ichev VA, Mangushev RA, Nikiforova NS (2012) Development of underground space in large Russian cities. Bases Found Soil Mech 2:17–20
2. Ulitsky VM, Shashkin AG, Shashkin KG (2010) Geotechnical maintenance of urban development. Georeconstruction, St. Petersburg
3. Ter-Martirosyan ZG (2009) Mechanic of soil. ASV, Moscow
4. Ulickiy VM, Shashkin AG, Shashkin KG (2015) Guide to geotechnical engineering (Guide to the grounds, foundations and underground structures), 2nd edn, additional. Saint Petersburg
5. Mangushev RA, Gursky AV (2016) Assessment of the impact of sheet pile indentation on additional precipitation of neighboring buildings. Geotechnics 2:2–7
6. Mangushev RA, Sapin DA (2015) Taking into account the rigidity of the "wall in the ground" on the draft of neighboring buildings. Hous Constr 9:3–7
7. Mangushev RA, Gursky AV, Sapin DA (2017) Taking into account the technological sediment of existing structures in the construction of new buildings around them with a developed underground expanse. In: Engineering-geological surveys, design and construction bases, foundations and underground structures. Sat. Tr. All-Russian scientific. tech. conf. 1–3 Feb 2017, pp 9–22. St. Petersburg
8. Mirsayapov IT, Hasanov RR, Safin DR (2017) Results of geotechnical monitoring of carriers building structures during reconstruction. Engineering-geological surveys, design and construction of foundations and underground structures. In: Sat. Tr. all-Russian scientific. tech. conf. 1–3 February 2017, pp 164–169. St. Petersburg
9. Nikiforova NS, Vnukov DA (2011) Geotechnical cut-off screens for the protection of buildings in the device of communication collectors. III Academic reading them. Professor A. A. Bartholomew. In: The foundations of the deep laying and problems of underground expanse development. Mat. intl. conf. Perm, October 18–19 2011, pp 413–42. Perm
10. Sokolov NS, Ryabinov VM (2016) The technology of appliance of continuous flight augering piles with increased bearing capacity. Hous Constr 9:11–14
11. Sokolov NS (2017) Criteria of economic efficiency of use of drilled piles. Hous Constr 5:34–38
12. Sokolov NS (2018) Determination of the type of buried structure of the reinforcement of the base under the embankment of high-speed rail line 9:62–66

Experimental Study and Computer Simulation of the Work of Combined Bolt and Weld-In Connections

N. V. Solodov and N. V. Vodyakhin

Abstract When operating load-bearing building structures, there is often a need to strengthen the bolted connection, if the connection has a lack of load-bearing capacity. This may be due to increased loads on the structure or damage to the connection (for example, as a result of corrosion). The method of strengthening the bolted connection with welds, therefore, has obvious practical significance, and the combined bolt and weld-in connection has its own field of application. Many authors devoted their works to studies of combined connections, which use the joint work of friction bonds in the form of prestressed high-strength bolts and bonds in the form of welds. However, connections on conventional, non-prestressed bolts, which are affected by shear forces, are still of interest for research. The subject of research in them is the joint work in one connection of two types of bonds—bolts and welds, each of which has a significantly different deformability. The paper presents the results of testing physical samples of bolt and weld-in connections and computer research of their numerical models. The impact of preloading on the load-bearing capacity of combined bolt and weld-in connections is revealed. The load-bearing capacity criterion for numerical models was adopted on the basis of the deformation compatibility of numerical simulation samples and physical tests.

Keywords Bolted connections · Welds · Combined connections · Multiformative connections · Reinforcement · Load-bearing capacity · Deformability · Numerical models · Computer simulation

1 Introduction

When reconstructing buildings and structures, it is often necessary to strengthen bolted connections or restore their load-bearing capacity. In this case, it is possible to use classical reinforcement options by replacing existing bolts with bolts of a

N. V. Solodov (✉) · N. V. Vodyakhin
Belgorod State Technological University named after V.G. Shukhov, Belgorod, Russia
e-mail: partietz5000@bk.ru

S. V. Klyuev (ed.), *Digital Technologies in Construction Engineering*,
Lecture Notes in Civil Engineering 173, https://doi.org/10.1007/978-3-030-81289-8_41

larger diameter and (or) strength class, or increasing the load-bearing capacity of the nodes by applying welds. In the latter case, we get combined bolt and weld-in connections.

In bolted connections, it is possible to use bolts of ordinary strength or high-strength. Depending on this, the value of the deformability of the bolted connection will be different [1]. Welds have a lower deformability compared to connections on bolts of normal strength and are comparable to the deformability of friction connections [2] on high-strength prestressed bolts. The combination of welds in combination with prestressed high-strength bolts, as shown by studies [3–6] is possible. However, when combining different-formative connections in a single connection, it becomes necessary to determine the parameters of each of the connections. This paper presents the results of experimental studies of physical samples of combined bolt and weld-in lap connections that perceive shear forces, as well as their numerical modeling.

2 Methods and Materials

Numerical simulation of the operation of the combined bolt and weld-in connections was performed in the ANSYS computing complex.

To confirm the reliability of the obtained results of numerical modeling, tests of physical samples were performed.

The test sample is an lap connection of three sheet parts connected by two bolts and welds. The results of field tests, as well as the analysis of the combined bolt and weld-in connection operation, are described in detail in [7].

As the load-bearing capacity of a combined bolt and weld-in connection is not the arithmetic sum of each of the combined connections [8, 9], a study of bolted, welded and combined connections was performed to assess the contribution (share) of each type of connection. All samples were divided into 3 groups: group C345B—bolted connection samples, group C345W—welds samples, group C345Comb—combined bolt and weld-in connection samples.

The application of welds for the samples of the third group was performed under load in order to simulate the actual operation of bolted connections of real structures [1]. In this case, preloading is understood as loading a sample of a bolted joint before it is reinforced with welds and then loaded as a sample of a combined connection. The preload value was 85% of the calculated load-bearing capacity of the bolted connection.

3 Results and Discussion

Figure 1 shows a general view of the sample of the first group—C345B, the sample of the second group—C345W, as well as their numerical models.

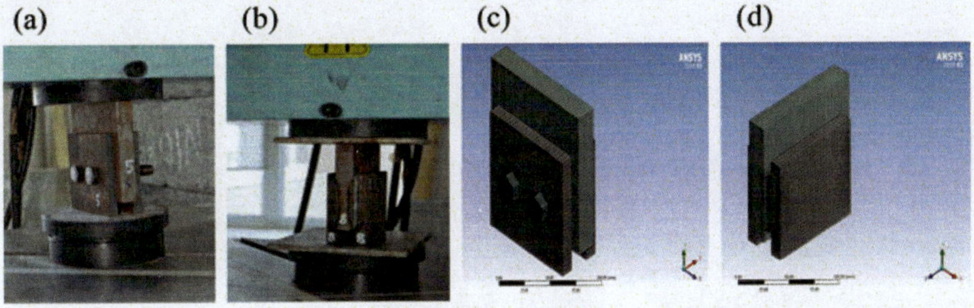

Fig. 1 General view of the samples under study, **a** general view of the sample of group C345B; **b** general view of the sample of group C345W; **c** numerical model of the sample of group C345B; **d** numerical model of the sample of group C345W

The average value of the actual load-bearing capacity for the physical model group C345B was 177.4 kN. During the numerical simulation, the load-bearing capacity of the C345B group samples was 165.0 kN. The total deformation of the bolted connection samples in the direction of loading was determined using an extensometer, its value was 3.3 mm. According to the results of numerical simulation—3.7 mm. Figure 2 shows a photo of a sample of the C345B group during and after the tests, as well as the results of numerical modeling.

The average actual load-bearing capacity of the samples of the second group (C345W) according to the results of field tests was 426.2 kN. The actual average load-bearing capacity (calculated resistance) of one cm^2 of the cross-section area of the weld is 35.41 kN/cm^2. Therefore, with the design value of the weld area in the C345W samples equal to 10.2 cm^2, the load-bearing capacity of the sample will be 361.2 kN. The load-bearing capacity of the C345W group samples according to the results of numerical simulation was 390 kN. Figure 3 shows a photo of the sample of the second group after the tests, as well as the result of numerical modeling.

Figure 4 shows a general view of the sample of the third group—C345Comb during the tests, as well as its numerical model.

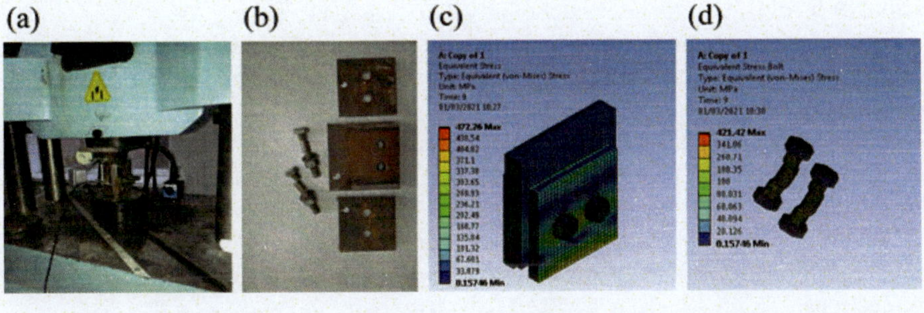

Fig. 2 Photo of the test sample of group C345B: **a** photo of the sample during the tests; **b** photo of the sample after the tests; **c**, **d** results of numerical modeling of the sample of the first group

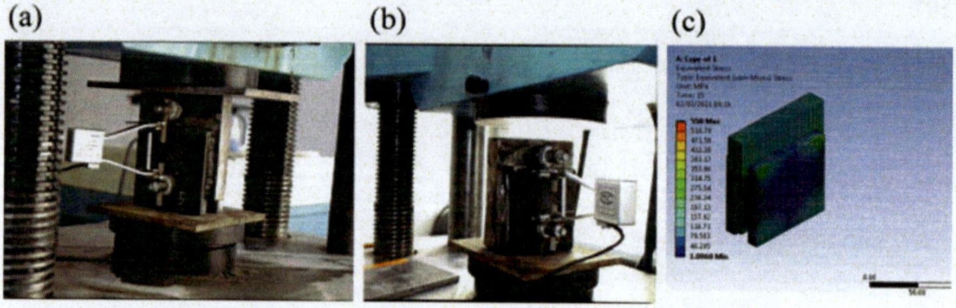

Fig. 3 Photo of the test sample of group C345W: **a** photo of the sample during the tests; **b** photo of the sample after the tests; **c** the result of numerical modeling of the sample of the second group

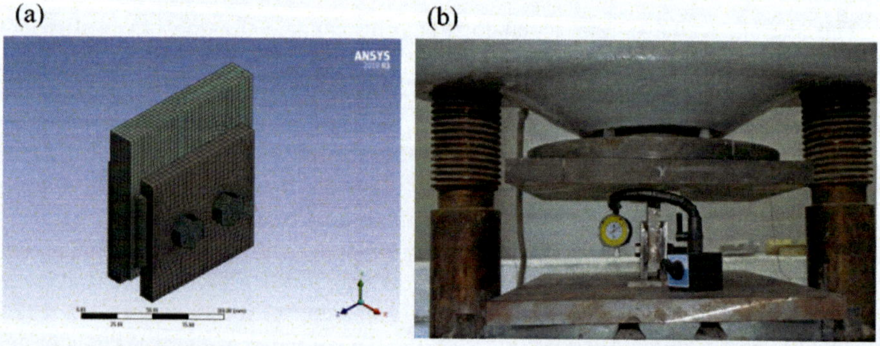

Fig. 4 General view of the samples of the C345Comb group: **a** numerical model of the sample; **b** general view of the sample

According to the results of field tests of two samples of the third group, the average load-bearing capacity was 525.0 kN with a strain value of 0.88 mm. The load-bearing capacity of the samples of the combined connection in the course of numerical simulation at a strain value of 0.88 mm was 490 kN. The photo of the sample of the third group after the tests as well as the results of numerical modeling is shown in Fig. 5.

Figure 6 shows the deformation diagrams of numerical models of combined bolt and weld-in samples, as well as the deformation of welds in the samples under study.

In the numerical simulation, the bolts were put into operation immediately when the load was applied, as in the case of the "positive" position of the bolt body, described in the works of Manuel and Kulak [2], Kulak an Grondin [10].

For the analysis, we compare the quantitative parameters that characterize the load-bearing capacity and deformability of bolted, welded and combined connections.

The design values of the bearing capacity of physical samples are equal to: bolted—P_B = 118.9 kN; welded—P_W = 361.2 kN; combined bolt and weld-in

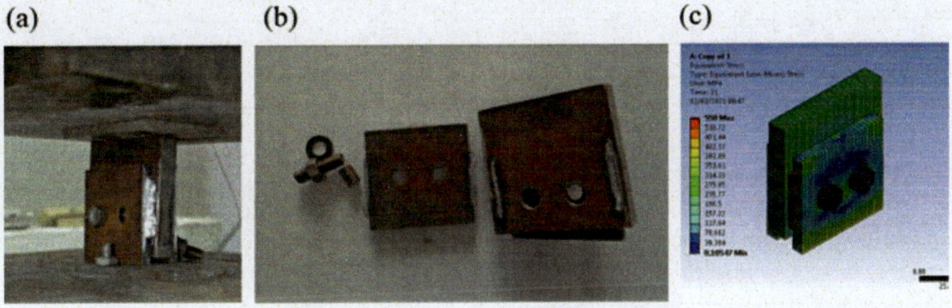

Fig. 5 Photo of the test sample of the C345Comb group: **a**, **b** photo of the sample after the tests; **c** the result of numerical simulation of the sample of the third group

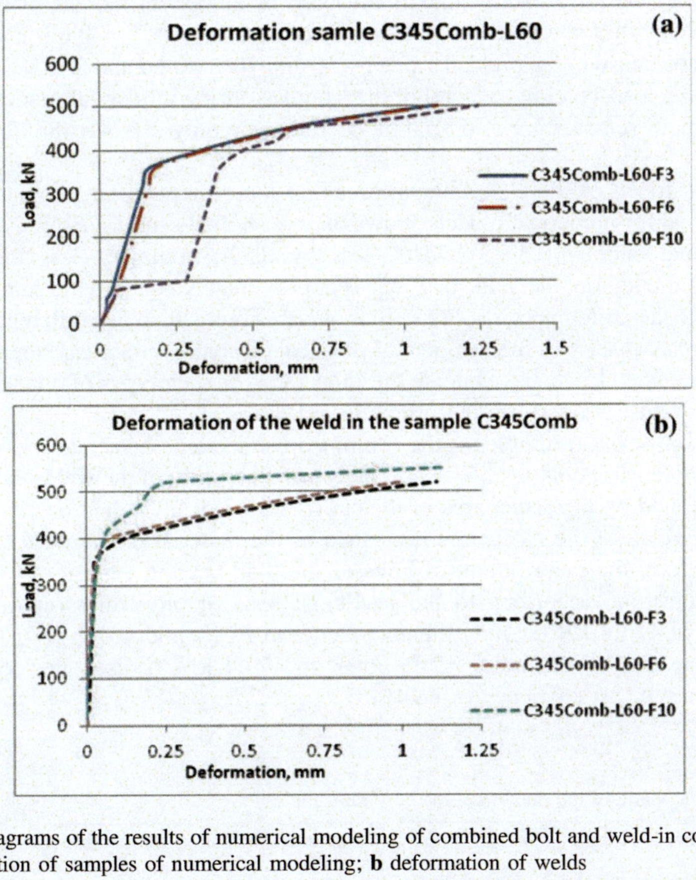

Fig. 6 Diagrams of the results of numerical modeling of combined bolt and weld-in connections: **a** deformation of samples of numerical modeling; **b** deformation of welds

—P_{comb} = 480.1 kN. For the analysis, the design load-bearing capacity of the welded samples is the value obtained taking into account the actual cross-section area of the welds and the actual value of the calculated resistance of the material of the welds. The arithmetic sum of the design values P_B and P_W is taken as the design load-bearing capacity of the combined connection.

The maximum load for the test samples is shown in Table 1.

The experimental (actual) load-bearing capacity of the combined connection samples is 15.0% less than the arithmetic sum of $P_B^{эксп.}$ and $P_W^{эксп.}$. This confirms the fact that the different deformability of bolts and welding does not allow these connections to work 100% effectively in one combined connection. We note, however, that the bolted connection was pre-loaded with a force of 100 kN before it was reinforced with welds, which corresponds to 84.1% of its design load-bearing capacity.

The load-bearing capacity (maximum load) of numerical models of combined connections with a strain value of 0.88 mm is equal to $P_B^{числ.}$ = 490.0 kN.

With a decrease in the preload value by 40 and 70% (to 60 and 30 kN, instead of 100 kN), the load-bearing capacity of the samples during numerical modeling, also with a sample deformation of 0.88 mm, decreases by only 2%—to 480 kN, instead of 490 kN.

A comparison of the experimental breaking load for physical samples and the limiting load for numerical models shows that: for bolted samples $P_B^{числ.}$ is less than experimental one by 9.3%; welded—less by 8.5%; combine—less by 9%. In numerical simulation, the value of $P_{comb}^{числ.}$ is approximately equal to the sum of $P_B^{числ.}$ and $P_W^{числ.}$, the difference is 14.6%. It is worth noting that the difference in the load-bearing capacity in the samples of physical tests and numerical simulations is almost the same (15%). To evaluate the joint work of such types of connections as bolts and welds, we perform the following comparison.

In tests of physical samples of the combined connection, the actual breaking load on the sample was equal to 525.0 kN. The bearing capacity of the welds in the same samples, found by the actual area of the cut of the welds after their destruction and the actual value of the calculated resistance of the material of the welds, is 407.2 kN. Therefore, the contribution of bolts to the load-bearing capacity of the combined connection, according to the results of tests of physical samples, can be estimated at 117.8 kN (as the difference between $P_{comb}^{эксп.}$ and 407.2 kN). Thus, the bearing capacity of the bolts $P_B^{эксп.}$, equal to 177.4 kN, is used in a combined connection with an efficiency of 66.4%.

Table 1 The value of the maximum load on the samples

№	Group of samples	Design (kN)	Actual experimental (kN)	Limit value for numerical models (kN)
1	C345B	118.9	177.4	
2	C345W	361.2	426.2	
3	C345Comb	480.1	525.0	

4 Conclusion

The results can be presented as follows.

(1) The computer simulation of the combined connection in the ANSYS computing complex as a whole reflects reliably the operation of the elements of the bolt and weld-in connection, including in terms of modeling the operation of the bolted connection reinforced under load by welds. (2) The results of testing physical samples and the study of numerical (computer models) showed that the load-bearing capacity of the combined bolt and weld-in connection is greater than the load-bearing capacity of welded samples. The joint operation of different-formative connections (bolts and welds) in a combined connection is possible after preliminary study of the crease deformations in the bolted connection before it is reinforced with welds. (3) The efficiency of the joint work of bolts and welds as connections in a combined bolt and weld-in connection can be estimated at 66.4% of the bearing capacity of the bolted connection (for the range of parameters of the sample studied in the paper). (4) Changing the preload value of a bolted connection from 84 to 30% of its load-bearing capacity has little effect on the efficiency of joint work of bolts and welds in a combined connection (in the studied range of parameters of experimental samples).

Acknowledgements This work was realized in the framework of the Program of flagship university development on the base of the Belgorod State Technological University named after V.G. Shukhov, using equipment of High Technology Center at BSTU named after V.G. Shukhov.

References

1. Solodov NV (2017) Strength and deformability when collapsing in a bolted connection. Bulletin of BSTU named after V.G. Shukhov 1:82–87
2. Manuel TJ, Kulak GL (2000) Strength of joints that combine bolts and welds. J Struct Eng 3:279–287
3. Zhang DX, Hou ZX, Sun L (2011) Nonlinear finite element analysis of shear bearing capacity of joint with combined bolts and welds. Steel Constr 8:9–13
4. Shi YJ, Wang L, Wang YQ (2013) Finite element analysis of the combined connection with bolts and welds. Build Sci Res Sichuan 2:100–104
5. Chen Q, Li FX, Le JQ et al (2015) Tensile capacity and design method of combined connections with bolts and welds. Eng Mech 33(1):112–121
6. Lan T, Ju J, Liu M (2019) Analysis of ultimate load-carrying capacity of combined connection with bolts and welds. Mechanika 426–433
7. Solodov NV, Vodyakhin NV (2021) Evaluation of the load-bearing capacity of the combined bolt-welded joint. Lect Notes Civil Eng 151:174–180
8. Wang JJ, Shi YJ, Wang YQ (2013) Finite element analysis on the bearing capacity for connection of sharing on a shear load by friction-type high-strength bolts and side welds. Build Sci Res Sichuan 1:15–19
9. Liu MS, Huang CA, Li JR, Ju JS (2017) Numerical modeling and mechanical analysis of combined connection with bolts and welds. Strength Mater 48(6):862–869
10. Kulak GL, Grondin GY (2003) Strength of joints that combine bolts and welds. Eng Second Q J 40(2):89–98

Thermolysis Technology and Technical Means for Processing Organic Technogenic Materials

V. S. Sevostianov, N. T. Shein, R. U. Shamgulov, and V. V. Obolonsky

Abstract Scientific and technical developments were carried out to create resource-saving technology and patent-protected devices for low-temperature thermolysis processing of municipal solid waste (MSW) with various physical and mechanical characteristics. Pilot tests of a thermolysis reactor with a spiral conveying body of combined action were carried out and design, technological and thermal engineering parameters of low-temperature processing of organic MSW, as well as rubber waste (RW) were obtained. Samples of recovered carbon black (rCB) were obtained, corresponding to industrial products in terms of their physicochemical characteristics and properties. The possibility of using carbon black obtained by low-temperature thermolysis in the paint and varnish industry has been confirmed. Presented are the test results of paint and varnish compositions made on the basis of three types of polymer binders (alkyd, epoxy, acrylic) carbon black grades P-803, K-354 and rCB, obtained by low-temperature thermolysis technology. It has been established that these compositions, as well as coatings based on them, in terms of their performance, meet the requirements for carbon black grades P-803 and K-354.

Keywords Ecology · Low-temperature thermolysis · Solid municipal waste · Carbon black · Recovered carbon black · Paint and varnish industry

1 Introduction

Carbon black (CB) is a highly dispersed carbonaceous material that is formed during incomplete combustion or thermal decomposition of hydrocarbons contained in natural or industrial gases, as well as in liquid products of petroleum or

V. S. Sevostianov · R. U. Shamgulov (✉)
Belgorod State Technological University named after V.G. Shukhov, Kostyukov St., 46, Belgorod 308012, Russia

N. T. Shein · V. V. Obolonsky
LLC «Transport Company Ecotrans», Serafimovicha St., 72, Belgorod 308017, Russia

S. V. Klyuev (ed.), *Digital Technologies in Construction Engineering*,
Lecture Notes in Civil Engineering 173, https://doi.org/10.1007/978-3-030-81289-8_42

coal origin [1]. The world's carbon black output is currently more than 13 million tons per year. Among the world's manufacturing firms, which produce more than 40% of the world's CB volume, two companies predominate—Cabot (USA) and Evonik (formerly Degussa, Germany) [2]. In Russia and the CIS countries, the largest manufacturer of CB—Omsk Carbon Group, is one of the world's top ten producers of carbon black.

Carbon black is widely used in various industries. The main share (more than 80%) of the produced CB is used in the tire industry, as well as in the preparation of rubber mixtures as filler for their reinforcement. In the electrical industry, a mixture of graphite, carbon black, and a binder (coal peck) is used to produce DC motor brushes [3]. For painting electrical insulation materials, gas channel carbon is used, which has a relatively high content of volatile substances (oxygen-containing groups—lactane, quinone, phenolic and carboxyl), which contributes to low electrical conductivity [4]. In the electrical industry, a mixture of graphite is used for the production of brushes of DC electric motors, which is a technical component in metallurgy, and is a mandatory component of most steel grades. By changing the ratio between the amount of ferrite and perlite in the structure of a solid metal, carbon makes it possible to obtain a metal with a wide range of mechanical properties [5]. Carbon black is also included in the composition of polymer materials for color change, and is also a stabilizer when used in polyolefines and other thermoplastic polymers to increase their resistance to the influence of the atmosphere, as this substance has the ability to protect polymers from the action of ultraviolet radiation on them [6]. In addition, CB is used in the production of abrasives, in printing, as well as an adsorbent in wastewater treatment. CB is not unimportant in the construction industry as a dye in the production of cement, concrete, self-leveling floors, asbestos cement, paving slabs and silicate bricks [7]. Carbon black plays a significant role in the paint industry as a pigment in the production of black and gray weather-resistant paints and enamels. Having high temperature resistance, resistance to light and many chemical compounds, the ability to absorb ultraviolet, infrared radiation and light of the visible part of the spectrum, CB gives paint and varnish materials significant performance qualities [8].

The above-mentioned areas of use of technical equipment, its wide application for the production of various types of technical products indicates the relevance of the development of this scientific and technical direction and the search for unused technological reserves, design and technological improvement of equipment.

2 Materials and Methods

In the industry, CB is obtained in various ways: furnace, channel thermal, lamp, acetylene. The feedstock can be liquid or gaseous. Carbon black is distinguished by its dispersion (high-, medium-, low-dispersed), by its activity in the rubber mixture (high-, medium-, semi-, low-active), and by its structure (high-, medium-, low-structure). Other methods of obtaining recovered carbon black (rCB) from organic raw materials are also possible, for example, from solid municipal waste (MSW), in particular - from rubber waste (RW), with low-temperature thermolysis of the feedstock [9]. The main advantages of the proposed method:

- effective solution of environmental problems in the complex processing of MSW;
- creation of resource-saving technologies and implementation of recycling of secondary (technogenic) materials;
- solving state problems of import substitution and obtaining innovative materials;
- implementation of national projects of the Russian Federation aimed at the rational use and processing of MSW to produce various types of marketable products, etc.

The author's team of researchers of BSTU named after V.G. Shukhov and engineering and technical employees of Ecotrans TC LLC, conducts scientific and technical developments and the introduction of resource-saving technology and special equipment for low-temperature thermolysis of organic waste into real production.

The method allows the use of recyclable organic MSW of various morphological and physical composition: small-scale polymer, wood, pulp and paper, rubber and other waste in a loose or compacted state during their heat moisture treatment. The proposed method of low-temperature processing of organic MSW at a temperature of up to 500 °C makes it possible to obtain high-quality products: recovered carbon black, liquid hydrocarbon fuel and synthetic hydrocarbon gas.

In the presented study, we used various patent-protected technological solutions: sealed material loading and unloading units, structural and technological solutions for the working elements of a thermolysis reactor, aspiration systems, rectification, etc., patent RU 2744225 C1.

A spiral thermal reactor is a combined-action transport body made in the form of a cylindrical body with spiral (or helical) elements placed inside [10]. The housing is equipped with pipes for the input and output, respectively, of the initial raw materials and the resulting products—gaseous reagents, followed by the production of liquid hydrocarbon fuel, carbon black. The spiral-shaped conveying body ensures the movement of the heat-treated material. The developed resource-saving technology and special equipment are currently undergoing pilot testing and implementation at the company "Ecotrans TC" LLC.

The estimated capacity of the low-temperature thermolysis process line for incoming raw materials is 400–500 kg/h. Resource- and energy-saving and design-technological solutions of the implemented developments are based on many years of experience in modeling technological processes in production conditions, testing thermolysis technologies for processing various MSW: rubber waste, wood and pulp and paper waste, polymer technogenic materials, etc.

One of the final products obtained by the developed technology for processing MSW, in particular RW, is carbon black. According to its characteristics, it is a combined meso-macro powder with a predominance of macropores, its specific surface area is 37.9 m^2/g, pore volume is 0. 23 cm^3/g, and the average pore size is 23.9 nm.

3 Results and Discussions

Due to the scale of the scientific and technical tasks to be solved, a wide range of processed MSW, we present the results of research on pilot tests of the technology for obtaining and using CB in the production of paint and varnish materials. At the same time, rubber crumb obtained from MRG waste and decommissioned tires were used as a raw material. Carbon of P-803 and K-354 grades was adopted as the reference recovered carbon black. Table 1 shows the physical and chemical characteristics of rCB.

To test the powder of rCB obtained by low-temperature thermolysis technology and used in paint materials, paint compositions were made based on the most common binders: alkyd (varnish PF-053), epoxy (resin solution E-41r) and acrylic (acrylic resin AR). The composition of the compounds is shown in Table 2.

Paint and varnish compositions were made by dispersing pigments and fillers in a film-forming solution on a laboratory bead mill. The technological parameters of the prepared paint and varnish compositions, as well as the properties of coatings based on them, are presented in Table 3. All the resulting paint and varnish compositions were black in color and after drying formed a homogeneous, even, free of extraneous inclusions surface.

The conditional viscosity, the mass fraction of non-volatile substances, and the degree of grinding were determined in undiluted compositions. To determine the remaining indicators, the paint and varnish material was applied with a paint

Table 1 Physical and chemical characteristics of the rCB sample obtained in the process of low-temperature thermolysis of rubber crumb from RW and decommissioned tires

Parameters	ASTM	Result
Surface area according to BET (m^2/g)	D6556	84.3
Surface area according to STSA (m^2/g)	D6556	55.1
Iodine number (mg/g)	D1510	124.5
pH-Volume (UZ suspension)		5.96
Moisture loss at ignition (125 °C, 1 h) (%)	D1509	1.01
Ash content at ignition (550 °C, 16 h) (%)	D1506	14.74
Sulphur content (%)	D1619	2.69
Fine particle content (%)	D1508	7.5
Toluene light transmission (%)	D1618	98.9

Table 2 Composition of paint and varnish compounds based on polymer binders and carbon black of various brands

Compound, №	1	2	3	4	5	6	7	8
Binder, wt%	89.2	89.2	35.61	35.61	35.61	15.0	15.0	15.0
Pigments and fillers, wt%	–	–	3.5	3.5	3.5	1.5	1.5	1.5
P-803	5.04	–	7.0	–	–	3.7	–	–
K-354	–	–	–	7.0	–	–	3.7	–
rCB (thermolysis)	–	5.04	–	–	7.0	–	–	3.7
Functional additives	2.52	2.52	–	–	–	1.8	1.8	1.8
Hardener	–	–	30.0	30.0	30.0	–	–	–
Solvent	2.52	2.52	23.89	23.89	23.89	78.0	78.0	78.0

Table 3 Properties of paint compositions and coatings based on them

Composition, №	1	2	3	4	5	6	7	8
Conditional viscosity at a temperature of (20.0 ± 0.5) °C according to a viscometer with a nozzle diameter of 4 mm (s)	35.0	27.0	19.0	25.0	18.0	58.0	61.0	64.0
Mass fraction of non-volatile substances (%)	38.9	45.1	46.5	46.0	45.3	27.4	25.3	21.5
Degree of grinding, Microns	60.0	60.0	40.0	40.0	40.0	40.0	40.0	40.0
Drying time to degree 3, at temperature (20 ± 2) °C (h)	24.0	24.0	6.0	6.0	6.0	2.0	2.0	2.0
Coverage of the dried film (g/m^2)	–		18.52	12.04	14.81	17.0	16.67	17.0

sprayer to the corresponding plates prepared according to GOST 8832-6 "Paint and varnish materials. Methods for obtaining a paint coating for testing". Table 4 shows data on the conditional light resistance of coatings. The tests were carried out in accordance with the requirements of GOST 21903-76 "Paint and varnish materials. Methods for determining the conditional light resistance". The DRT-400 lamp was used for testing. Samples of coatings applied to white drawing paper according to GOST 597-73 with a size of 100 × 200 mm were subjected to UV irradiation and after 2, 6, 12 and 24 h, the brightness coefficient was measured on a OUF device with a correcting absorber. The test results are shown in Table 4.

Table 4 Conditional light resistance of coatings based on polymer binders and carbon black of various brands

Compound, №	1	2	3	4	5	6	7	8
Brightness coefficient before UV irradiation (%)	–	0.10	0.13	0.10	0.10	0.18	0.10	0.17
2 h after UV irradiation (%)	–	0.10	0.13	0.10	0.10	0.19	0.10	0.17
6 h after UV irradiation (%)	–	0.10	0.13	0.10	0.10	0.21	0.10	0.18
12 h after UV irradiation (%)	–	0.10	0.13	0.10	0.10	0.28	0.11	0.18
24 h after UV irradiation (%)	–	0.16	0.13	0.13	0.17	0.29	0.12	0.18

4 Conclusion

1. A complex of scientific and technical developments and research on the creation of resource-saving technology and patent-protected devices for the processing of MSW with various physical and mechanical characteristics was carried out.
2. According to the results of pilot tests, the design, technological and thermal parameters of the thermolysis technology for processing various MSW were worked out. The technical, economic and environmental feasibility of using the completed developments was confirmed.
3. The possibility of using CB samples obtained by low-temperature thermolysis technology in the formulations of paint and varnish materials is investigated.
4. Paint and varnish compositions are made on the basis of polymer binders of three types (alkyd, epoxy, acrylic) of carbon black grades P-803, K-354 and rCB, obtained by low-temperature thermolysis technology.
5. It is established that paint and varnish compositions made using the technical technology of low-temperature thermolysis, as well as coatings based on them, meet the requirements for technical carbon grades P-803 and K-354. The obtained results meet the requirements for obtaining paint and varnish coatings (GOST—9.032-74).
6. Recovered carbon black obtained by low-temperature thermolysis technology can be used in the production of black and gray paint materials.

Acknowledgements The work was carried out within the framework of project No. 10089447, the program of the REC “Innovative solutions in the agro-industrial complex”, the scientific and production platform “Rational use of natural resources”.

References

1. Ivanovsky V (2019) Carbon black. Processes and apparatus schoolbook, 2nd edn. Printing house Blankom, Omsk
2. Gulmisaryan T, Kapustin V, Levenberg I (2017) Technical carbon black: morphology, properties, production. Caoutchouc and rubber, Moscow

3. Machulin L (2018) Cinderella of the carbon world. Sci Life 7:40–47
4. Rothon R (2017) Fillers for polymer applications. Springer International Publishing, Switzerland
5. Kushner V (2008) Materials science schoolbook. OmGTU, Omsk
6. Liu M, Horrocks A (2002) Effect of Carbon Black on UV stability of LLDPE films under artificial weathering conditions. Polym Degrad Stab 3(57):485–499
7. Perfilov V, Zubova M (2015) The impact of carbon additives on the properties of fine-grained fibrous concrete. Internet-Vestnik VolgGASU 1(37):1–7
8. Leong CK, Chung DDL (2006) Improving the electrical and mechanical behavior of electrically conductive paint by partial replacement of silver by carbon black. J Electron Mater 35:118–122
9. Glagolev S, Shein N, Sevostianov V, Obolonsky V, Shamgulov R (2020) Technologies for integrated processing of solid municipal waste. Ecol Indus Russ 12(24):11–15
10. Bocaver K (2018) Spiral reactors in heterogeneous technological processes. LAP LAMBERT Academic Publishing, Beau Bassin

Simulation of the Cement Loading Movement in the Chamber of a Pneumatic Chamber Pump

S. Yu. Lozovaya, Yu. M. Fadin, A. V. Gavrilenko, and Yu. V. Gavrilenko

Abstract The production of building materials is a complex technological process associated with the processing of raw materials with different physical and mechanical properties, as well as with the use of various degrees of complexity of technological equipment and auxiliary mechanisms. High-pressure pneumatic transport has found application in the enterprises of construction, refractory, chemical and other industries. High performance of pneumatic chamber pumps can be achieved by organizing effective pseudo-fluidizing of the material, i.e. uniform mixing of the transported material with the air. The continuous Eulerian approach used in the paper allows describing the motion of air and material environment using the same type of continuity equations and momentum transfer equations. The unloading of the pump chamber is simulated, the velocity distributions of the carrier air flow, the concentration fields of cement particles are determined for various stages of this process, and the mass flow rates of the phases at the inlet to the discharge pipe are calculated. In this case, the excess pressure in the pump chamber, the diameter of the discharge pipe, the distance of the inlet to the discharge pipe and the aeration device to the bottom of the pump chamber, as well as the distribution of compressed air between the upper free zone of the chamber and the multi-nozzle aeration device varied. Computer simulation of the process of unloading the pump allowed considering the operation of the pump at various stages.

Keywords Pneumatic chamber pump · Pseudo-fluidizing of the material · Computer simulation · Aeration device · Consumable concentration of the material

S. Yu. Lozovaya · Yu. M. Fadin · A. V. Gavrilenko (✉) · Yu. V. Gavrilenko
Belgorod State Technological University Named After V.G. Shukhov,
Kostyukov St., 46, Belgorod 308012, Russia
e-mail: Andrej.402@mail.ru

S. V. Klyuev (ed.), *Digital Technologies in Construction Engineering*,
Lecture Notes in Civil Engineering 173, https://doi.org/10.1007/978-3-030-81289-8_43

1 Introduction

Pseudo-fluidizing is a transition state between a stationary layer of bulk material and the carry-over of the material by a stream of compressed air, which occurs in a certain range of air flow rates through the material layer, which depends on the design of aerating devices. It should be noted that an increase in the concentration of the cement-air mixture leads to an increase in the performance of the pneumatic chamber pump. One of the main directions of reducing energy consumption for cement transportation using pneumatic chamber pumps is to improve the process of cement aeration in the pump chamber, which allows obtaining more concentrated cement-air mixtures at the inlet to the discharge pipe while reducing the unproductive costs of compressed air [1–4].

2 Methods and Materials

The high efficiency of a pneumatic chamber pump equipped with a new multi-nozzle aeration device protected by the Russian Federation patent for utility model No. 153059 was experimentally demonstrated [5]. To reduce the volume of laboratory and industrial tests, a computer simulation of the pump operation was performed, which allows identifying the influence of the design and technological parameters of the aeration device on the pump performance.

The turbulent gas-dispersed flow of the cement-air mixture, which is formed by the aeration device in the lower part of the pump chamber and is pumped by compressed air into the cement pipeline, will be studied on the basis of a two-liquid model. In the framework of this model, the dispersed phase is considered as a pseudo-continuous medium—a pseudo-gas of particles with their specific pressure, viscosity and temperature (kinetic theory of granular (gas-dispersed) flows [6–9, 10]. Kinetic theory allows applying a continuous Eulerian approach to the dispersed phase, so we can use the same type of continuity equations and momentum transfer equations to describe the motion of both phases.

The continuity equations for the air and dispersed phases have the form

$$\frac{\partial(\varepsilon\rho)}{\partial t} + \frac{\partial(\varepsilon\rho u_i)}{\partial x_i} = 0, \tag{1}$$

$$\frac{\partial(\beta\rho_m)}{\partial t} + \frac{\partial(\beta\rho_m u_j)}{\partial x_j} = 0, \tag{2}$$

there ε and β —volume concentrations of air and particles in the flow of cement air mixture; ρ, ρ_M—air and particle densities, kg/m^3; u_i, u_j—components of the average velocity of the air and dispersed phases, m/s.

To find out the turbulent characteristics of gas-dispersed flows in the kinetic theory, the turbulence model is used.

When cement is aerated due to its large specific surface area, the temperature of the air and the particles are almost instantly aligned, so the process of unloading the pump can be considered isothermal and the energy conservation equation is not considered when modeling it.

The system of Eqs. (1 and 2) applied to the operating conditions of the TA-29 pneumatic chamber pump is implemented in the Maple 13 package [11].

3 Results and Discussion

As a result of simulation of the unloading of the pump chamber for various stages of this process (Fig. 1), the velocity distributions of the carrier air flow, the concentration fields of cement particles, and the mass flow rates of the phases at the inlet to the discharge pipe are determined. During the computational experiment, the excess pressure in the pump chamber, the diameter of the discharge pipe, the distance of the inlet to the discharge pipe and the aeration device to the bottom of the pump chamber, as well as the distribution of compressed air between the upper free zone of the chamber and the multi-nozzle aeration device were varied. Computer simulation of the pump unloading process confirms its characteristic features observed in laboratory and production conditions: the pulsating nature of the flow of the cement-air mixture in the discharge pipe (Fig. 1b, c), the break under certain conditions of air jets without capturing the material (Fig. 1c), the formation of a material residue in the pump chamber (Fig. 1d), and others.

Figure 1a shows a pump with a material-loaded chamber. When air is supplied to the pump chamber, the material in the lower part of the chamber is pseudo-fluidized due to the air coming out of the nozzles of the multi-nozzle aeration device. Also, air was supplied to the upper part of the chamber above the material layer (the air supply pipe is not shown). At the same time, there is an effective unloading of the pump chamber with a high concentration of cement-air mixture (Fig. 1b, c). There is also an uneven movement of cement towards the discharge pipe due to the friction of the cement layers against the walls of the pump chamber and the discharge pipe. There is a decrease in the concentration of the mixture, which leads to an increase in air consumption (Fig. 1c). At the end of unloading the chamber, the formation of air bubbles (gaps) in the discharge pipe is observed, which leads to a decrease in the concentration of the cement-air mixture, which is accompanied by a sharp decrease in productivity and an increase in the consumption of compressed air (Fig. 1d).

Processing of the results of computational experiments allowed establishing the approximate dependence of the cement consumption concentration on the design and technological parameters of the pump

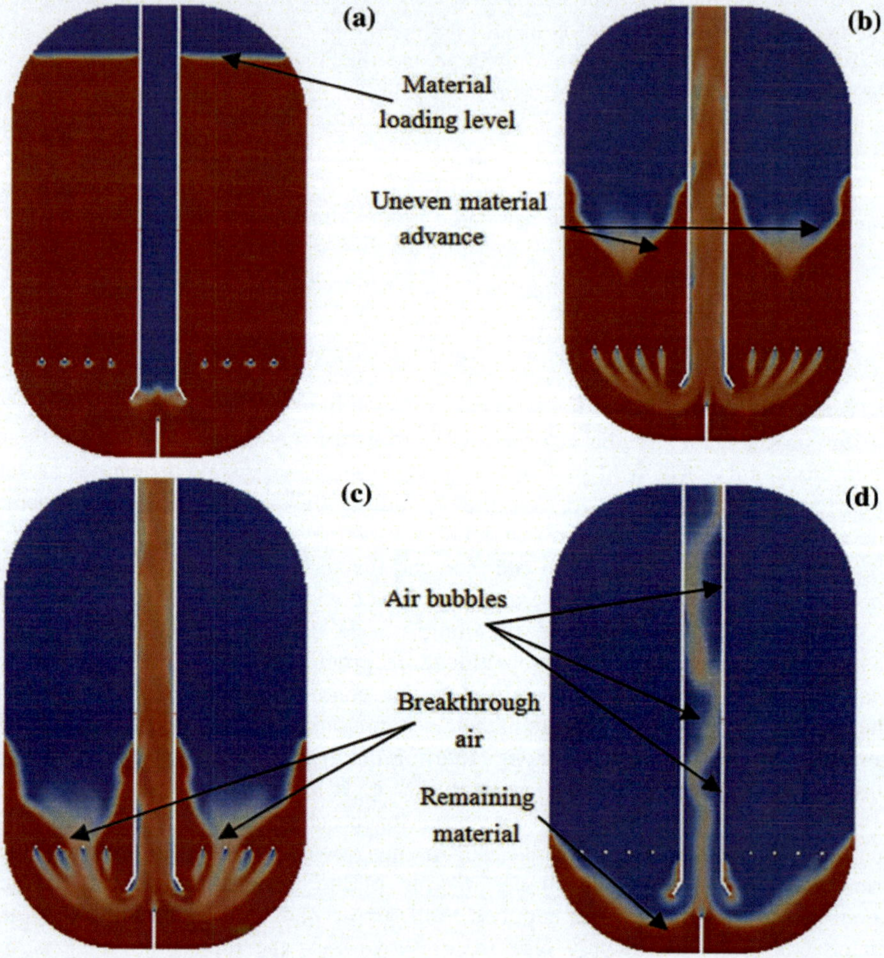

Fig. 1 Stages of unloading the pump chamber

$$\mu = \frac{G_y}{G_B} = \frac{2.8}{\left(\frac{\rho}{\rho_0}\right)^{2.6}\left(\frac{D}{D_\kappa}\right)^{0.8}}, \tag{3}$$

there G_y—pump capacity, kg/s; G_B—compressed air mass flow rate, kg/s; D, D_κ—discharge pipe and pump chamber diameters, m; ρ, ρ_0—air densities under operating and normal conditions, which are set by the dependence

$$\frac{\rho}{\rho_0} = \frac{PT_0}{P_0 T}, \tag{4}$$

there P—operating pressure, Pa; T—operating air temperature, K; $P_0 = 101$ kPa, $T_0 = 293$ K.

Equation (3) confirms the experimentally established increase in the consumption concentration of cement during unloading of the pneumatic chamber pump with a decrease in the diameter of the discharge pipe.

Based on the known consumption concentration of cement and the working pressure in the pump chamber, the specific energy consumption for pumping cement can be found using the empirical dependence

$$E = \frac{0.18 k_z k_n T}{\eta_K \mu} \lg\left(\frac{P}{P_0}\right), Vt \cdot ch/t, \tag{5}$$

there k_z—reserve coefficient, $k_n = 1.15$—coefficient that takes into account unproductive losses of compressed air, $\eta_K \approx 0.75$—the efficiency of the compressors.

It follows from Eq. (5) that the energy consumption for transporting cement decreases with an increase in the concentration μ and pressure of air P in the pump chamber, however, these values cannot be set arbitrarily, as they depend on the characteristics of the pneumatic conveying system as a whole.

The operating pressure varied in the range from 3 to 6 atm. (0.3–0.6 MPa), the total compressed air flow rate—from 0.5 to 1.2 nm^3/sec, the operating temperature from 120 to 140 °C. The highest concentration of cement at the inlet to the discharge pipe $\mu = 18$ kg occurred at a total compressed air flow rate of 1 nm^3/s and its following distribution: to the upper zone of the chamber 20%, through the aeration device—70% and the central nozzle—10% of the total flow rate. The air velocity at the inlet to the loading nozzle took values from 4 to 4.5 m/s.

It is also found out that the lowest residue of unloaded cement is achieved when the following conditions are met

$$h_{rt} = (0.5 - 2)D, h = (1 - 1.7)D, \tag{6}$$

there h_{rt}, h_a—distances from the bottom of the chamber to the entrance to the discharge pipe and to the level of the nozzles of the aeration device, m.

Let us consider the dependence of the productivity on the change in the excess pressure and the height of the discharge pipe from the bottom of the chamber over the entire range of their variation at fixed values of the height of the aeration device $h_a = 40$, 46, 55, 64, 70 mm (Fig. 2). Analysis of the graphs showed that they are increasing and decreasing in nature.

We get the highest performance for each h_a:

- at $h_a = 40$ mm and pressure $P_{izb} = 1.25$–1.5 atm., $G_y = 5.8$–8 kg/s at $h_{rt} = 34$ mm; $G_y = 6$–7.5 kg/s at $h_{rt} = 55$ mm (Fig. 2a);
- at $h_a = 46$ mm and pressure $P_{izb} = 1.25$–1.5 atm., $G_y = 6.5$–8.6 kg/s at $h_{rt} =$ 34 mm; $G_y = 7$–8.3 kg/s at $h_{rt} = 55$ mm (Fig. 2b);

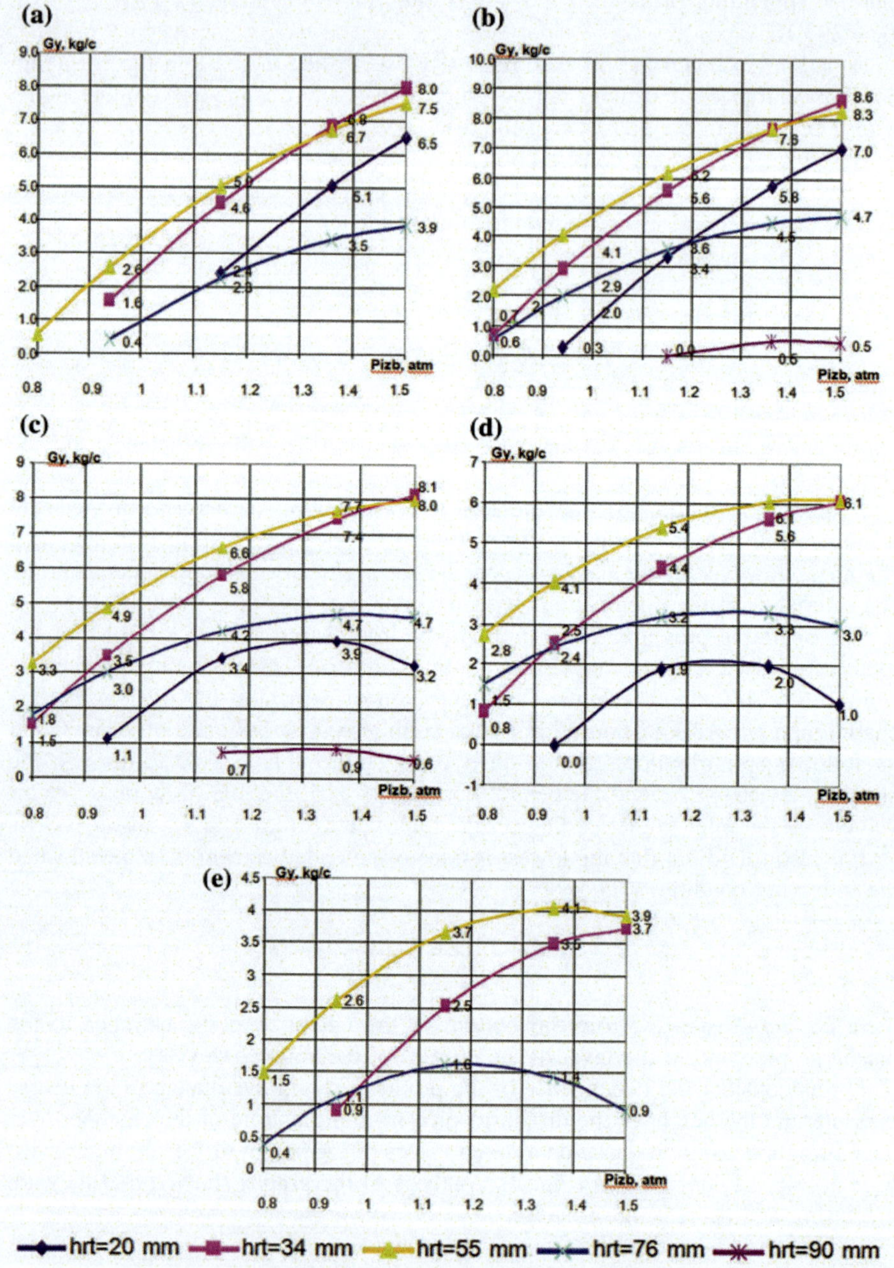

Fig. 2 The dependence of productivity on the change in excess pressure and the height of the discharge pipe when the height of the aeration device varies from the bottom of the chamber h_a: **a** h_a = 40 mm; **b** h_a = 46 mm; **c** h_a = 55 mm; **d** h_a = 64 mm; **e** h_a = 70 mm

Table 1 Maximum performance from changes in excess pressure (P_{izb} = 0.8–1.5 atm.)

Height of the discharge pipe from the bottom of the chamber h_{rt} (mm)	Height of the aeration device from the bottom of the chamber h_a (mm)	G_y for $P_{izb\ min}$ = 0.8 atm.	G_y for $P_{izb\ max}$ = 1.5 atm.
34	40	5.4	6.0
	46	7.2	8.4
	55	7.2	8.0
55	40	6.5	7.4
	46	7.4	8.1
	55	7.5	7.9

- at h_a = 55 mm and pressure P_{izb} = 1.25–1.5 atm., G_y = 6.6–8.1 kg/s at h_{rt} = 34 mm; G_y = 7/2–8 kg/s at h_{rt} = 55 mm (Fig. 2c);
- at h_a = 64 mm and pressure P_{izb} = 1/25–1/5 atm., G_y = 5–6.1 kg/s at h_{rt} = 34 mm; G_y = 5.8–6.1 kg/s at h_{rt} = 55 mm (Fig. 2d);
- at h_a = 70 mm and pressure P_{izb} = 1.25–1.5 atm., G_y = 3.1–3.7 kg/s at h_{rt} = 34 mm; G_y = 3.9 kg/s at h_{rt} = 55 mm (Fig. 2e); at pressure P = 1.36 atm., G_y = 4.1 kg/s.

According to the obtained results, the values of the performance of the pneumatic chamber pump G_y were studied, but at a pressure in the pump chamber close to the pressure at the production of P_{izb} = 1.5 atm.

The analysis of the results summarized in Table 1 showed that for the height of the discharge pipe from the bottom of the chamber h_{rt} = 20–34 mm, and for the values of the height of the aeration device from the bottom of the chamber h_a = 40 mm and h_a = 46–55 mm with a minimum P_{izb} = 0.8 atm. the productivity is G_y = 5.4 kg/s and G_y = 7.2 kg/s, respectively. Also, at the maximum value of P_{izb} = 1.5 atm. the maximum productivity is G_y = 8.4 kg/s at h_{rt} = 34 mm, h_a = 46 mm, and at h_{rt} = 55 mm, h_a = 55 mm, P_{izb} = 0.8 atm. the productivity is G_y = 7.5 kg/s. It should be noted that with an increase in pressure of almost 2 times, the productivity will increase by about 1.2 times.

4 Conclusion

The computer simulation of the movement of the cement-air mixture in the pump chamber showed that the aeration device provides effective aeration of cement in the discharge part of the pump chamber, and due to the geometric arrangement of the nozzles of this device, a homogeneous pseudo-fluidized layer is formed in the pump chamber. Based on this simulation, the ratio for the consumption concentration of cement in the cement-air mixture is obtained, which depends on the design and technological parameters of the pump.

Acknowledgements This work was realized in the framework of the Program of flagship university development on the base of the Belgorod State Technological University named after V.G. Shukhov. The work was realized using equipment of High Technology Center at BSTU named after V.G. Shukhov.

References

1. Bogdanov VS, Fadin YM, Shaptala VV, Gavrilenko AV (2015) Calculation of parameters of a pseudo-fluidized layer in a pneumatic chamber pump. Bull BSTU named after V.G. Shukhov 2:65–67
2. Bogdanov VS, Fadin YM, Shaptala VV, Gavrilenko AV (2016) Characteristics of cement-air mixture flows during cement pneumatic transportation. Bull BSTU named after V.G. Shukhov 2:110–112
3. Bukhmirov VV, Rodionov GA (2013) Mathematical modeling of two-phase flows in installations of pneumatic transport of fine materials. Bull Ivanovo State Univ 5:24–28
4. Zarnitsyna EG, Terekhova ON (2011) Ventilation installations and pneumatic transport: a textbook. AltSTU Publishing House, Barnaul
5. Bogdanov VS, Fadin YM, Shaptala VV, Gavrilenko AV (2015) Patent 153059 Russian Federation, IPC B65G53/40. Pneumatic chamber pump for the transport of bulk materials. applicant and patent holder BSTU named after V.G. Shukhov №2014140286
6. Broks M, Zimmer D (2015) Pneumotransport technologies for the cement industry. Cem Appl 5:1–4
7. Bogdanov VS, Lozovaya SYu, Fadin YuM, Gavrilenko AV (2016) Research of influence of the main parameters on the capability of the pneumatic chamber pump with multijet aeration unit. Int J Pharm Technol 8–4:24669–24680
8. Shaptala VG, Shaptala VV, Gavrilenko AV (2015) Numerical simulation of the movement of a cement-air mixture in a pneumatic chamber pump. Bull BSTU named after V.G. Shukhov 2:159–161
9. Gavrilenko AV (2015) Determination of pressure losses in the installation of pneumatic transport of materials. Bull IrSTU 4:23–26
10. Urban Y (1967) Pneumatic transport. Ed. by L.M. Shvedov. Mechanical engineering, Moscow
11. Shaptala VG, Shaptala VV, Suslov DY (2013) Problems of modeling and calculation of bubbling reactors. Bull BSTU named after V.G. Shukhov 5:189–192

Impact of Fasteners on the Heat Engineering Uniformity of Accessible Roof Area

L. A. Suleymanova, I. S. Ryabchevskiy, and I. N. Ziyatdinova

Abstract In modern megacities, the cost of building land increases every year, and therefore the problem of rational use of the territory available for construction, including the use of space on the roof of an existing building, becomes more and more urgent. Accessible roof areas allow changing almost completely the appearance of the building and using actively the roof space for the construction of terraces, recreation areas, parking lots, and more. As a rule, along with the increased requirements for appearance, requirements for the applied materials and technologies of installation of layers of the roof structure are imposed on the accessible roof area. Taking into account the climatic features of the construction region, special requirements are imposed on the hydro and thermal insulation layer. The authors consider the effective coating structures of the accessible roof areas, and also calculate the reduced heat transfer resistance of the inhomogeneous part of the structure, taking into account linear and point heat engineering uniformities. Based on this calculation, it is revealed that point mechanical fasteners, fixing in addition to the insulation layer of waterproofing to the supporting base, reduce significantly the heat engineering uniformity of the thermal insulation layer of the accessible roof area, while increasing the material consumption, labor intensity of installation and reducing the service reliability and durability of the roof.

Keywords Accessible roof area · Thermal insulation · Waterproofing · Fastening · Installation

1 Introduction

Currently, in industrial and civil construction, flat roofs with a slight slope of 0.5… 3% are becoming increasingly popular. This is due to the fact that the absence of a slope allows increasing the functionality of the roof by placing technical and utility rooms, recreation areas and landscaping, parking areas, etc. on it [1].

L. A. Suleymanova (✉) · I. S. Ryabchevskiy · I. N. Ziyatdinova
Belgorod State Technological University named after V.G. Shukhov, Belgorod, Russia

S. V. Klyuev (ed.), *Digital Technologies in Construction Engineering*,
Lecture Notes in Civil Engineering 173, https://doi.org/10.1007/978-3-030-81289-8_44

The structure of the accessible roof area is multilayered (Fig. 1). Alternating materials provide protection of the roof from loss of thermal energy, water penetration on the underlying layers of the structure, accumulation of condensate [2, 3].

In the version of the traditional multi-layer roof (Fig. 1a) of the accessible roof area, first of all, a vapor barrier is laid, trapping household fumes, insulation, which is protected from the outside by a waterproofing coating, after which the insulating

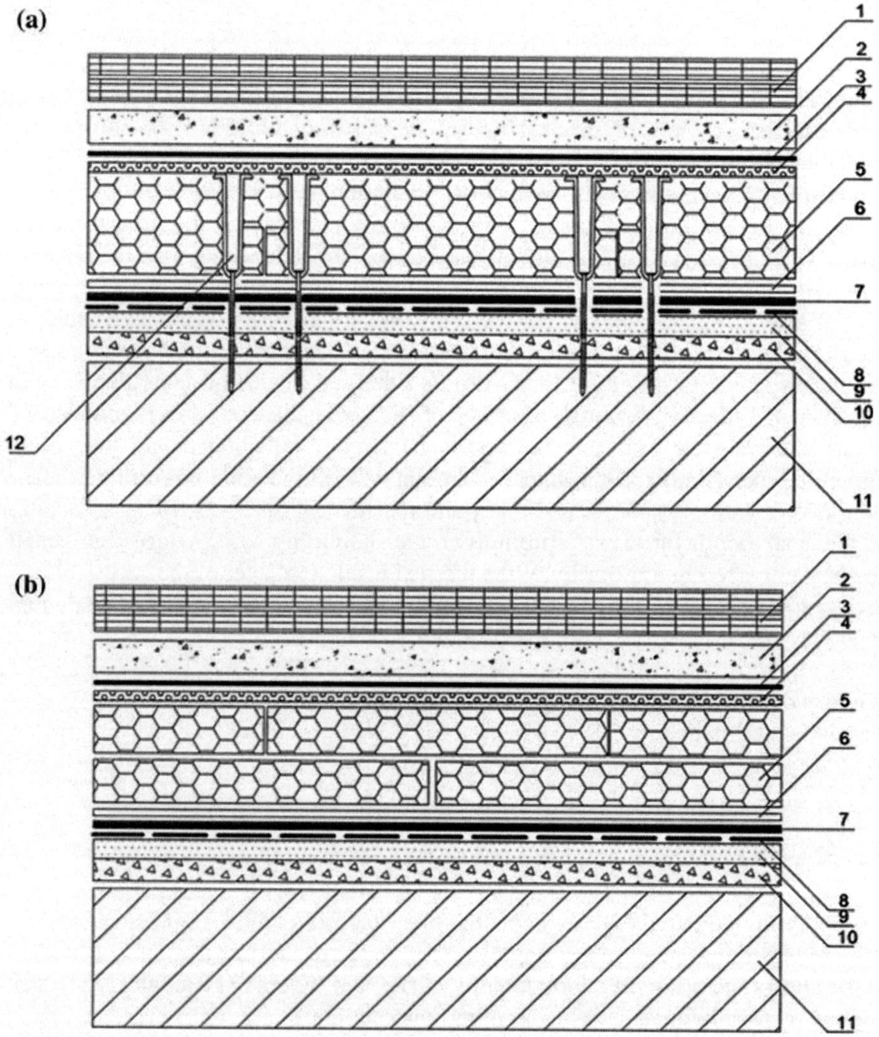

Fig. 1 Roof system designs: **a** roof with mechanical fastening; **b** inversion ballast roof: 1—paving slab; 2—cement-sand reinforced covering; 3—separation layer; 4—drainage layer; 5—thermal insulation layer; 6—protective and drainage layer; 7—waterproofing in two layers; 8—bitumen primer; 9—reinforced cement-sand covering; 10—roof sloping; 11—concrete slab; 12—fired pin

part of the structure is fixed to the supporting base by a point fixing system made in the form of metal or polymer fired pins [4]. The external waterproofing layer of a standard roof, made of polymer, bituminous or bituminous-polymer solid roll, exposed to atmospheric influences, reduces its service reliability and the entire roof structure as a whole.

In this connection, the inversion ballast scheme for laying roof layers is gaining popularity (Fig. 1b), the feature of which is the arrangement of a waterproofing layer under the insulation. In this case, the insulation part of the structure is fixed to the bearing base by the load from the weight of the overlying layers, and the thermal insulation arrangement in two layers provides the shear rigidity of this layer [5–7].

There is no vapor barrier in the inversion structure, as household vapors are captured by a layer of thermal insulation made in the form of slabs made of extruded polystyrene foam. Due to the use of this technological solution, the cost of installing a ballast accessible roof area is further reduced, speeding up the construction process [8, 9]. In addition, the advantages of this type of roof include increased wear resistance, especially in regions with expressed aggressive climatic conditions in relation to building materials, reduced material consumption and labor intensity due to the reduction of materials used and their laying processes, and environmental safety due to the preservation of natural balance and the absence of the release of volatile toxins when heating the roof.

As the point fixing system in the form of fired pins used in the traditional accessible roof area is heat-conducting inclusions when fixing the insulation layers to the load-bearing base, the study of the effect of fasteners on the heat engineering uniformity of the insulation will allow justifying the choice of an inversion ballast roof as the most energy-efficient.

2 Methods and Materials

To determine the impact of the point fixing system of the thermal insulation layer of the accessible roof area, the authors calculated the reduced heat transfer resistance of the inhomogeneous part of the structure, taking into account linear and point heat engineering uniformities [10, 11].

For the calculation, we consider a fragment of the thermal insulation layer of an inversion ballast roof with a thickness of 100 mm, attached to the supporting base by a load from the overlying layers of the structure (Fig. 2a), consisting of two layers of thermal insulation in the form of slabs of extruded polystyrene foam with a thickness of 50 mm each, arranged in a staggered order, and a fragment of the thermal insulation layer of the roof, attached to the supporting base by a point fastening system (Fig. 2b), consisting of a layer of thermal insulation in the form of a slab of extruded polystyrene foam with a thickness of 100 mm, a metal dowel with a height of 70 mm and a thickness of 10 mm, and a metal fired pin, fixed at the

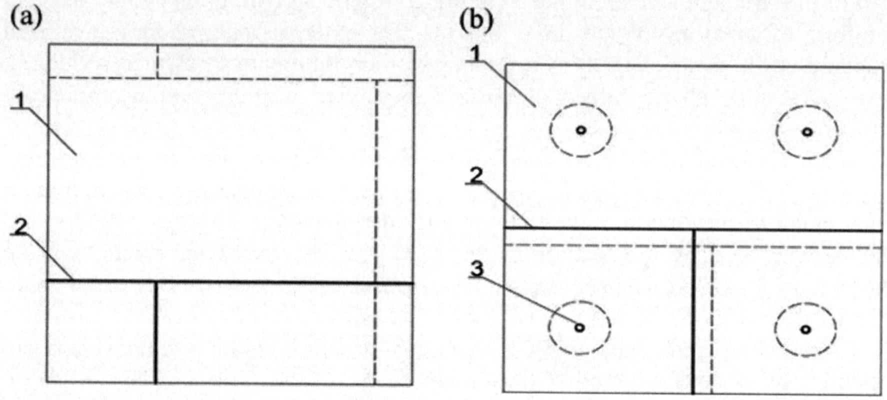

Fig. 2 General view of the fragments of the thermal insulation layer made of extruded polystyrene foam for determining the reduced heat transfer resistance: **a** the thermal insulation layer of the inversion ballast roof; **b** the thermal insulation layer of the roof with a point fixing system; 1—the section of the extruded polystyrene layer 2—the section of the binding of the seams of the thermal insulation slabs; 3—the section of the fastening of the thermal insulation layer to the load-bearing base

corners of the insulation plate at a distance of 300 mm from the edges of the plate. The area of the fragments is assumed to be 1 m^2.

3 Results and Discussion

The fragments under consideration are represented by three types of a homogeneous part of the structure:

- the section of the thermal insulation layer made of extruded polystyrene foam (Fig. 2, pos. 1);
- the section of binding of seams of thermal insulation slabs (Fig. 2, pos. 2);
- the section of fixing the thermal insulation layer to the bearing base with a point fixing system (Fig. 2, pos. 3);

The reduced heat transfer resistance of the structure of the calculated fragment is determined by the formula (1) [9, 10]:

$$R_o^r = \sum A_i / \left(\sum A_i / R_{o,i} + \sum L_j \psi_j \right) \quad (1)$$

where A_i—the area of the structure of the *i*th type in the fragment under consideration, m^2;

L_j—the length of all joints of the *j*th type in the fragment under consideration, m;

$R_{o,i}$—heat transfer resistance of the homogeneous part of the structure of the *i*th type, (m^2·°C)/Вт; ψ_j—additional specific linear heat loss through the joint of the *j*th type, W/(m·°C).

To determine the additional specific linear heat losses through the joints, the authors adopted values based on the calculation of the two-dimensional temperature field of the joint of the structure [12]. The temperature field of the considered area is shown in Fig. 3.

The reduced heat transfer resistance is determined by a fragment of the heat-insulating layer of extruded polystyrene foam attached to the bearing base by a linear fastening system and a fragment of the heat-insulating layer of extruded polystyrene foam attached to the bearing base by a point fastening system. The calculation data is presented in Table 1.

From the data presented in the table, it follows that the thermal uniformity of the thermal insulation layer of the inversion ballast accessible roof area is 4.6% higher than that of the roof with the fixing of the insulation layers with a point fixing system due to the absence of through heat-conducting inclusions.

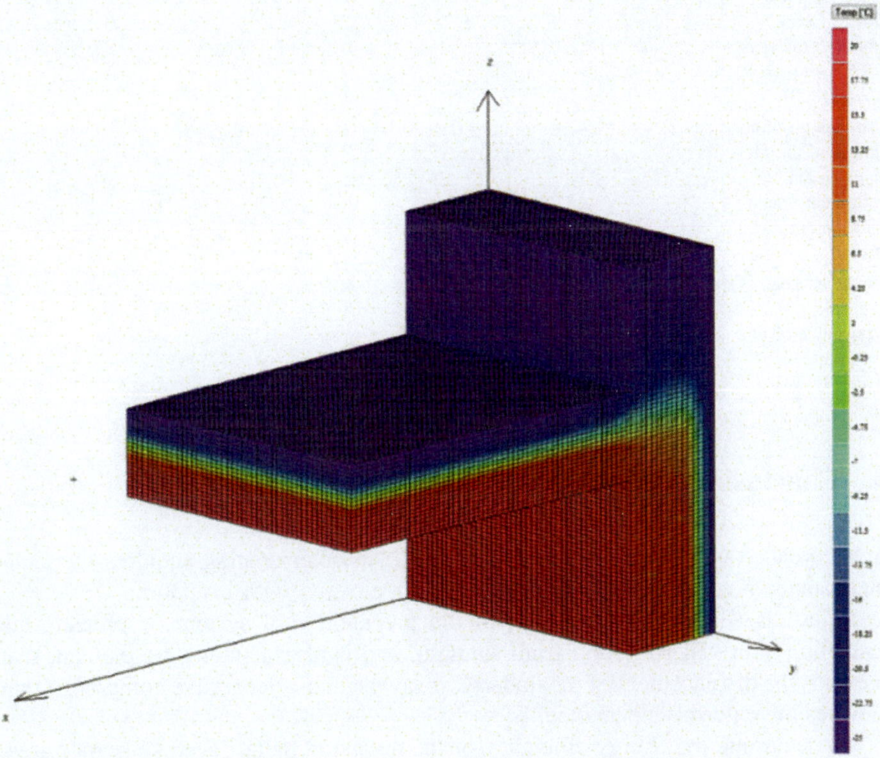

Fig. 3 Temperature fields of the junction of the inversion ballast roof to the parapet

Table 1 Heat engineering indicators for determining the reduced heat transfer resistance of fragments of a thermal insulation layer

Indicators	Inversion ballast roof		Roof with point fixing system			
	Binding of seams of thermal insulation slabs	Section of thermal insulation layer	Binding of seams of thermal insulation slabs	Section of thermal insulation layer	Section of fixing the thermal insulation layer with a fired pin	Section of fixing the thermal insulation layer with a fired pin
Fragment thickness (m)	0.1		0.1			
Fragment section area A_i (m^2)	0.003	0.997	0.0030	0.99669	0.00008	0.00024
Heat transfer resistance of the homogeneous part of the structure $R_{o,i}$ ((m^2·°C)/W)	3.52339	2.79	3.52339	2.79	4.13838	1.38539
Additional specific linear heat loss through the joint of the *j*th type ψ_j (W/(m·°C))	0.02		0.02543		0.0154	
The length of all joints of the *j*th type in the fragment under consideration L_j (m)	3.0		3.0		0.1256	
Reduced heat transfer resistance of the structure $R^r{}_o$ ((m^2·°C)/W)	2.389		2.284			

4 Conclusion

In accordance with the inversion technology, the outer coating acquires the technical status of a waterproofing layer, which is covered with insulation.

Ballast is a mandatory element of the inversion roof system. It protects the insulation with waterproofing from sunlight, and is also a stabilizing element that prevents the displacement of the underlying layers, and a decorative component that elevates the exterior appearance.

To determine the energy efficiency of the inversion ballast accessible roof area, the authors calculated the reduced heat transfer resistance of the inhomogeneous part of the structure, taking into account the linear and point thermal uniformities of the thermal insulation layer from the extruded layer fixed by the fastening system and the ballast, and the conclusions were established.

The use of the ballast system allows, due to the absence of mechanical fasteners, reducing the material consumption and labor intensity of the installation of the thermal insulation layer of the accessible roof area, increasing the durability of the structure, while ensuring high thermal uniformity.

Acknowledgements The study is implemented in the framework of the Flagship University Development Program at Belgorod State Technological University named after V.G. Shukhov, using the equipment of High Technology Center at BSTU named after V.G. Shukhov.

References

1. Yang W (2015) Sustain Cities Soc 16:1–12
2. Tam VW, Wang J, Le KN (2016) Build Environ 110:46–54
3. Brenneisen S, Gedge D (2013) Green roof green roof planning in urban areas green roof in urban areas. Sustainable built environments. Springer, New York
4. Mechanical Fastening, Joining, and Assembly (2015) Second Edition. CRC Press, pp 199–230 (9 feb)
5. Francke B, Geryło R (2018) Inverted roof insulation kits and their durability. In: MATEC web of conferences. EDP sciences 163. 08005
6. Egorov AN, Tugushev AA (2021) System of criteria and analysis appraisals for green inverted roof. Contemporary problems of architecture and construction. CRC Press, pp 377–380
7. Misar I, Novotný M (2016) Defects and behaviour of inverted flat roof from the point of building physics. In: MATEC web of conferences. EDP sciences 93. 02002
8. Pedrosa GA, Río MM, Fonseca VC (2014) Durability analysis of PVC-P membrane in inverted flat roof. Construction and building research. Springer, Netherlands, pp 515–521
9. Pedrosa A, Río M, Fonseca C (2014) Interaction between plasticized polyvinyl chloride waterproofing membrane and extruded polystyrene board, in the inverted flat roof. Materiales de Construcción Editorial CSIC 64(316):037
10. Suleymanova LA, Pogorelova IA, Marushko MV, Ryabchevsky IS (2020) Energy-efficient double-row masonry of the exterior walls in the buildings made of cellular concrete blocks. In: IOP conf. series: materials science and engineering 913, 022044
11. Suleymanova LA, Pogorelova IA, Suleymanov KA (2020) Energy efficiency improvement of aerated concrete block wall fences. In: IOP conf. series: materials science and engineering 945, 012006
12. Almeida F, Beyrichen H, Dodamani N, Caps R, Müller A, Oberhoffer R (2020) Thermal conductivity analysis of a new sub-micron sized polystyrene foam. J Cell Plast. 0021955X2094310 (SAGE Publications)

Determination of the Residual Service Life of the Operated Buildings and Structures in Terms of the Margin of Resistance to Technogenic Impacts

M. A. Aleynikova, N. Yu. Soytu, N. A. Maslennikov, and A. V. Novozhilova

Abstract The indicators of safe operation of buildings and structures are the technical condition, which is determined on the basis of the survey results, and the residual service life. Determination of the residual service life of building structures is carried out in order to establish the timing of their further reliable operation and to justify unscheduled repair work in accordance with the revealed patterns of change in the controlled parameters of the technical state of structures and building elements obtained from the results of calculations, which are the basis for making a decision on further operation, reconstruction or demolition of the house. The calculation of the residual life of buildings is of particular importance in the context of the influence of various kinds of technogenic impacts. In this regard, the article deals with the specifics of determining the residual service life of operated buildings and structures in terms of the margin of resistance to technogenic impacts. In the course of the study, detailed attention was paid to the calculation of reliability index, as well as the assessment of the risk of influence of various technogenic factors. Also, the article touches upon the methodology for calculating the stability margins of building structures of operated buildings exposed to the influence of technogenic impacts of a natural and technological nature. The procedure for calculating the coefficients of structures resistance to technogenic impacts according to certain parameters is presented on the example of earthquakes.

Keywords Residual Service Life · Building · Operation · Technogenic Impacts · Loads · Stability · Reliability

1 Introduction

The task of trouble-free operation of buildings and structures, which provides for compliance with safety requirements for building structures, is to calculate the residual service life based on the results obtained in the process of technical

M. A. Aleynikova (✉) · N. Yu. Soytu · N. A. Maslennikov · A. V. Novozhilova
St. Petersburg State University of Architecture and Civil Engineering, St. Petersburg, Russia

S. V. Klyuev (ed.), *Digital Technologies in Construction Engineering*,
Lecture Notes in Civil Engineering 173, https://doi.org/10.1007/978-3-030-81289-8_45

inspection and assessment of the current state of objects. Safety requirements, that is, requirements for the structural safety of buildings, mean compliance with an appropriate level of reliability of all loads and impacts during the entire period of their operation [1].

The modern concept of the functioning of complex engineering systems provides for the transition from the concept of "absolute safety" to the concept of "acceptable risk" [2]. This, in turn, necessitates the analysis of the residual service life of building objects, as well as identifying the risks of their operation in order to develop a management system for them, that is, to reduce the risks to an acceptable level.

The calculation of the residual service life is carried out by specifying the existing loads, the design scheme, the identified defects and damages (DD), taking into account their development over time. However, there is always the possibility that during the survey some part of the DD will not be found, and their number, distribution and location will be unknown. In addition, even for the predicted DD, their spontaneous uncontrolled development is possible in the process of unplanned subjective change in the design parameters of the functioning of "unit—structure—building—load—material—design scheme" system under the influence of various kinds of technogenic factors.

In this regard, in the process of determining the residual service life, it is necessary to take into account how the resistance of a structure to natural and technological influences changes over time, considering the changes in loads, their intensity and the technical state of structures and foundations.

In light of the above, according to the author, the following situations should be considered:

- natural design basis technogenic impact that can lead to a partial or complete loss of the performance of a building, structure or foundation;
- design basis accident of a technological nature that can lead to a partial or complete loss of the operational qualities of a building, structure or foundation;
- the maximum calculated natural technogenic action, which can cause global destruction of the base, building, structure or part thereof or a threat to people.

In modern methods and regulatory documents, the main approaches to strengthening and restoring individual elements of building objects are systematized with various options for design solutions. In addition, in these methods and documents models for calculating the residual bearing capacity of structures and design solutions are given and organizational and technological requirements and recommendations for their implementation are indicated.

However, these documents do not contain a scientifically grounded, reliable methodology for assessing, predicting and regulating the technical state of buildings and structures in specific conditions of operational impacts, taking into account the influence of various kinds of technological factors.

Thus, these circumstances predetermine the choice of the topic of this article, and also confirm its theoretical and practical significance.

Recently in Russia a lot of attention has been paid to the problem of determining the residual service life of buildings and structures. In the works of such domestic authors as: Zolina T.V., Shutova M.N., Evtushenko S.I., Dudorov V.E., Kilyazova E.A. the problems of conceptual assurance of reliability and determination of the residual service life of building structures were systematized.

Such foreign scientists as Dhawan, Sushil Kumar; Bindal, Abhinav; Bhalla, Suresh; Bhattacharjee, Bishwajit have developed a practical methodology for determining the risks of operating construction sites using the example of reinforced concrete structures.

However, despite the available works, it should be noted that most of the developments concern either directly substantiating the relevance of the survey as such, or the problems of calculating the residual service life of individual building objects. Also, the works often focus on the economic aspects of assessing possible damage.

In addition, it is necessary to emphasize that the methods for determining the technical condition and residual service life of structures subject to various influences have not been sufficiently developed, which requires additional, more in-depth studies.

2 Purpose of the Study

The purpose of the study is to consider the specifics of determining the residual service life of operated buildings and structures in terms of the margin of resistance to technogenic impacts.

3 Materials and Methods

The article uses modern methods and techniques of visual and instrumental research in field surveys of buildings and structures; analytical techniques for studying the stress–strain state of structures using the theory of elasticity and plasticity; numerical finite element method for modeling the stress–strain state; statistical methods for processing identified defects and damage; comparative analysis of the results of analytical calculations and mathematical modeling.

For the experiments, mathematical statistics and probability theory were used, thanks to which the results of field surveys of building structures in operation were processed, and their reliability indicators were evaluated, the technical condition of construction objects was diagnosed, supplemented by an analysis of their physical and mechanical characteristics.

Monitoring of building structures was carried out on the basis of the approved program with the help of technical means of non-destructive testing and with a certain frequency. The frequency of work was determined taking into account the classes of responsibility of building structures. At the same time, the minimum frequency of observations was set in accordance with the design and regulatory documentation or special requirements for a specific object. The result of monitoring is information about the state of the object and the territory of development, which was recorded in the interim and final reports. Reports are drawn up on the basis of registration, accumulation and primary analysis of information from sensitive elements that were installed on building structures subject to control.

4 Results and Discussion

So, according to the author, as a basis for determining the residual service life of the operated buildings and structures in terms of the margin of resistance to technogenic impacts it is advisable to use a reliability index, a detailed description of which is presented in the Eurocodes and harmonized national standards [3–5]:

$$\beta = \frac{\bar{s}}{\sigma(S)} = \frac{\bar{R} - \bar{F}}{\sqrt{\sigma^2(R) + \sigma^2(F)}}$$

where $\bar{R}$ is the average value of the bearing capacity, that is, the generalized strength of the element (structure);

$\bar{F}$ is an average value of the effect of loading (actions) on an element (structure);

$\bar{s}$ is safety margin for all distribution laws of R and F;

σ^2 (*R*) *u* σ^2 (*F*) is variance of bearing capacity and load, respectively.

To assess the risk *Y* of the influence of technogenic impacts on the residual service life of operated buildings, it seems appropriate to use the functional *H*, which connects the probability *P* of an adverse event and the mathematical expectation of damage *U* from this adverse event:

$$Y = H_y\{U, P\} = \sum_i [H_{y_i}(U_i, P_i)] = \int C(U)P(U)dU = \int C(P)U(P)dP$$

where *i* is a type of adverse event;

C is a weight function taking into account the mutual influence of risks.

The minimum permissible values of the controlled parameters are established according to the results of calculations of building structures by the well-known methods of structural mechanics and resistance of materials to determine the bearing capacity and compare it with the maximum acting forces:

$$F_{cr}[x_1(t),\ x_2(t), \ldots x_m(t)] > F$$

where $F_{cr}[x_1(t),\ x_2(t), \ldots x_m(t)]$ is a function of the bearing capacity of the elements.

The parameters $x_1(t),\ x_2(t), \ldots x_m(t)$ are the dimensions of the cross-section and the strength of materials as a function of time. The calculation of the value of the acting force F for statically determined structures is not difficult from a conceptual point of view. For statically undetectable structures, the value of F is calculated from the results of mathematical modeling of the technical condition established during the survey [6].

In general, for the quantitative and qualitative analysis of risks and the calculation of the residual service life according to the presented formulas, based on studies of complex dynamic nonlinear hazardous processes (the occurrence of damage, failures, destruction and accidents as a result of technogenic impact), the construction of physical and mathematical models of objects is carried out. In the models, differentiation of threats constituting separate types of hazard (local risks) and threats of a complex hazard should be carried out according to the corresponding combinations and types of hazard (global risk). These models and scenarios of the occurrence and development of adverse events take into account both specified and postulated technogenic hazardous processes that develop in time t. This approach uses risk timelines $R(t)$. The safety condition in this case can be presented in the following form [7]:

$$\{R,\ R_{F,t}\} \leq \{[R],\ [R_{F,t}]\}$$

where $R_{F,t}$ is risk value for a given point F and time t;

$[R],\ [R_{F,t}]$ is maximum allowable risk value.

We will carry out the initial calculation of the reliability index for the truss structures built in 1966. The truss structures were surveyed in 1988, 2005 and 2017. At the time of the last survey, the building had been in operation for 51 years with a design service life of $T_0 = 60$ years. Table 1 shows the calculation results.

Let us consider the calculation of the residual service life and the margins of truss structure resistance to technogenic impacts according to certain parameters of their intensity (F), using the example of an earthquake.

First of all, it is necessary to determine the effective peak value of technogenic impact (load or impact, F_c) for a given probability of non-exceedance (P_H) and residual service life (T). If a sufficient probability of non-exceedance (P_H) of the effective peak value of the load or effect (F) is determined for a given class of consequences for the structure, then this non-exceedance must be maintained for the design and any residual or reassigned service life of the structure [8]. In other words, P_H *is const*. Then, with a change in the residual service life (T) of the structure the required period of excess of the action (influence) parameter (Tc) changes according to the formula:

Table 1 Results of calculating the truss structure reliability index

Element	Load	$\Sigma_{av} = F$ (MPa)	Variance	Standard deviation	Coefficient of variation $V\bar{F}$	$\bar{R}$	Coefficient of variation $V\bar{R}$	$\xi = \bar{R}/\bar{F}$	β	Reliability probability P_f	Reliability probability P_S
Upper belt	1	118.20	43.60	6.60	0.06	205	0.024	1.73	10.54	≈0	≈1
	2	127.12	52.50	7.25	0.057	215	0.024	1.69	9.88	≈0	≈1
	3	137.54	236.39	15.37	0.11	215	0.024	1.56	4.78	≈0	≈1
	4	151.59	964.31	31.05	0.20	215	0.024	1.42	2.01	0.0222	0.9778
	5	172.19	3169.37	56.30	0.327	215	0.024	1.25	0.76	0.2265	0.7735
Bottom belt	1	162.91	67.26	8.20	0.050	205	0.025	1.26	4.35	≈0	≈1
	2	175.22	92.24	9.60	0.0548	223	0.025	1.27	4.30	≈0	≈1
	3	182.15	176.63	13.29	0.073	223	0.025	1.22	2.83	0.0023	0.9977
	4	186.52	241.04	15.53	0.083	223	0.025	1.20	2.21	0.0135	0.9865
	5	193.02	381.57	19.53	0.101	223	0.025	1.16	1.48	0.0706	0.9294

$$T_c = 1/\left[1 - (P_H)^{1/T}\right]$$

For these periods, an excess of an action parameter is defined as the effective peak value of that action. Thus, the value of the parameter (*Fc*) for the residual service life (*T*) is calculated by the formula:

$$F_{C,T} = F_{C \cdot 50} \cdot K_E$$

where $F_{C \cdot 50}$ is a value of the parameter of technogenic impact (load or impact) for the service life of 50 years;

$F_{C,T}$ is a value of the parameter of technogenic impact (load or impact) for the residual service life of T years;

K_E is coefficient of the effective peak value of the technogenic impact parameter in relative units.

Table 2 shows an example of a change in the coefficient of the effective peak value of the parameter of technogenic impact (K_E) in relative units for a given probability of non-exceedance (P_H = 0.9 or 90%) in case of an earthquake that has an intensity of 9; 8; 7 points according to the MSK-64 scale.

Thus, the residual service life of the truss structure ΔT in terms of resistance to technogenic impact can be determined by the formula:

$$\Delta T = T_e \times |(FS_T - FS_K)/|(FS_O - FS_T)$$

where T_e is time from the start of operation of the structure (or the start of construction) to the moment of determining the safety factor for technogenic impact according to a certain parameter FS_T;

FS_T T is a value of the safety factor for technogenic action according to a certain parameter for the time T_e;

FS_K is the minimum or maximum permissible value of the safety factor for technogenic action according to a certain parameter.

FS_O O is the design value of the safety factor for technogenic action according to a certain parameter.

Table 2 K_E coefficient value for an earthquake with a given probability of non-exceedance (P_H = 0.9 or 90%)

Residual service life of the structure (T), years	Period of exceeding the parameter or average repeatability (Tc), years	Intensity 9 points	Intensity 8 points	Intensity 7 points
1	10	0.25	0.185	0.18
5	50	0.45	0.35	0.25
10	100	0.625	0.5	0.5
50	500	1	1	1

Table 3 Parameters for determining the residual service life of the structure

Structures consequence class	Construction criticality rating	Safety factor of technogenic action FS, %				
		Technical conditions				
		1	2	3	4	5
CC3	A	100–95	95–90	90–85	–	85 and less
	Б	100–95	95–90	90–85	–	80 and less
	B	100–95	95–85	85–75	–	75 and less
CC2	A	100–95	95–90	90–80	–	80 and less
	Б	100–95	95–85	90–85	–	75 and less
	B	100–95	95–85	85–85	75–50	50 and less
CC1	A	100–95	95–90	90–80	80–75	75 and less
	Б	100–95	95–90	95–90	80–70	70 and less
	B	100–95	95–85	95–85	75–25	25 and less

If the considered technogenic impact is described by several $(1 \leq i < n)$ controlled parameters, then the smallest of the calculated values of residual service life T_e is accepted.

The technical condition of the structure in accordance with the margin of resistance to technogenic actions is determined according to Table 3.

5 Conclusion

During the construction and operation of buildings and structures, defects and damage can form and accumulate in structural elements and their joints. In addition, aging and corrosion of materials, rheological changes in the base can occur. This leads to a decrease in the resistance of structures to loads, the intensity of which increases under the influence of various technogenic impacts.

Taking into account the above, the article reveals the specifics of determining the residual service life of operated buildings and structures in terms of the margin of resistance to technogenic impacts. These specifics must be taken into account in verification calculations for the resistance of building structures to technogenic impacts.

The methodological approach outlined in the article can be used in practice to determine the parameters of the residual service life of industrial structures and buildings that are under the influence of various technogenic factors. The application of this approach will improve the safety of operation of production facilities operated in difficult engineering and geological conditions.

Prospects for further research are to carry out a statistical assessment of the magnitude and nature of damage; to determine the hazard category of the main types of defects and damages; to calculate the maximum permissible values of damage depending on the category of their hazard and the state of structures. Also of considerable interest are methods for calculating the residual and reassigned service life of construction objects with accumulated damage based on the results of calculating the margins of resistance to technogenic loads.

References

1. Shutova MN, Evtushenko SI (2019) Application of an integrated approach to determining the residual service life of an industrial building with deformed trusses. Constr Archit 7(3):30–35
2. Dhawan SK et al (2019) Expected residual service life of reinforced concrete structures from current strength considerations. Adv Struct Eng 22(7):1631–1643
3. ISO 13822:2010. Bases for design of structures—assessment of existing structures. ISO 44 (2010)
4. Eurocode 0: ENV 1990:2002+A1. Basic of structural design. CEN, Brussels 116 (2002)
5. Eurocode 3: ENV 1993-1-1, Part 1.1, Design of steel structures, CEN, Brussels (2005)
6. Ortega NF (2016) Assessment of residual life of concrete structures affected by reinforcement corrosion. HBRC J 12(2):114–122
7. Kelasev NG, Kodysh EN (2020) Determination of the service life of buildings and structures. Ind Civil Constr 2:12–17
8. Rad R (2019) Zahra: Probabilistic seismic collapse and residual drift assessment of smart buildings equipped with shape memory alloy connections. Eng Struct 197:18–23

Application of Phosphate Slag for the Production of Heavy Concrete

D. S. Dyussembinov, R. E. Lukpanov, D. V. Tsygulyov, and S. B. Yenkebayev

Abstract The article presents the results of research on the effect of phosphorus slag in combination with post-alcoholic bordeaux and slaked lime on the quality of heavy concrete. The complex application of these components contributes to significant improvement of strength, frost resistance, and water absorption. To determine the optimal composition of the additives used 5, 10, 15, 20% phosphorus slag in relation to the cement binder was taken. The concentration of post-alcohol bard (alcoholic bard) and caustic lime was taken in proportion to the concentration of phosphorus slag. Studies have shown that the use of post-alcoholic bard and slaked lime in relation to the phosphorus slag allows getting the best quality indicators of concrete at all stages of its preparation. Alcoholic bard adds plasticity to the concrete mixture, and slaked lime due to its high alkalinity normalizes the slightly acidic environment after alcoholic bard and increases the activity of phosphorus slag. Thus, obtained complex additive from industrial waste and slaked lime will not only increase the physical and mechanical properties of concrete but also reduce its cost, as the used filler, phosphorus slag in combination with the alcohol bard and slaked lime, can replace part of the cement binder without loss of strength characteristics. However, it is necessary to consider the optimal amount of all components, as all the characteristics can be preserved only at the optimum ratio with the cement binder.

Keywords Phosphorus slag · Concrete · Cement · Post-Alcoholic bard · Slaked lime · Strength · Frost resistance

---

D. S. Dyussembinov · R. E. Lukpanov · D. V. Tsygulyov · S. B. Yenkebayev (✉)
Eurasian National University of L. N. Gumilyov, Nur-Sultan, Kazakhstan
e-mail: yenkebayev-serik@mail.ru

S. V. Klyuev (ed.), *Digital Technologies in Construction Engineering*,
Lecture Notes in Civil Engineering 173, https://doi.org/10.1007/978-3-030-81289-8_46

1 Introduction

The use of industrial waste in the production of building materials is widespread, and new technologies allow us to expand the range of waste applications. With the development of science, many of the simplest technologies of waste application in the production of building materials have received a new leap in technological development, namely, the effectiveness of the integrated use of waste has been substantiated [1].

The qualitative complex use of components is often accompanied by a synergistic effect. For the first time this term in the production of building materials in the post-Soviet space was conducted by Professor V.I. Solovev [2]. His research consisted in the integrated use of food industry waste and man-made waste. The results showed the effectiveness of the optimal modes and ratios of all components [3]. This effect allows all the components used in the additive to strengthen the qualitative indicators of each other.

In practice, it is widely known the use of hydraulic ash, which in a certain amount and with the use of plasticizing additives can improve some qualitative indicators of concrete, as well as reduce its cost [4]. An important fact is the use of microsilica in combination with plasticizing additives of the different genesis of polycarboxylates and naphthalene formaldehyde [5]. These additives increase the density of the concrete mixture, as well as its mobility, but using them in combination with microsilica it is possible to obtain high-strength concrete using ordinary M400 cement [6].

The use of industrial waste and recycled materials in the production of building materials is an important aspect of waste management all over the world [7]. The use of complex additives from industrial waste in the production of concrete is not new [8]. However, new research in the field of complex application of industrial waste is relevant because it solves not only the problems of the construction industry but also the problems of ecology [9]. We have considered the issues of complex application of industrial wastes of the phosphorus slag, and after alcoholic bard, according to the developed MPE and the results of waste research on the specific effective activity of natural radionuclides, the wastes do not exceed the permissible norms [10].

The purpose of the article research of complex application of phosphorus slag, after alcohol bard and slaked lime on qualitative indicators of concrete. The results made it possible to determine the optimum ratio of all components added to the concrete mixture to obtain the best qualitative characteristics.

2 Materials and Methods of Research

The production process of the additive must be made in the following sequence [11]: alkali is combined with post-alcohol bard at a temperature, not below 60 °C; after the resulting solution is passed through the rotary pulsation machine (RPM dispersant) to accelerate the chemical reaction. The resulting emulsion is readily soluble in water, thus providing a set of properties of both hydrophobization and plasticization. The technological process makes it possible to obtain a homogeneous emulsion easily soluble in water without oil staining, which cannot be achieved by mixing the same components by mechanical mixing. Full solubility of the additive in water makes it possible to react effectively with the cement binder, without the formation of micropores with a reduced water-cement ratio and thus improve the physical–mechanical and construction-technical properties of concrete. The technological scheme of additive production is shown in Fig. 1.

To study the effect of the complex use of industrial waste on the quality of concrete, the following compositions were adopted:

- Type 1—Control sample;
- Type 2—Sample with 5% finely ground phosphate slag of the cement weight;
- Type 3—Sample with 10% finely ground phosphate slag of the cement weight;
- Type 4—Sample with 15% finely ground phosphate slag of the cement weight;
- Type 5—Sample with 20% finely ground phosphate slag of the cement weight.

The amount of post-alcoholic bard was up to 30% of the weight of phosphorus slag and slaked lime up to 2%. Table 1 shows the compositions for the study of concrete for physical and mechanical properties. The mineralogical composition of phosphate slag is presented in Table 2.

The concrete samples for the strength were tested on an automatic press machine —Controls (Pilot) 500 kN. The strength of one series was determined as the arithmetic average of the tested specimens. Eighteen specimens were tested, nine from each series. In order to increase the contact area when evaluating water absorption, large specimens were used: height h = 80 mm, width b = 350 mm, and

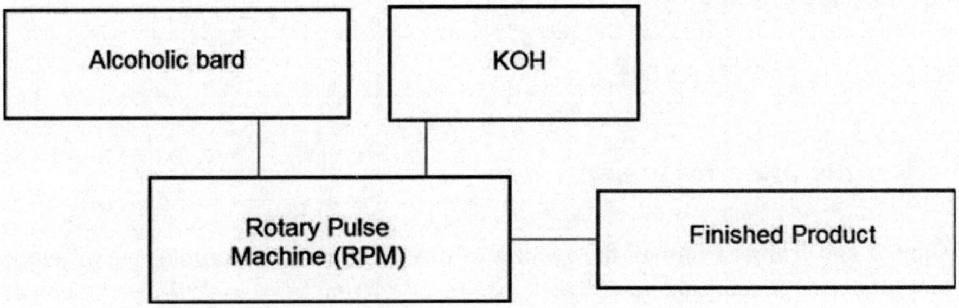

Fig. 1 Production technology of concrete additive

Table 1 Composition for the study

Type of sample	Cement, kg	Sand, kg	Crushed stone, kg	Water, l	Complex additive, kg	Density, kg/m^3
Type 1	300	800	1150	150	–	2250
Type 2	280	800	1150	112	19.8	2310
Type 3	260	800	1150	91	39.6	2375
Type 4	240	800	1150	72	59.4	2398
Type 5	220	800	1150	75	79.2	2359

Note the density was determined by the mass of finished samples

Table 2 Composition of phosphorus slag

Item	Component content %										
	SiO_2	CO_2	Fe_2O_3	Al_2O_3	CaO	MgO	K_2O	Na_2O	P_2O_5	SO_3	F
Quantity	41	0.3	5.6	1.3	36	8	1	0.8	1	1	4

length l = 550 mm. Frost resistance was determined on 100 × 100 × 100 cube specimens which reached 28 days (12 samples of each series).

Freezing time was 4 h at minus 18 °C, and thawing time was the same 4 h at plus 18 °C, at a humidity of 95%. The main criterion for the end of frost-resistance tests was visual changes in the surface of the sample (cracks, chips) as well as the loss of weight.

To determine the water absorption of concrete, the samples were weighed on a calibrated scale with an accuracy of 0.1% and then placed in a container with water. The water level in the tank was 60 mm above the water level of the samples and the water temperature was 21 °C. The samples were weighed every 24 h until constant weight.

Water absorption by mass was calculated using the equation:

$$W_M = \frac{m_w - m_d}{m_d} * 100, \quad (1)$$

where

W_M—water absorption by weight, g;
m_w—the weight of saturated sample g;
m_d—the weight of dry sample g.

3 Results and Discussion

Figure 2 and Table 3 shows the results of strength tests on samples of different concentrations. According to the test results, the strength of type 1 specimens is lower than type 2, 3, 4, 5 specimens in which industrial waste additives were used. However, it should be noted that samples with an additive of 20% begin to lose

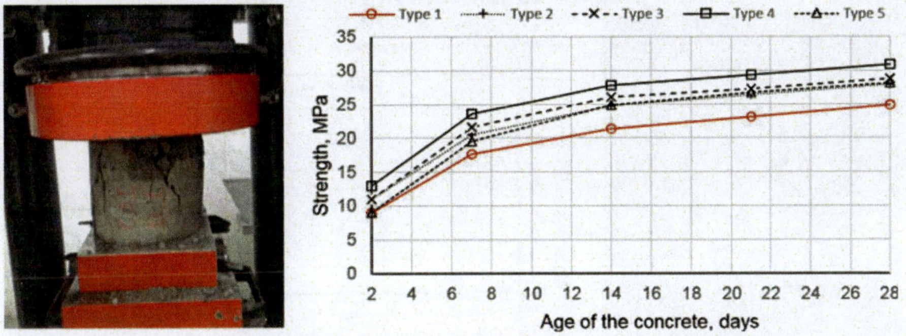

Fig. 2 Results of strength measurements

Table 3 Strength of concrete measurements results

Type of sample	Age of concrete, days				
	2	7	14	21	28
Type 1	9.1	17.7	21.4	23.1	24.8
Type 2	10.9	20.5	24.7	26.4	27.9
Type 3	11.2	21.6	26.1	27.2	28.8
Type 4	13.1	23.5	27.8	29.3	30.9
Type 5	8.9	19.5	24.8	26.7	28.1

strength, therefore, the use of an additive more than 20%, in terms of improving the strength properties is not recommended.

According to the analysis of partial values, the maximum increase in strength is observed for type 4 specimens. All data points have high reliability, convergence and close relationship, as evidenced by the statistics: thc standard deviation of type 1 does not exceed 1.09, at an average value of 24.8 MPa coefficient of variation is 4.33%; the standard deviation of type 2 does not exceed 1.05, at an average value of 27.5 MPa coefficient of variation is 3.76%; the standard deviation of type 3 does not exceed 1.06, with an average value of 28.8 MPa the coefficient of variation is 3.67%; the standard deviation of type 4 does not exceed 1.51, with an average value of 30.9 MPa the coefficient of variation is 4.89%; the standard deviation of type 5 does not exceed 1.63, with an average value of 28.1 MPa the coefficient of variation is 5.82%. The correlation coefficient for the increase of strength characteristics by increasing the percentage of the additive has a direct proportional pattern in the range from 0 to 15%, is 0.98. The same coefficient in the range up to 20% decreases to 0.69, indicating that the 15% additive concentration is optimal. Specimen Type 5 has a low set strength in the first two days, which further confirms the rationality of using a 15% solution.

The results of water absorption are shown in Fig. 3.

According to the results of the research presented in Fig. 3, the effect of plasticizer (post-alcohol bard) on the water-cement ratio and density of concrete is

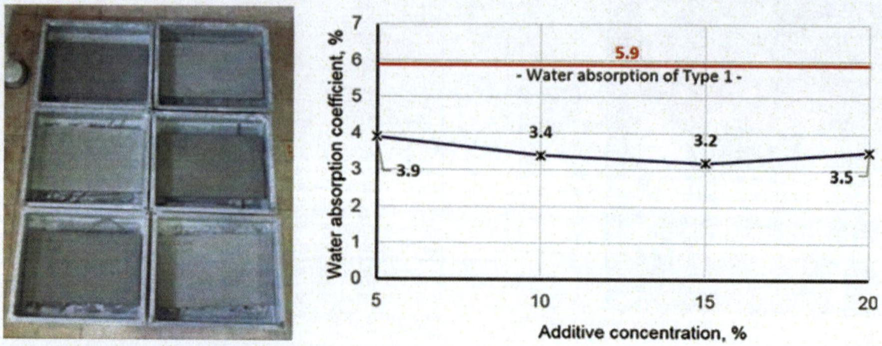

Fig. 3 Results of water absorption measurements

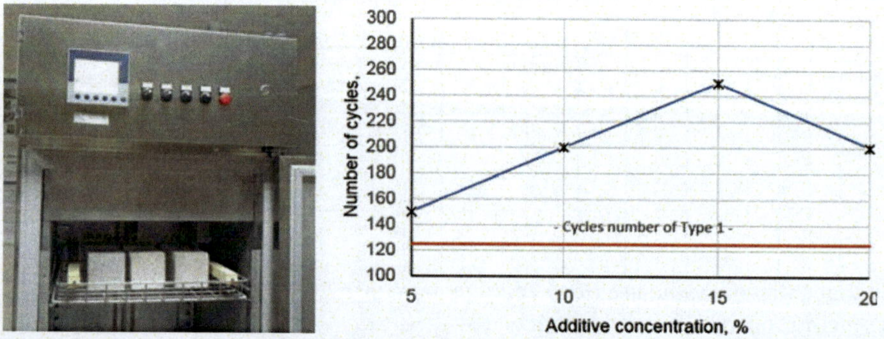

Fig. 4 Results of frost resistance measurements

observing. In this case, there is a synergistic effect, where there is an increase in the density of concrete. Due to the action of the three components is achieved a positive result—an increase in density reduction, water-cement ratio, and reduction of water absorption. Studies of water absorption also showed a sharp increase with the addition of 20% additive similar to the results of strength. The results of data points also have a medium to close relationship as evidenced by the statistics: the standard deviation varies from 0.105 to 0.421, the coefficient of variation from 2.7 to 12.4%.

The results of frost resistance measurements of comparable concrete samples are shown in Fig. 4, and Table 4.

The results of data points also have a medium to close relationship, as evidenced by the statistics: the standard deviation varies from 1.15 to 2.25, the coefficient of variation from 1.7 to 5.9%. According to studies, the use of additives from industrial waste increases the frost resistance of concrete. This effect is obtained through the use of hydrated lime. In the process of hydration free lime is formed, which migrates in the structure of concrete and in the presence of a catalyst, that is, water reacts with sulfates to form gypsum. The latter may explain the increase in

Table 4 Frost resistance of concrete

Types of samples	Weight loss of samples, %						
	F 50	F 75	F 100	F 125	F 150	F 200	F 250
Type 1	0	0	1.2	Over 2%	–	–	–
Type 2	0	0	0.2	0.9	Over 2%	–	–
Type 3	0	0	0	0.5	1.1	Over 2%	–
Type 4	0	0	0	0.1	0.3	0.9	1.3
Type 5	0	0	0	0.3	1.4	Over 2%	–

frost resistance. However, the Type 5 sample with 20% phosphate slag loses its frost resistance. Thus, the optimal concentration is also type 5 with a percentage of slag of 15% (5%) strength loss, on average, decreases 4 times to specimens without impregnation.

4 Conclusion

The results of the study confirm the effectiveness of the integrated use of industrial waste. The results showed that the maximum increase in physical and mechanical parameters corresponds to 15% of phosphorus slag concentrate in the composition of concrete. All obtained partial values of the estimated parameters have good convergence, and hence the reliability of the results. The increase in a strength corresponding to the optimum concentration of the additive is 24.6%, the increase in water absorption is 49.2%, and the increase in frost-resistance is 50%. Given the fact that the use of an additive occurs with the replacement of the binding component (cement), it can be assumed that the technological result of the use of an additive also contributes to reducing the cost of production of heavy concrete. If it is a question of reducing the cost (rather than increasing the characteristics) of concrete without reducing its strength properties, it is possible to use the maximum concentration of an additive equal to 20% by the weight of cement.

References

1. Orazova DK, Zhussupbekov AZ, Lukpanov RE, Yenkebayev SB (2016) Evaluation of wind power unit reliability according to the results of field studies on the example of Ereymentau wind power station. In: Challenges and innovations in geotechnics—proceedings of the 8th Asian young geotechnical engineers conference, 8AYGEC 2016, pp 65–69
2. Lukpanov RE (2016) Laboratory modeling of soil dam reinforced by geosynthetic material. In: Challenges and innovations in geotechnics—proceedings of the 8th Asian young geotechnical engineers conference 8AYGEC 2016, pp 159–162

3. Chen X, Wang H, Najm H, Venkiteela G, Hencken J (2019) Evaluating engineering properties and environmental impact of pervious concrete with fly ash and slag. J Clean Prod 237. https://doi.org/10.1016/j.jclepro.2019.117714
4. Zhussupbekov AZ, Tulebekova AS, Lukpanov R, Zhumadilov IT (2016) Comparison analysis of features in Eurocode and Kazakhstan norms requirements. In: Challenges and innovations in geotechnics—proceedings of the 8th Asian young geotechnical engineers conference 8AYGEC 2016, pp 251–255
5. Zhang C, Hu M, Yang X, Amati A, Tukker A (2020) Life cycle greenhouse gas emission and cost analysis of prefabricated concrete building façade elements. J Ind Ecol 24(5):1016–1030. https://doi.org/10.1111/jiec.12991
6. Manjunath R, Narasimhan MC, Umesha KM (2019) Studies on high performance alkali activated slag concrete mixes subjected to aggressive environments and sustained elevated temperatures. Constr Build Mater 229. https://doi.org/10.1016/j.conbuildmat.2019.116887
7. Ksonshkevych LM, Barabash IV, Krantovska OM, Synii SV, Sunak PO (2019) Disperse reinforced concrete with polycarboxylate additive on a mechanically activated binder. IOP Conf Ser Mater Sci Eng 708(1). https://doi.org/10.1088/1757-899X/708/1/012092
8. Keskin SB, Özlem KK, Yıldırım G, Şahmaran M, Anıl Ö (2018) Determination of self-healing performance of cementitious composites under elevated CO2 concentration by resonant frequency and crack opening measurements. https://doi.org/10.1007/978-3-319-64349-6_47
9. Hrbek V, Zobal O, Prošek Z, Dureje J, Tesárek P (2018) Effect of activation recycled concrete using different admixtures on the strength and modulus of elasticity of the resulting cement composite. In: Experimental stress analysis—56th international scientific conference, EAN 2018—conference proceedings, pp 118–125
10. Cho B, Kim Y, Kim D, Choi S (2018) Effect of ferronickel slag powder on microhydration heat, flow, compressive strength, and drying shrinkage of mortar. Adv Civil Eng. https://doi.org/10.1155/2018/6420238s
11. Lukpanov R, Dusembinov D, Sabitov Y (2017) Komlex modified additive for heavy concrete 1047/1

Kinetics of Changes in the Physical and Mechanical Properties of High-Strength Cement Concretes of Different Types in the Long-Term Hardening Process Under Normal Temperature and Humidity Conditions

T. A. Nizina, A. S. Balykov, D. I. Korovkin, and V. V. Volodin

Abstract It is known that the process of hydration and hardening of cement concretes is not limited to 28 days, but can last for years and even decades. The strength characteristics of concrete significantly depend on the temperature and humidity of the environment, which requires strict control of the hardening conditions. At the same time, the kinetics of concrete strength gain can vary significantly depending on the operating conditions and the concrete composition. Currently, one of the priority directions of building materials science is the development of modified cement concretes, characterized by a complex of high performance characteristics. At the same time, there is evidence that due to the structural features, high performance multicomponent concretes can be potentially more non-equilibrium and active in relation to environmental influences compared to traditional three-or four-component concretes with ordinary strength. The aim of the work was to establish the regularities of changes in the physical and mechanical properties of high-strength cement concretes of different types (fine-grained, lightweight, fibre-reinforced) during the long-term hardening process under normal temperature and humidity conditions. The kinetics of increasing the strength of concrete under compression and bending, the elastic modulus under compression are revealed taking into account the complex modifiers used in their formulation based on polycarboxylate superplasticizer and mineral additives with different chemical and mineralogical composition (siliceous, sulfoaluminate, carbonate).

Keywords High-strength fine-grained concrete · Lightweight concrete · Fibre-reinforced concrete · Complex additive · Normal temperature and humidity conditions · Hardening kinetics

T. A. Nizina · A. S. Balykov (✉) · D. I. Korovkin · V. V. Volodin
National Research Mordovia State University, Saransk, Russia

S. V. Klyuev (ed.), *Digital Technologies in Construction Engineering*,
Lecture Notes in Civil Engineering 173, https://doi.org/10.1007/978-3-030-81289-8_47

1 Introduction

Long-term strength gain caused by the time-consuming hydration processes of clinker minerals is the most important indicator of cement concrete, which distinguishes it from other types of building materials. Compared with structures of metal, wood and plastic the load-carrying capacity of reinforced concrete structures under favorable operating conditions tends to increase over a long period of time—1–3 years, and in some cases even more [1].

It is known that, depending on operating conditions, the kinetics of concrete strength gain can vary significantly [2, 3]. The logarithmic dependence of the following form is widely used for describing the kinetics of the heavy weight cement concrete strength increasing in normal temperature and humidity conditions:

$$R_t = R_{28} \cdot \frac{\lg t}{\lg 28}, \quad (1)$$

where R_t and R_{28} are the compressive strength of concrete at age t and 28 days.

However, when considering numerous compositions of cement concretes, as a rule, there are significant discrepancies between the experimentally obtained data and model (1), which requires the search for more accurate approximating functions.

Currently, much attention is paid to the development of modified concretes with high performance properties, characterized by a complex multicomponent composition with a wide use of various types of modifiers that allow to directed manage the structure formation and properties of the material [4–11]. In this case, to increase the strength characteristics of cement systems the most effective is the combined use of active mineral additives and superplasticizers, which levels the high water demand of powdered mineral component, reduces the water-cement ratio without compromising the rheological properties of mixtures and growth rate of concrete strength in the early stages of hardening [12–15].

The use of new binders and modifiers significantly changes the basic characteristics of concretes. This does not exclude the ambiguity and inconsistency of the manifestation of the properties of high-strength concretes under the influence of various external and internal factors, which may be a problem in their practical application. The authors [16] note that, due to the structure features, high performance modified concretes can be potentially more non-equilibrium and active in relation to environmental influences compared to traditional three- or four-component concretes. And first of all, it concerns the problem of the influence of temperature-humidity state of such concretes on manifestation and change of their construction-technical and operational properties. The relevance of this problem is due to the fact that in the structure of all physical and climatic environmental influences temperature and humidity factors are constantly acting and it depends on them the manifestation of almost all the basic properties of concrete which are strength, frost resistance, deformability, thermal conductivity, etc.

The aim of the work was to establish the regularities of changes in the physical and mechanical properties of high-strength cement concretes of different types (fine-grained, lightweight, fibre-reinforced) during the long-term hardening process under normal temperature and humidity conditions.

2 Methods and Materials

The main component of the binder was Portland cement 500-D0-N (PC) according to the Russian State Standard GOST 10,178–85 produced by Mordovcement PJSC. Polycarboxylate superplasticizer Melflux 5581 F (SP) and the following types of mineral additives (MA) were used as modifiers: MS-85 condensed non-compacted microsilica (MS), expanding sulfoaluminate modifier (ESAM) and carbonate filler microcalcite (MCC). As a fine aggregate, we used natural quartz sand of the Novostepanovsky quarry (Smolny village, Ichalkovsky region, the Republic of Mordovia) with a fraction less than 0.63 mm (sand fineness modulus M_{sf} = 1.5–1.6; content of clay particles—1.8–2.0%). Dispersed reinforcement of concretes was provided by the introduction of wave fiber from steel wire with a cutting length of 15 mm and a diameter of 0.3 mm. For lightweight concrete, glass microspheres with a true density of 0.53 g/cm^3 produced by trading company ForeSphere were used as a porous filler.

The water-binder ratio of modified concrete mixtures was selected on the basis of providing equivalent values of the water-binder ratio of cement paste and the so-called true (effective) water-binder ratio of the concrete mixture (water-binder ratio of cement paste in the concrete mixture), taking into account the thickening ability of used mineral additives and water demand of fillers and aggregates, which allowed to achieve close mobility in the respective groups of cement concretes: 0.500 for control unplasticized composition; 0.285–0.325 for plasticized compositions, including those with 20% content of MA (MS, ESAM, MS + ESAM + MCC); 0.27 for lightweight concrete; 0.31 for fibre-reinforced concrete. The Hagermann cone spread diameter of concrete mixes according to the Russian State Standard GOST 310.4 was: for unplasticized control composition of 106 mm; plasticized compositions of fine-grained and fibre-reinforced concretes—220–230 mm; lightweight concrete—160 mm.

The studied compositions of cement concretes are presented in Table 1.

After hardening under normal temperature and humidity conditions for 28 days, the specimens were additionally exhibited in the same environment for 6 months. The physical and mechanical characteristics (ultimate bending and compressive strength, compressive modulus of elasticity) were determined both for specimens at project age (28 days) and after 2, 4 and 6 months after the beginning of the test cycle. Experimental studies were carried out on specimen-prisms with dimensions of 40 × 40 × 160 mm.

The bending and compressive strength of cement concretes was determined using the WilleGeotechnik® press (13-PD/401 model). The primary parameters

Table 1 The studied compositions of cement concretes

Composition number	Portland cement	MS	ESAM	MCC	Melflux 5581 F	Natural quartz sand (fraction less than 0.63 mm)	Porous glass microspheres (true density 0.53 g/cm^3)	Steel wave fiber (d = 0,3 mm; L = 15 mm)
	% by weight of the binder (Portland cement + mineral additives)							
0	100	0	0	0	0	300	0	0
1	100	0	0	0	1	175	0	0
2	80	20	0	0	1	175	0	0
3	80	0	20	0	1	175	0	0
4	80	5	5	10	1	175	0	0
5	80	5	5	10	1	104	13	0
6	80	5	5	10	1	175	0	24

were configured and obtained experimental results were recorded using the GEOSYS 8.7.8 software.

3 Results and Discussion

The values of physical and mechanical characteristics of the concretes under study in the control condition and after 2, 4 and 6 months of exhibiting under normal temperature and humidity conditions are given in Table 2.

Analysis of strength parameters of concretes at project age (28 days of hardening in normal temperature and humidity conditions after making samples; in Table 2 corresponds to the duration of exposure 0 months) indicates a significant variation in properties depending on the composition of the cement system. The lowest strength values were obtained for concrete of the unplasticized control composition (No. 0), with compressive and bending strengths of 52.5 and 4.5 MPa, respectively. The next level of strength indicators, ranging in the range of 74.4–87.8 MPa in compression and 6.8–7.9 MPa at bending is fixed for high strength fine-grained and lightweight concretes of compositions Nos. 1–4 and No. 5, respectively. At the same time the highest compressive strength values are observed for concretes of compositions Nos. 2 and 4 with 20% content of MS or complex additive MS + ESAM + MCC, respectively.

Table 2 Test results of cement concrete samples exposed under normal temperature and humidity conditions

Duration of exposure, months	Composition number						
	0	1	2	3	4	5	6
Bending strength, MPa							
0	4.5	7.6	7.9	6.8	7.9	7.4	12.1
2	4.8	7.9	10.1	5.5	9.2	7.4	12.6
4	4.9	9.8	10.2	6.8	10.6	9.8	12.6
6	5.0	10.0	8.9	7.0	10.7	9.9	12.8
Compressive strength, MPa							
0	52.5	82.7	87.3	74.6	87.8	74.4	95.6
2	59.1	86.8	92.4	78.8	92.3	76.6	103.2
4	64.2	97.2	99.9	86.5	98.1	78.5	107.2
6	68.0	100.0	103.1	88.8	98.6	78.8	118.0
Compressive modulus of elasticity, MPa							
0	28,223	37,390	38,733	35,132	38,743	30,177	43,204
2	29,371	37,647	39,281	36,314	39,332	30,582	43,503
4	31,227	39,877	40,167	38,580	40,380	30,807	43,722
6	32,681	40,805	41,034	38,705	40,775	31,380	45,844

Introduction into the concrete mixture composition containing complex mineral additive MS + ESAM + MCC additional 2.3% of steel wave fiber (by volume of the concrete mixture) leads to an increase in the bending and compressive strength of fibre-reinforced by 52 and 9%, respectively. The strength characteristics of lightweight concrete are similar to those of high-strength fine-grained concrete containing a sulfoaluminate additive ESAM, introduced in place of 20% Portland cement. It should be noted that in obtaining lightweight concrete was carried out replacing 71% (by weight of the binder PC + MA) of the fine aggregate (natural quartz sand of fraction less than 0.63 mm) with 13% of hollow glass microspheres with a true density of 0.53 g/cm^3. This composition of lightweight concrete, which has a compressive strength of 74.4 MPa, a density of 1889 kg/m^3 and a specific strength of 39 MPa, can undoubtedly be classified as high performance modern concrete.

Thus, when evaluating the strength increasing processes of the concrete samples of the studied compositions exposed for 6 months in normal temperature and humidity conditions, the following groups of cement concretes will be analyzed: control fine-grained concretes without MA—unplasticized (No. 0) and plasticized (No. 1); high-strength fine-grained concretes with MA—No. 2 (MS), No. 3 (ESAM), No. 4 (MS + ESAM + MCC); high-strength lightweight (No. 5) and fibre-reinforced (No. 6) concretes.

Let's carry out the analysis of change of strength indicators of the abovementioned groups of cement concretes, based on estimation of the relative change of properties (Fig. 1), as well as the absolute increase of indicators (Table 2) in the process of normal temperature and humidity conditions exposure over a period of 6 months. It was to be expected that with increasing duration of exposure for all compositions there is an increase in the ultimate strength and modulus of elasticity of concretes in compression (Fig. 1). In general, for the entire study period was a fixed increase in the compressive strength of the concrete studied compositions by 5–22 MPa, which in relative units (relative to the initial strength of the samples of different compositions) is 5–30%. The greatest relative gain, reaching 30% for compressive strength and 16% for compressive modulus of elasticity, is observed (Fig. 1) for the concrete of the control unplasticized composition (No. 0); in absolute values, the level of increase in compressive strength of concrete of this composition was 16 MPa (Table 2). For lightweight concrete, after 6 months of exposure in normal temperature and humidity conditions, there is the smallest increase in strength indicators in compression, estimated both in absolute and relative values (Table 2, Fig. 1).

Similar character of the strength gain of cement concrete of plasticized compositions without mineral additives (No. 1), as well as those containing 20% MS (No. 2) or ESAM (No. 3) is established (Fig. 1). After 6 months of exposure of samples in normal temperature and humidity conditions, there is an increase in strength by 14–17 MPa (Table 2), which is 18–21% of the strength of samples in the control state. Strength gain of high-strength fine-grained concrete containing complex mineral additive (MS + ESAM + MCC) does not exceed 11 MPa (12% of the initial strength, see Fig. 1). Obviously, this decrease in strength gain rate is associated with a two-fold decrease in the proportion of active mineral additives

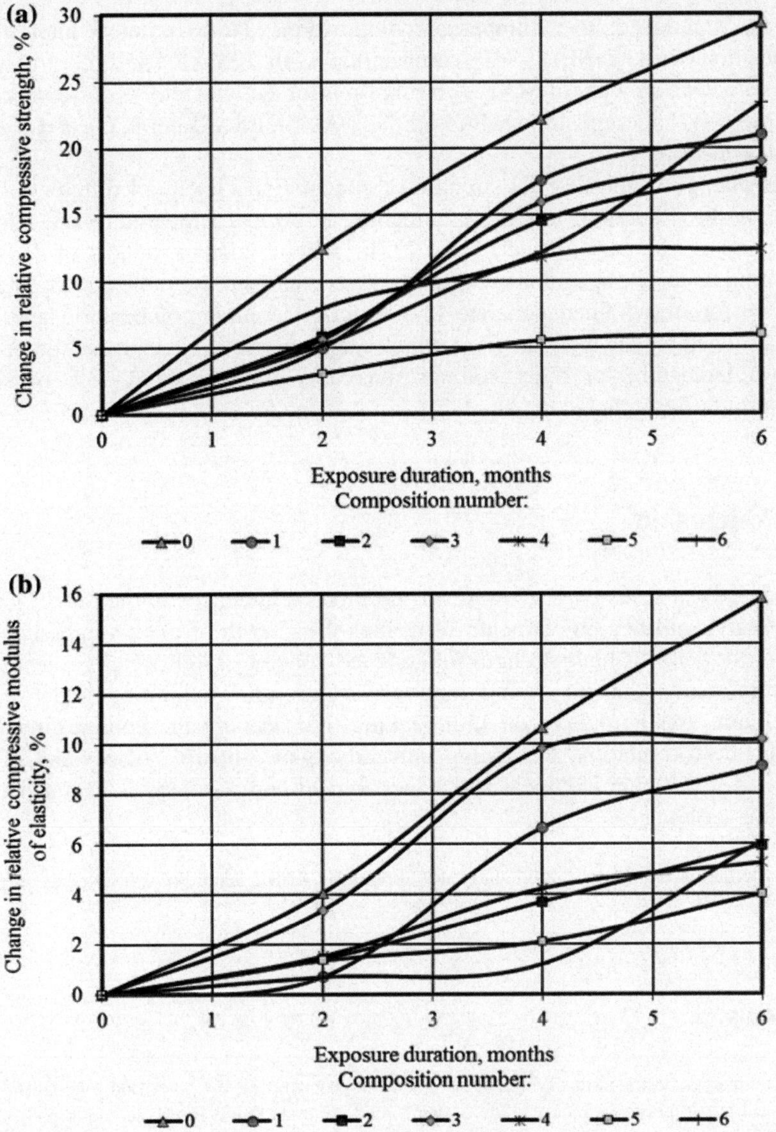

Fig. 1 Changes in the relative compressive strength (**a**) and compressive modulus of elasticity (**b**) of cement concretes depending on the exhibition duration under the normal temperature and humidity conditions

(MS + ESAM) in the modified cement binder when used them in combination with an inert carbonate filler named microcalcite.

The positive effect associated with the increase in the compressive modulus of elasticity of research concretes, both in absolute (Table 2) and in relative values

(Fig. 1), is reduced in a number of compositions: No. 0 (control unplasticized composition without MA) → 3 (composition with ESAM) → 1 (control plasticized composition without MA) → 6 (fibre-reinforced concrete) → 2 (composition with MS) → 4 (composition with complex (MS + ESAM + MCC) and 5 (lightweight concrete).

The change in the flexural strength of concrete specimens of different compositions in the process of long-term hardening in normal temperature and humidity conditions is less pronounced (Table 2). In general, after 6 months of additional exhibiting under the specified conditions, there is an increase in flexural strength by 31–33% for fine-grained concrete of plasticized control composition and lightweight concrete, as well as fibre-reinforced concrete. An increase in bending strength indicators for other compositions does not exceed 3–13% of similar indicators in the initial state.

4 Conclusion

The carried out researches allowed us to estimate differences in the rate of gaining compressive and bending strength as well as the growth of the compressive modulus of elasticity of high strength fine-grained concretes, lightweight concrete and fibre-reinforced concrete during long-term (6 months) exhibiting under normal temperature and humidity conditions taking into account the complex modifiers used in the formulation, containing polycarboxylate superplasticizer and mineral additives of different chemical and mineralogical compositions (siliceous, sulfoaluminate, carbonate).

Acknowledgements The reported study was funded by RFBR according to the research project № 18-29-12036.

References

1. Pyatnica AI, Murashkin GV (2009) Features of determining the wear and evaluation of the resource of operated objects made of reinforced concrete. Mod Probl Sci Educ 6–1:32–33
2. Kudaibergenova NA, Chumadova LI, Vatin NI, Bakirova IG, Bratashov AA, Kabanov AV (2016) The kinetics of curing of concrete at an early freeze. Constr Unique Build Struct 2 (41):7–17
3. Toshin DS, Rovenskaya EA (2020) Influence of long-term hardening conditions on the strength of heavy concrete. Akademicheskij Vestnik Uralniiproekt RAASN 2(45):71–75. https://doi.org/10.25628/UNIIP.2020.45.2.012
4. Shin HO, Yoo DY, Lee JH, Lee SH, Yoon YS (2019) Optimized mix design for 180 MPa ultra-high-strength concrete. J Market Res 8(5):4182–4197. https://doi.org/10.1016/j.jmrt.2019.07.027

5. Rassokhin AS, Ponomarev AN, Figovsky OL (2018) Silica fumes of different types for high-performance fine-grained concrete. Mag Civil Eng 78:151–160. https://doi.org/10.18720/MCE.78.12
6. Nizina TA, Balykov AS, Korovkin DI, Volodin VV (2019) Modified fine-grained concretes based on highly filled self-compacting mixtures. IOP Conf Ser Mater Sci Eng 481:012048. https://doi.org/10.1088/1757-899X/481/1/012048
7. Kocak Y (2020) Effects of metakaolin on the hydration development of Portland–composite cement. J Build Eng 31:101419. https://doi.org/10.1016/j.jobe.2020.101419
8. Balykov AS, Nizina TA, Volodin VV, Kyashkin VM (2021) Effects of calcination temperature and time on the physical-chemical efficiency of thermally activated clays in cement systems. Mater Sci Forum 1017:61–70. https://doi.org/10.4028/www.scientific.net/MSF.1017.61
9. Nedunuri SSSA, Sertse SG, Muhammad S (2020) Microstructural study of Portland cement partially replaced with fly ash, ground granulated blast furnace slag and silica fume as determined by pozzolanic activity. Constr Build Mater 238:117561. https://doi.org/10.1016/j.conbuildmat.2019.117561
10. Carballosa P, García Calvo JL, Revuelta D, Sánchez JJ, Gutiérrez JP (2015) Influence of cement and expansive additive types in the performance of self-stressing and self-compacting concretes for structural elements. Constr Build Mater 93:223–229. https://doi.org/10.1016/j.conbuildmat.2015.05.113
11. Celik K, Hay R, Hargis CW, Moon J (2019) Effect of volcanic ash pozzolan or limestone replacement on hydration of Portland cement. Constr Build Mater 197:803–812. https://doi.org/10.1016/j.conbuildmat.2018.11.193
12. Tarakanov OV, Kalashnikov VI (2017) Prospects of application of complex additives in new generation concretes. News KSUAE 1(39):223–229
13. Nizina TA, Balykov AS, Korovkin DI (2020) Influence of polycarboxylate superplasticizer and mineral additives of various nature on the kinetics of early hardening stages of cement systems. Lect Notes Civil Eng 95:184–190. https://doi.org/10.1007/978-3-030-54652-6_28
14. Kirsanova AA, Kramar LY (2013) Organomineral modifiers based on metakaolin for cement concretes. Build Mater 11:54–56
15. Nizina TA, Ponomarev AN, Balykov AS, Korovkin DI (2018) Multicriteria optimization of the formulation of modified fine-grained fibre concretes containing carbon nanostructures. Int J Nanotechnol 15(4/5):333–346. https://doi.org/10.1504/IJNT.2018.094790
16. Chernyshov EM, Slavcheva GS, Kim LV (2015) About constructional potential of high performans concretes structures with due regard for temperature-humidity operational conditions. Build Mater 9:3–17

Influence of the Loading Mode on the Strength and Deformability of Reinforced Carbon Plastics

S. I. Merkulov, S. M. Esipov, and D. V. Esipova

Abstract This article presents a method for evaluating the effect of loading parameters on the strength of an anisotropic unidirectional carbon fiber composite material under tension in the fiber orientation plane. The mechanisms of deformation and destruction of samples are experimentally investigated, and the dependences of the mechanical properties of reinforced carbon plastics are given. The purpose of the study is to determine the applicability of carbon fiber plastics for external reinforcement of reinforced concrete elements that have lost strength in stretched zones, as well as to analyze the degree of influence of the loading parameters of structures on their strength after reinforcement. The results of the study include tabular and graphical dependences of the time resistance and elastic modulus as a function of the load application rate over a wide range. Creep diagrams were also constructed at various stress levels, from the average experimental to the maximum calculated, taking into account the system of reducing coefficients. The article presents the results of long-term tests of bent elements with plastic elements, taking into account their creep. The results of the study can potentially be used for verification calculations of reinforced concrete elements reinforced with external composite reinforcement on the impact of dynamic short-term loads.

Keywords External reinforcement · Carbon fiber · Composite material · Load

1 Introduction

In the modern construction industry, composite materials—reinforced plastics-are widely used in the selection of materials for strengthening bent and compressed reinforced concrete building structures. Among the variety of precursors of composite materials, its carbon fiber-based composites are widely studied and available [1].

S. I. Merkulov
Kursk State University, Kursk, Russia

S. M. Esipov (✉) · D. V. Esipova
Belgorod State Technological University V. G. Shukhov, Belgorod, Russia

S. V. Klyuev (ed.), *Digital Technologies in Construction Engineering*,
Lecture Notes in Civil Engineering 173, https://doi.org/10.1007/978-3-030-81289-8_48

Based on the listed tasks of composites for reinforced concrete reinforced concrete reinforcement, it can be concluded that the applicability and effectiveness of reinforced plastics depends on their mechanical properties during tensile work. To determine these properties and mechanisms of deformation, it is necessary to use experimental methods, since analytical methods for describing the stress–strain state of a composite materials are complex and operate with a large number of independent parameters [2]. The high degree of anisotropy of the fiber and binder, the absence of plastic deformations of the fiber, the influence of scale coefficients and stress concentrators create difficulties in processing experimental data and obtaining objective characteristics even in cases of uniaxial stress–strain state [1, 2]. In the work described in [3], it was experimentally shown that a 10% reduction in the cross-sectional area of a bent element made of a composite material leads to a drop in the load-bearing capacity by at least 30%. For cases of stretching, this dependence is also true. In experiments [4] it was found that the ultimate strength for the reinforcing elements of composite materials is not a constant value, a is a function of the distribution of defects in the body of the fiber. A similar situation is observed for structural steels that are similar in properties to isotropic materials. The scientific community explains the difference in theoretical and practical results by the presence of microdefects of the fiber—damage caused by external factors. The larger the diameter of the fiber, the more defects are observed and the more they affect the strength. Experimental studies are described in detail in [5]. The situation is similar with the dependence of the strength on the loading speed. However, the specific relationship between the static loading rate and the ultimate strength can only be determined by experiments, and theoretical analysis is not sufficient. Determining the influence of the speed and loading mode (long-term or short-term) on the mechanical properties of reinforced plastics is the problem of the article, since after the installation of external reinforcement, the reinforced concrete element and the composite begin to work together. It follows that we need to understand how the nature of the load applied to the reinforced element (the amount of unloading before reinforcement, the amount of temporary short-term and long-term loads, the proportion of constant loads, pre-stress) affects the load-bearing capacity after reinforcement. In addition, it is worth noting that composites behave like viscoelastic materials when deformed [6], i.e. creep deformations and creep limit are also of interest among the material characteristics. The dotted curve is constructed from the points that are the intersection of the normal from the axis σ and the transition point from the 1st phase of creep to the 2nd, at the angle with which the plot section corresponding to the 2nd phase for each of $\sigma 1 \ldots \sigma 4$ is inclined. The values $\sigma 1 \ldots \sigma 4$ mean some stress levels at which the graphs are plotted. At the same time, there is a noticeable change in the nature of creep with increasing stresses—the appearance of the 3rd phase of creep and the acceleration of growth rates in the 2nd phase. However, when understanding the issue of creep, there are 2 problematic factors: (1) To determine the desired values, it is necessary to apply the Holpin equation and the Findley dependence between the total strain and the creep strain [6], which contain a set of material constants. (2) The problem of choosing a point corresponding to the creep limit. From the point of view of practical use, it is considered [3, 6] that the creep limit is the stress that occurs when the creep strain is 0.1% in 104 h. The real time of the tests described in the article was 1025 h. This duration limit was set based on the working conditions of the laboratory. Based on these factors, it is not possible to correctly determine the desired values that fall under the generally accepted definitions in the framework of this study. An additional series of long-term static tests was carried out to record the creep phenomenon of the carbon composite.

2 Methods and Materials

To obtain practical data, 6 series of static tensile tests of carbon composite were carried out. Compression tests were not carried out, because as part of the tasks of strengthening reinforced concrete structures with carbon composite, the elements of external reinforcement do not experience compression forces [2]. Static tests were characterized by a small amount of acceleration of the moving parts of the test machine, including the resulting inertia forces could be ignored. In series #1–#5, 3 samples were tested, in series #6–3 samples. As the main criterion for the separation of series, the loading speed was chosen: for series No. 1—0.05 kN/s, for series No. 2—0.1 kN/s, for series No. 3—0.15 kN/s, for series No. 4—0.2 kN/s, for series No. 5—0.25 kN/s. The tests of the series No. 6 were carried out for 102.5 h at a constant load value with an interval of 10 h for registering deformations. To determine the mechanical characteristics, a strip single-layer carbon fiber composite with a width of 75 mm and a length of 600 mm was used. The dimensions of the sample were assumed based on the condition of creating a field of uniform deformations, since The San-Venant principle determines the increase in the zones of action of the edge effect. The orientation of the carbon fibers is parallel to the longitudinal axis of the sample. The diameter of the fiber is 0.131 mm. The thickness of the binder—impregnated composite is 1 mm. For anchoring in the wedge grips of the WEW-600D test machine, the samples are provided with glass-fiber linings 75 mm wide and 100 mm long [2], which means that the length of the working part of the sample was 400 mm. As the strain gauge equipment, pasted foil sensors with a base of 20 mm, a digital strain gauge Terem-4.1, an extensometer IDN-10–50 are used.

To assess the applicability of the obtained values, long-term tests of reinforced concrete beams with a single-layer composite material glued to the lower face, identical to the previously described one, were carried out. The calculated span of the elements was 1200 mm, the cross-section height was 150 mm, and the width was 100 mm. The test procedure was generally identical to the short-term tests described in [5]. A force traverse with a controlled compression force was used to simulate a constant load. The time of testing and registration of deformations coincided with those shown in Fig. 5. The general view of conducting long-term tests is shown in Fig. 1.

Fig. 1 General view of conducting long-term tests

3 Results and Discussion

1. The loading rate affects the value of the ultimate strength of the material (Fig. 2). In a certain range of initial values of the loading rate (Vн), the increment of parameters is not observed. At a certain value of Vk, the diagram begins to bend, and a break appears. Gradually, moving into the area of shock loads (Vy), the diagram tends to rectify.

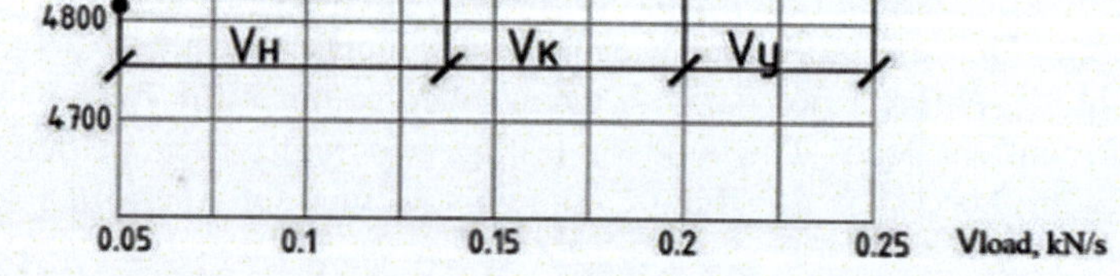

Fig. 2 Dependence of the composite strength on the rate of load change

This leads to the following assumptions: (a) the work of the boundary layers separating the fiber and the matrix at different loading rates is different, (b) the impact strength of the composite exceeds the impact strength of its components [3, 7].
The loading speed affects the value of the elastic modulus of the 1st kind (Fig. 3). The point of curvature of the diagram is located in the same range of values of the loading speed as in Fig. 3. In the area of shock loads, the value of the elastic modulus begins to grow with a character close to hyperbolic. This is due to the simultaneous nature of two factors: (a) there is an increase in the strength of the material (by 2.5–3%), (b) its specific elongation decreases (by 21%). Comparing these data with other studies [3], we can conclude that the strain rate has an influence on the elastic modulus of the 1st kind and does not affect the elastic modulus of the 2nd kind.
Carbon-based reinforced plastics behave like viscoelastic materials. There is an increase in deformations under static load (Fig. 4). In the framework of the experiment, deformations of the initial compression of the samples and accidental slippage were excluded from the results. With an increase in the initial stress in the sample, the growth of creep deformations accelerates. The growth of deformations is linear up to a certain time limit (100 h), after which the acceleration of the creep increase is noticeable [8, 9]. The creep phenomenon probably occurs during stress relaxation in the reinforcing fiber. The results obtained are summarized in Table 1. It was found that due to the presence of creep, especially at high stress levels, the load-bearing capacity of the bent elements decreases. This is most typical when the load levels are close to the limit. The characteristic parameters of the stress–strain state are described in Table 2 and Figs. 5, 6 and 7. As can be seen from the presented materials, the level of the load-bearing capacity drop in the area of 0–75% of the breaking moment is close to 0.

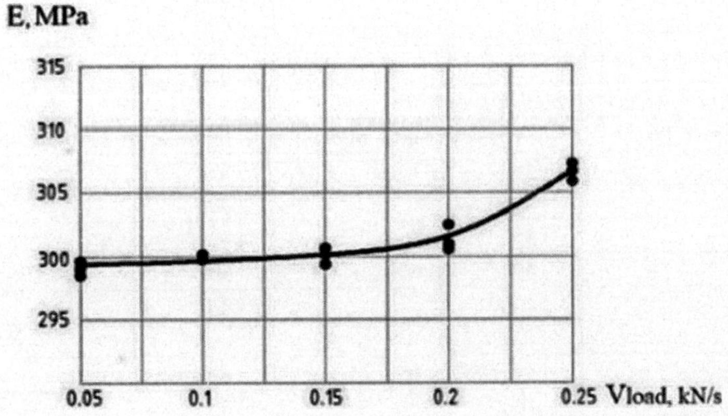

Fig. 3 Dependence of the elastic modulus of the 1st kind on the rate of change of the load

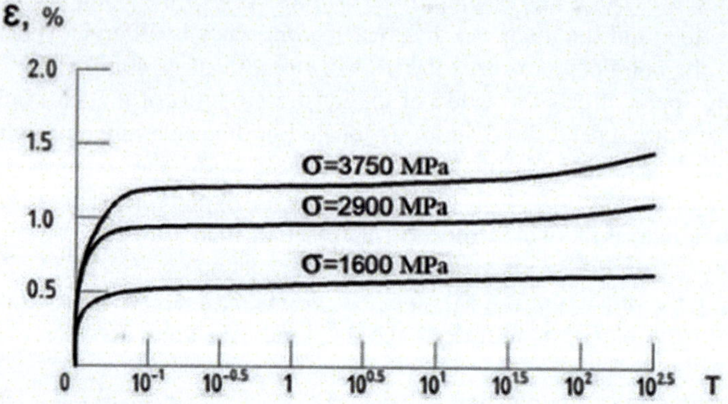

Fig. 4 Creep of carbon composites

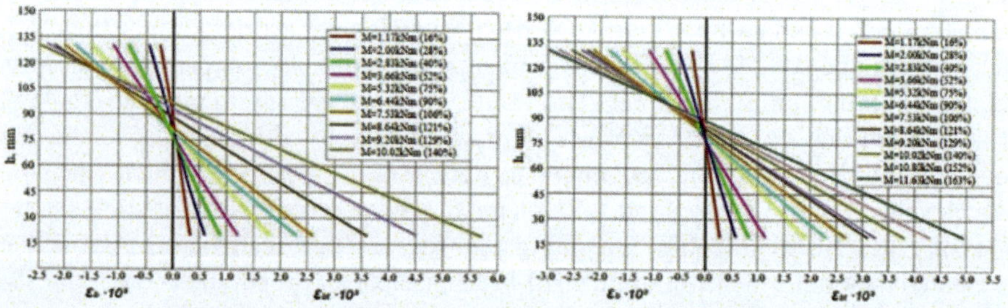

Fig. 5 Diagrams of relative deformations of sample B1 at loading time of 1 h (top) and 1000 h (bottom)

Table 1 Average research results about carbon composites

Studies of the influence of loading speed		
Loading speed, kN/s	Tensile strength, MPa	Modulus of elasticity, GPa
0.05	4830	298
1.00	4840	300
1.5	4865	301
2.0	4925	302
2.5	4960	307
Studies of the influence of loading time		
Loading time, h	Tensile strength, MPa	Relative strain, %
0.1	1600	0.5
	2900	0.9
	3750	1.18

(continued)

Table 1 (continued)

1	1600	0.54
	2900	0.94
	3750	1.22
10	1600	0.6
	2900	0.97
	3750	1.26
100	1600	0.66
	2900	1.04
	3750	1.33
320	1600	0.7
	2900	1.08
	3750	1.43

Table 2 Average research results about concrete beams

8.5	Test features	Loading time, h	Relative deformations of the stretched zone, %
B1	Without preloading	1	0.360
		1000	0.377
B2	With preloading (0.5)	1	0.342
		1000	0.378
B3	With preloading (0.7)	1	0.350
		1000	0.365

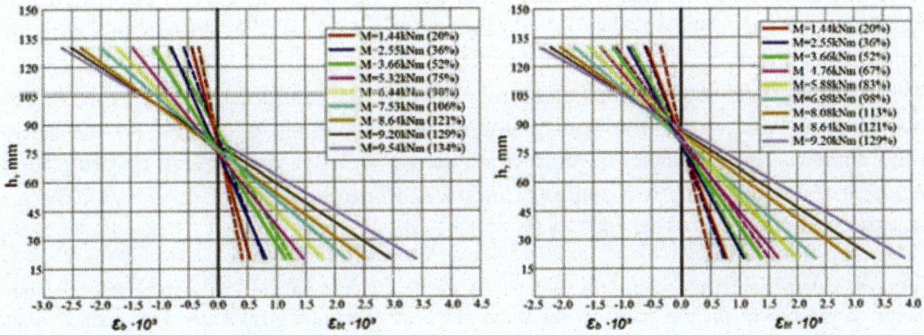

Fig. 6 Diagrams of relative deformations of sample B2 at loading time of 1 h (left) and 1000 h (right)

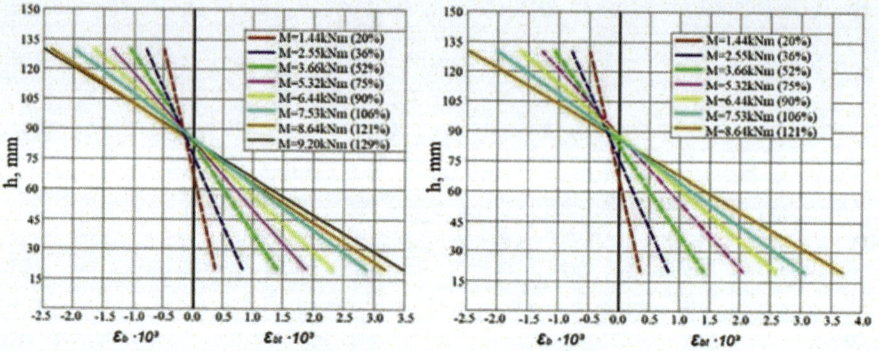

Fig. 7 Diagrams of relative deformations of sample B3 at loading time of 1 h (left) and 1000 h (right)

4 Conclusion

When using polymer composite reinforcement, attention should be paid to the calculated values of the strength and deformation characteristics. Within the framework of long-term loading of the structure, it is unacceptable to use the strength values of the composite obtained on the basis of short-term tests, since they overestimate the actual strength. On the basis of the above, it is possible to determine the tasks, as a result of which it is possible to improve certain provisions of the normative methods of designing reinforcement of reinforced concrete structures by external reinforcement with composite materials: (1) Develop a methodology for determining correction factors that take into account the proportion of the duration of operation of a reinforced or projected structure under load. (2) Enter into the calculation of reinforcement the possibility of viscoelastic deformation of reinforced plastics and the increase in creep deformations. (3) Take into account in the calculation of the loss of the pre-stress value of structures with composite reinforcement due to the creep of composite materials. (4) The creep of composite elements of external reinforcement reduces the load-bearing capacity of the bent elements by an average of 5–8%. At the same time, there is an increase in deformations by 4.3–10.5%. Creep deformations manifest themselves with a loading duration of more than 500 h, previously no negative effects were observed.

Acknowledgements This work was realized in the framework of the Program of flagship university development on the base of the Belgorod State Technological University named after V. G. Shukhov. The work was realized using equipment of High Technology Center at BSTU named after V. G. Shukhov.

References

1. Nevolin DG, Smerdov DN, Smerdov MN (2015) Experimental studies of the bearing capacity of reinforced concrete structures of mining buildings and structures. Proceedings of higher educational institutions. Mining J 8:138–142
2. Fib Bulletin 14. Externally Bonded FRP reinforcement for RC structures (2001)
3. Lu XZ, Teng JG, Ye LP, Jiang JJ (2005) Bond–slip models for FRP sheets/plates bonded to concrete. Eng Struct 27:920–937
4. ACI 440.2R-08 (2008) Guide for the design and construction of externally bonded FRP systems for strengthening concrete structures. Michigan/American concrete Institute, ACI Committee 440
5. Merkulov SI, Esipov SM (2018) Increase in the bearing capacity of reinforced concrete structures by strengthening the external reinforcement composite material BST. Bull Constr Equip 2(1002):56–57
6. Rimshin VI, Merkulov SI, Esipov SM (2018) Concrete structures reinforced with composite material. Bull Eng School Far East Fed Univ 2(35):93–100
7. Yushin AV, Morozov VI (2014) Experimental studies of two-span reinforced concrete beams reinforced with composite materials on an inclined section. Bull Civil Eng 5(46):50–57
8. Bokarev SA, Smerdov DN (2010) Experimental studies of bent reinforced concrete elements reinforced with composite materials. Proc High Educ Inst Constr 2(614):112–124
9. Smerdov DN, Selivanova EA (2017) The study of the creep properties in the elements of the system of external reinforcement during prolonged exposure to load. Polytransport system materials IX International scientific-technical conference. Siberian state University of railway engineering, pp 53–56
10. Dai J, Ueda T, Sato Y (2005) Development of the nonlinear bond stress–slip model of fiber reinforced plastics sheet–concrete interfaces with a simple method. J Compos Constr 9:52–62

Efficiency of Using Lightweight Concrete in the Construction of Buildings and Underground Structures in the Arctic

A. E. Mestnikov and A. L. Popov

Abstract The article discusses the features of the use of granular foamed aggregates from local mineral raw materials and lightweight concrete based on them in the Arctic. To expand promising technologies for the Arctic, the previous experience in the production of foamed aggregates and lightweight concrete obtained on their basis has been studied. It was revealed that in the conditions of Yakutia and the Arctic, technologies for the production of foamed aggregates for lightweight concrete based on unlimited reserves of natural aluminosilicate and siliceous rocks are relevant. The results of a study of the physical and mechanical characteristics of foamed aggregates produced in Yakutia in comparison with other Russian aggregates and lightweight concrete based on them are presented. It is shown that the shape of the grains plays an important role in achieving high brand strengths of lightweight concrete based on foamed aggregates. Thus, the efficiency of using granular foamed zeolite from local raw materials and lightweight concrete based on it in low-rise, underground and road construction in permafrost and cold climates has been substantiated. In this case, an increase in frost resistance can be achieved due to the low water absorption of foamed aggregates and thermal characteristics due to the low thermal conductivity of the resulting lightweight concrete.

Keywords Siliceous raw materials · Aluminosilicate raw materials · Granular foamed aggregate · Lightweight concrete · Construction · Energy efficiency

1 Introduction

The Arctic today is a center of increased interest of the international community, including due to the growing availability of natural resources to it [1, 2]. It is known that in 1996 the International Arctic Council was formed, whose work is aimed at

A. E. Mestnikov
Ammosov North-Eastern Federal University (NEFU), Yakutsk, Russia

A. L. Popov (✉)
Institute of Oil and Gas Problems of the Siberian Branch of the RAS, Yakutsk, Russia

S. V. Klyuev (ed.), *Digital Technologies in Construction Engineering*,
Lecture Notes in Civil Engineering 173, https://doi.org/10.1007/978-3-030-81289-8_49

protecting the natural resources of the region and the further development of the Arctic. The official document on the creation of the Arctic Council was signed by eight countries: the USA, Finland, Norway, Denmark, Canada, Iceland, Sweden and the Russian Federation.

For the Russian Arctic, 8 support zones have been identified in eight regions of the country: Kola, Arkhangelsk, Nenets, Vorkuta, Yamalo-Nenets, Taimyr-Turukhansk, North Yakutsk and Chukotka [3]. Along the coast of the Arctic Ocean, they can be conditionally divided into 3 parts: European, West Siberian and Northeast. Of these, in industrial terms, the European part is the most developed, followed by the level of development and development are the northern regions belonging to West Siberia. And the least developed territory is the Northeast part, which includes the North Yakutsk (with the center in the village of Tiksi) and the Chukotka (with the center in Anadyr) support zones.

Currently, due to the global change in the geopolitical situation, the Government of the Russian Federation pays close attention to the Northeast part of Arctic territories [4]. The development and use of new architectural and technological solutions for construction in the Arctic is due to the development strategy of the Arctic territories of the Russian Federation, first of all, the development and restoration of the infrastructure of seaports and the northern sea route [5]. The development of a set of solutions that take into account extreme climate conditions is observed in the construction of social infrastructure, airfields, seaports, as well as autonomous residential complexes, including shift camps and military camps. At the same time, difficulties arise in solving logistic problems, since the used building materials are completely delivered from industrialized centers.

Taking into account the world experience in the development of the Arctic and Antarctica, the basis of today's autonomous residential complexes in the hard-to-reach regions of the Arctic zone of the Russian Federation is made up of a lightweight prefabricated metal frame and a pile foundation [6–10]. An alternative option for metal structures is the Massiv-Holz-Mauer technology [11], which is the production of environmentally friendly solid wood walls. In the conditions of Yakutia, the MHM technology has found wide application in the construction of shift camps for diamond workers in the Arctic and subarctic regions [12]. During the Soviet period, the exploration and development of the Arctic territories, the basis of the construction of buildings was made of stone buildings made of large blocks and panels based on expanded clay aggregates. Arctic cities and large settlements (Murmansk, Arkhangelsk, Naryan-Mar, Norilsk, Vorkuta, Tiksi, Pevek, Anadyr, etc.) were built up with buildings of the same type, which are characterized by a fairly high durability.

In our opinion, for the construction of buildings and underground structures in the cold climate and permafrost of the Arctic regions, blocks, panels and slabs, as well as solutions for 3D printing and spray concreting from lightweight concrete with granular foamed aggregates from local natural and technogenic mineral raw materials.

Thus, expanded clay (vermiculite, azerite) was practically uncontested lightweight aggregate for lightweight concrete during the period of Soviet construction, including in Yakutia. In the works of the N.V. Chersky Mining Institute of the North of the Siberian Branch of the RAS, it was shown that it is promising to use expanded vermiculite as a foamed aggregate in the mines of South Yakutia, azerite for the mines of the Norilsk Combine, and crushed silicate foamed concrete for the mines of ALROSA company (Tables 1 and 2) [13–16].

In Table 2, the sign (*) marks lightweight concretes, which are used as heat-shielding shotcrete in the mine workings of the permafrost zone [16].

Features of the technology for the production of granular foamed glass determine its high cost, however, its high service life, a more familiar and simple technology for the production of building products based on them, as a result, can significantly increase the technical and economic indicators of its use.

Table 1 Characteristics of foamed aggregates

Foamed aggregates	Bulk density, kg/m^3	Compressive strength in the cylinder, MPa	Water absorption, %wt
Expanded clay	450–500	1.6–2.4	22.3–24.5
Azerite	350–400	3.4–4.6	23.2–24.5
Vermiculite	100–200	0.3–0.5	400–500
Foamed silicate granules	600–700	1.8–2.8	30–70

Table 2 Characteristics of lightweight concrete based on foamed aggregate

Foamed aggregates	Average density of concrete, kg/m^3	Compressive strength, MPa	Thermal conductivity coefficient, W/(m × °C)
Expanded clay	750	7.5	0.27
	900	10.1	0.24
	1200	13.1–16.4	0.41
Azerite	1270*	10.3	0.48
	1280*	15.7	0.60
	1720*	7.1	0.40
Vermiculite	500	1.3	0.11
	620*	2.4	0.20
	750*	3.7	0.22
	970*	4.4	0.30
Foamed silicate granules	910*	0.9	0.21
	1220*	2.55	0.24
	1360*	3.75	0.28

2 Methods and Materials

As part of the work, a granular foamed zeolite of fraction 5–20 produced by OOO Suntarzeolite (Russia, Suntar village) from zeolite-containing raw materials of Yakutia—zeolite-hongurin, was investigated (Fig. 1a).

For comparison, granular foamed glass of fraction 10–20 produced by ZAO Modis (Russia, Yaroslavl) was studied (Fig. 1b).

The characteristics of foamed aggregates according to the manufacturer's passport are given in Table 3.

To obtain lightweight concrete, Portland cement of the CEM I 32.5 N grade, produced by JSC Yakutcement, was used.

The main physical and mechanical properties of the samples of foamed aggregates and lightweight concrete were determined according to standard methods:

- bulk and true density, water absorption, strength of the aggregate when squeezed in the cylinder—GOST 9758–2012 "Non-organic porous aggregates for construction work. Test methods";

(a) (b)

Fig. 1 General view of foamed aggregates: **a**—granular foamed zeolite, **b**—granular foamed glass

Table 3 Characteristics of foamed aggregates according to the manufacturer's passport

Foamed aggregates	Bulk density, kg/m^3	Compressive strength in the cylinder, MPa	Thermal conductivity coefficient, W/(m × °C)
Granular foamed zeolite	200–300	0.7	0.075
Granular foamed glass	150–350	0.5–1.5	0.09–0.12

Table 4 Characteristics of foamed aggregates

Foamed aggregates	Average bulk density, kg/m^3	Average density of grain, kg/m^3	Intergranular space, %	Average compressive strength in the cylinder, MPa	Average water absorption, %wt
Granular foamed zeolite	190	315	39	0.88	13.1
Granular foamed glass	245	510	52	0.92	14.2

- density—GOST 12,730.1–78 "Concrete. Methods for Determination of Density";
- compressive strength—GOST 310.4–81 "Cements. Methods of bending and compression strength determination";
- thermal conductivity—GOST 7076–99 "Bullding materials and products. Method of determination of steady-state thermal conductivity and thermal resistance" (Table 3).

3 Results and Discussion

According to the obtained data on the main characteristics of foamed aggregates (Table 4), the studied granular foamed glass produced by CJSC "Modis" has lower indicators of bulk density, grain density and water absorption. At the same time, the compressive strength of granular foamed glass grains is comparable to the strength of granular foamed zeolite. It is within the permissible error, despite the lower density, which is associated with a more regular spherical shape of the foamed glass grains and a smaller intergranular space, due to which the stress distribution occurs during compression.

The granular foamed zeolite produced at Suntarzeolite LLC consists of a mixture of fractions 5–10 and 10–20, while the analytical sample contains fraction 3–5 in an amount of 14%. In this regard, the properties of individual fractions were studied (Table 5). It is shown that larger fractions have lower strength and are characterized by lower density and water absorption rates.

Thus, the comparatively low water absorption of granular foamed glass and foamed zeolite and lightweight concretes based on them can contribute to the expansion of their field of application in contact with frozen and thawed soil. For example, bulk thermal insulation from granules and lightweight concrete based on them can be used in road and underground construction, as well as in elements of insulated foundations in permafrost conditions.

Table 5 Characteristics of individual fractions of granular foamed zeolite

Fraction, mm	Fraction content, %	Average bulk density, kg/m^3	Average density of grain, kg/m^3	Average compressive strength in the cylinder, MPa	Average water absorption, %wt
3–5	14	286	625	1.02	15.2
5–10	44	240	510	0.95	14.1
10–20	42	230	480	0.91	13.2

Table 6 Characteristics of lightweight concrete based on foamed aggregate

Foamed aggregates	Average density of concrete, kg/m^3	Average compressive strength, MPa	Thermal conductivity coefficient, W/(m × °C)
Granular foamed zeolite	430	1.81	0.14
	530	1.87	0.16
	620	1.97	0.18
Granular foamed glass	400	2.2	0.12
	500	3.3	0.14
	600	4.6	0.16

Cement consumption for lightweight concrete samples was selected based on the target density corresponding to the class D500, D600, D700. The water-cement ratio was 0.4. The physical and mechanical characteristics of lightweight concrete, obtained in accordance with regulatory documents, are shown in Table 6.

Samples of lightweight concrete made using granular foamed glass have increased strength at low densities of 400–600 kg/m^3, about 2.2–4.6 MPa, compared to lightweight concrete based on granular foamed zeolite (1.81–1.97 MPa) at the same densities (Fig. 2). At the same time, the strength of lightweight concrete based on granular foamed zeolite hardly increases with an increase in its density, in contrast to lightweight concrete made using granular foamed glass. This fact shows that the shape of the grains has an important role in achieving high brand strengths of lightweight concrete based on foamed aggregates. It is the spherical shape of the grains that makes it possible to evenly distribute stresses over the composite matrix and create a framework for the strength of lightweight concrete.

Thus, from Tables 5 and 6, it can be seen that on the basis of granular zeolite foamed there is a possibility of producing products from lightweight concrete D400–D600 with strength characteristics typical of autoclaved aerated concrete. Such concretes can be used in low-rise construction of residential and industrial buildings in remote areas using various construction technologies—shotcreting, monolithic housing construction, 3D printing, panel and stone construction. It should be noted that the location of the Suntarzeolite enterprise and the huge reserves of raw materials predetermine the effectiveness of the use of granular foamed zeolite in low-rise construction in the Arctic and subarctic regions of Yakutia.

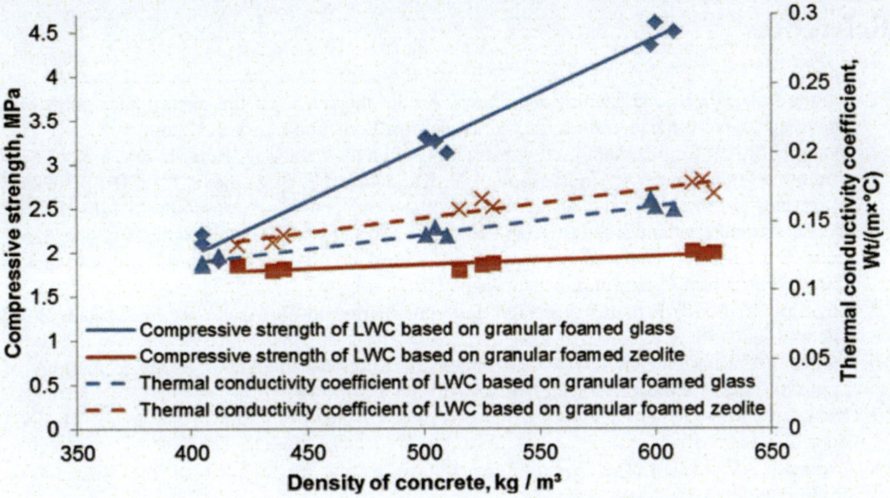

Fig. 2 Dependence of strength and thermal conductivity of lightweight concrete (LWC) on density

4 Conclusion

The studied foamed aggregates (granular foamed glass, granular foamed zeolite) are characterized by a sufficiently high strength (0.8–1.0 MPa) at a low density of 190–250 kg/m^3, as well as relatively low water absorption.

It is shown that the shape of the grains plays an important role in achieving high brand strengths of lightweight concrete based on foamed aggregates. It is the spherical shape of the grains that makes it possible to evenly distribute stresses over the composite matrix and create a framework for the strength of lightweight concrete.

Thus, on the basis of foamed aggregates, it is possible to obtain lightweight concrete for the manufacture of blocks, panels and slabs, including monolithic structures using traditional technologies, as well as 3D printing and shotcrete. The use of lightweight concrete in construction can significantly reduce the weight of the structure being erected, which is very important when performing construction work in the Arctic and permafrost. And also an increase in frost resistance can be achieved due to the low water absorption of foamed aggregates and thermal characteristics due to the low thermal conductivity of the resulting lightweight concrete.

References

1. Young OR (2019) Constructing the "new" Arctic: the future of the circumpolar north in a changing global order. Outlines Global Transform Polit Econ Law 5(12):6–24
2. Wang H (2019) Sino-russian joint construction of arctic waterway. Rep Sci Soc 1(21):19–24
3. Smirnova OS, Lipina SA, Kudryashova EV, Krejdenko TF, Bogdanova YN (2016) Creation of development zones in the Arctic: methodology and practice. Arct North 25:148–157
4. Decree of the President of the Russian Federation. Strategy for Developing the Russian Arctic Zone and Ensuring National Security until 2035. The Russian President's official website (2020). URL: http://kremlin.ru/acts/news/64274/
5. Savinova V (2018) Features of design and construction in the Arctic region. Sustain Build Technol 4:50–56
6. Puntus VA, Mjasepp KK (2017) 3D printer construction—innovation in the technology of improving the Architecture of the Arctic. Am Sci J 15:7–14
7. Nekrasova MA (2014) Greenproject management of green building in the arctic. RUDN J Ecol Life Saf 3:10–16
8. Zagorski AV (2018) The Arctic defense postures in the context of the Russia-West confrontation. Arct North 31:80–96
9. Elshina L, Yarmakovskiy V, Kirillov I, Panteleev V (2018) Scientific assistance of hazardous construction in Russian arctic region. ACI Spec Publ 326:97.1–97.10
10. Barenberg A (2018) The city as showpiece: arctic camp, arctic city: the GULAG and the construction of Vorkuta. The City in Russian Culture, pp 121–139
11. Santi S, Pierobon F, Corradini G (2016) Massive wood material for sustainable building design: the Massiv–Holz–Mauer wall system. J Wood Sci 62:416–428
12. Mestnikov AE, Antipkina TA (2020) Building materials and technologies for the North and the Arctic: monograph. ACB Publishing House, p 233
13. Lotov VA (2011) Production of foam glass from the natural and technogenic aluminum silicates. Steklo Keram 9:34–37
14. Davidyuk AN, Savin VI, Kostin AA, Fedoseev AV (2015) Lightweight concrete of a new generation on granular foamed glass. Beton i zhelezobeton 5:2–6
15. Alekseev KN, Kurilko AS (2017) Prospects for the application of light heat-protective fiber-reinforced concrete. Min Inform Anal Bull 24:254–263
16. Orlov A, Nezhikov A (2017) Foam glass-ceramics as an aggregate for high-technological lightweight concretes. Bull SRC Stroitelstvo 3(14):163–171

Influence of the Composition of the Soil on the Effectiveness of Its Strengthening with Cement

A. A. Bezrodnikh, V. V. Strokova, I. Yu. Markova, and M. A. Stepanenko

Abstract The paper shows the effectiveness of strengthening of members of various types of clayey soils with an inorganic binder of the hydration type of hardening—cement. The relationship between the complex of physical–mechanical and physical–chemical properties of the soil and the final physical and mechanical indicators of the hardened composite has been substantiated. A comparative analysis of the obtained values of the physical and mechanical characteristics of the compositions of hardened soils, where the amount of binder was determined experimentally and on the basis of calculation, was carried out. A calculation method for the selection of cement concentration was considered and applied, taking into account the hydration activity of alite and belite in the composition of cement, as well as the sorption capacity of soil determined by Zaporozhets method and its optimal moisture content. As a result, it was found that the values of the experimental and calculated data have a high convergence. However, the use of the calculation method allows to minimize labor costs for the selection of components. Thus, the effectiveness of the calculation method has been proven in view of providing a greater margin of safety for the considered types of clay soils with a calculated amount of binder. In addition, it was found that soils with a large proportion of the clay component require a slightly higher dosage of cement. At the same time, the strengthened system is distinguished by a higher density, which will provide a lower filtering capacity of the consolidated system and predict a longer resistence in the environment.

Keywords Soil · Cement · Strengthening · Strength · Activity

A. A. Bezrodnikh (✉) · V. V. Strokova · I. Yu. Markova · M. A. Stepanenko
Belgorod State Technological University Named After V.G. Shukhov, Belgorod, Russia
e-mail: nelubova@list.ru

S. V. Klyuev (ed.), *Digital Technologies in Construction Engineering*,
Lecture Notes in Civil Engineering 173, https://doi.org/10.1007/978-3-030-81289-8_50

1 Introduction

The problem of depletion of high-quality stone materials poses the problem of finding ways to use the base soil surface as the road basement. However, as a rule, soils do not meet the requirements for physical and mechanical characteristics, in particular, strength. In addition, the high filtering and sorption capacity of soils leads to an increase in their heaving when the ambient temperature changes sign. All this necessitates the search for effective ways to stabilize soils in order to give them physical stability under external influences [1–3].

Currently, a wide range of physical (mechanical), chemical and complex methods of soil strengthening is known [4–8]. Many years of experience in soil strengthening have proven the effectiveness of using cement as a component that provides the required soil characteristics during the design period. In this case, the concentration of cement is determined experimentally without taking into account the characteristics of the soil [9–11].

A significant territory of Russia is made up of clayey soils, characterized by high humidity and sorption capacity [12–16]. This leads to some difficulties in strengthening such soils with a cement binder, which is associated with overconsumption of the binder, its accelerated hydration and densifying of the system. In this regard, the purpose of this work is to select an effective dosage of cement as the main fixer by a calculation and experimental method based on data on the composition of cement and the properties of clay soil as a stabilized system.

2 Methods and Materials

Natural soils of various compositions were used for research (Table 1, 2). According to the data on physical and mechanical properties, the soils used according to the number of plasticity belong to loams; based on the data on the granulometric composition—sample No. 1 belongs to heavy silty loams, sample No. 2 belongs to light silty loams (Table 3).

The specific surface area of the dispersed soil was determined by air permeability methods using a PSKh-12 device and gas adsorption by the BET method using a device Sorbi M.

Table 1 Physical and mechanical properties of soils

Sample	Moisture level				Plasticity number	Density, kg/m^3
	Hygroscopic	At the liquid limit	At the plastic limit	Optimal		
Sample 1	4.6	36.8	24.2	15	12.6	1960
Sample 2	3.2	19.7	11.4	18.6	8.3	2140

Table 2 Chemical composition of soils

Sample	Oxides, mass. %								
	SiO_2	Al_2O_3	CaO	Fe_2O_3	MgO	K_2O	TiO_2	Na_2O	other
sample 1	63.2	16.5	8.9	4.6	2.7	2.1	0.8	0.7	0.8
Sample 2	75.3	14.7	1.9	2.7	2.3	1.6	0.6	0.5	0.4

Table 3 Granulometric composition of soil samples according to sieve analysis data

Sample	Sieve size, mm						
	0.1	0.2	0.5	1	2	5	10
Sample 1	2.97	18.63	27.06	14.41	16.43	18.93	0.74
Sample 2	11.36	25.46	19.63	12.51	15.71	13.88	0.33

The phase composition of the soils was determined using an ARL 9900 WorkStation X-ray fluorescence spectrometer.

The activity of the soils was assessed as an integral characteristic according to the indicators of the sorption capacity, determined by the Zaporozhets method, and the free energy of the surface according to the contact angle of wetting with test liquids using a KRUSS device.

The calculation of the minimum amount of cement for soil strengthening was carried out according to the formula in accordance with the work [17]:

$$C_{cem} = \frac{100}{S_{C_3S} \cdot 0.48677 + S_{C_2S} \cdot 0.15056} \cdot (E_p + 1.18 \cdot 10^{-3} \cdot W_{opt}), \quad (1)$$

S_{C_3S}, S_{C_2S}—amount of $Ca(OH)_2$, released during hydration of alite and belite;
E_p—absorption capacity (sorption capacity according to method of Zaporozhets);
W_{opt}—optimum moisture content of soil.

Portland cement CEM I 42.5 N brand produced by CJSC "Belgorodsky cement" was used for reinforcement (Table 4 and 5).

Evaluation of the effectiveness of measures to improve the performance of the soil was carried out on the basis of the characteristics of the final material: the maximum density of the hardened soil (method according to State Standard 22733–2002), compressive strength and tensile strength in bending (methods according to State Standard 10180–2012), as well as frost resistance (method according to State Standard 10060–2012).

Table 4 Mineralogical composition of cement clinker

Mineralogical composition of cement clinker, %				
C_3S	C_2S	C_3A	C_4AF	Other
61.7	17.2	5.5	13.6	2

Table 5 Properties of Portland cement CEM I 42.5 N

Characteristic	Value
Normal consistency, %	26
Sounding of cement, mm	0.12
Average activity of cement at the age of 28 days, not less, MPa	40

3 Results and Discussion

According to the obtained data (Table 6), divergence of values of specific surface area of soils (higher in the case of sample No. 2) is within 5%. The increase in the active specific surface area is associated with the high porosity of the layered structure of clay materials. In this case, the active specific surface area of sample 2 is somewhat higher, which is associated with a larger proportion of the finely dispersed component in its composition.

Light loam is distinguished by a higher free energy of the surface: by 30% compared to heavy loam (Table 6). This means that uncompensated bonds prevail on the surface of the particles of a given material, which provide its high chemical activity. However, unlike light soil, heavy loam is characterized by a higher sorption capacity: the total amount of absorbed CaO is 22% higher. The high absorptive capacity of clayey soils is explained by their high porosity and «packet» structure. In this regard, it is obvious that heavy loam (sample 1) with a greater proportion of the clay component in its composition demonstrates an increased adsorption capacity compared to light loam.

Substituting the numerical data of soil properties (moisture and sorption capacity) (Table 6), as well as taking into account the mineral composition of cement (Table 4) from formula (1), we obtain that to stabilize light silty loam 6–7% of cement is required, and for heavy silty loam—7–8%. This is due to the higher adsorption capacity of the latter due to the greater proportion of the clay component in its composition. However, the calculated data require practical confirmation.

The addition of cement to soils qualitatively changes their physical and mechanical characteristics (Table 7). In this case, the nature of the dependence is the same for density and strength: an increase in the cement dosage provides an increase in the numerical values of the characteristics.

Table 6 Physicochemical properties of soils

Sample	Free surface energy, mN/m	Sorption capacity, mg/g	Specific surface area	
			By air permeability, cm^2/g	By BET method, m^2/g
Sample 1	54.42	55.89	1654	12.9
Sample 2	72.25	45.65	1722	16.4

Table 7 Properties of reinforced soil depending on its type

Cement, %	Medium density, kg/m^3		Compressive strength, MPa	
	Light silty loam	Heavy silty loam	Light silty loam	Heavy silty loam
3	2050	2070	1.6	1.5
4	2060	2080	2.4	2.2
5	2080	2100	2.6	2.4
6	2090	2110	3.3	3.2
7	2110	2130	4.5	4.4
8	2120	2160	5.1	4.9

The density of the hardened soil grows insignificantly: from 2050 to 2160 kg/m^3 at the maximum dosage of cement (Fig. 1). At the same time, the density of the composite when using heavy loam is somewhat higher, which is associated with a larger share of the fine fraction in its composition, and, consequently, a higher packing density of particles in the volume.

Taking into account the fact that the strength characteristics of the hardened soil directly depend on its maximum density, it is fair to assume the correlation of the data. In general, the addition of cement to clay soil provides an increase in strength by more than three times with an increase in the cement dosage from 3 to 8% (Fig. 2). However, the density of heavy loam is higher than light loam. Then, when analyzing the strength, the opposite is noted—the strength of the strengthened light soil is slightly higher than that for heavy soil. Nevertheless, in the case of both soils, the addition of only 4% of cement provides the values required by regulatory documents: the compressive strength must be at least 2 MPa. However, an increase in cement concentration to the calculated values (7% for light and 8% for heavy loam) provides a significant margin of safety for the consolidated system.

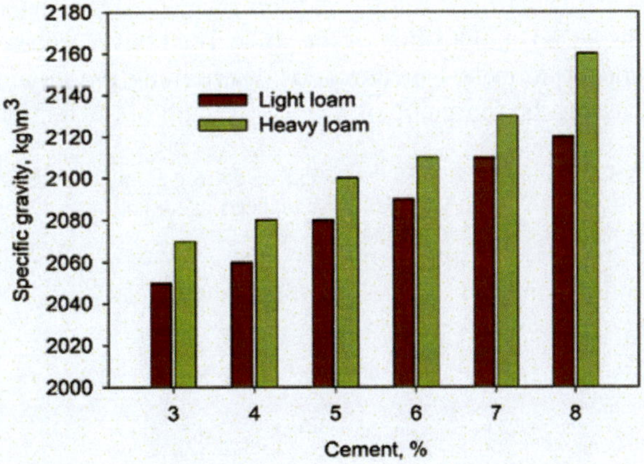

Fig. 1 Dependence of the density of hardened soil on the concentration of cement

Fig. 2 Dependence of the strength of the strengthened soil on the concentration of cement

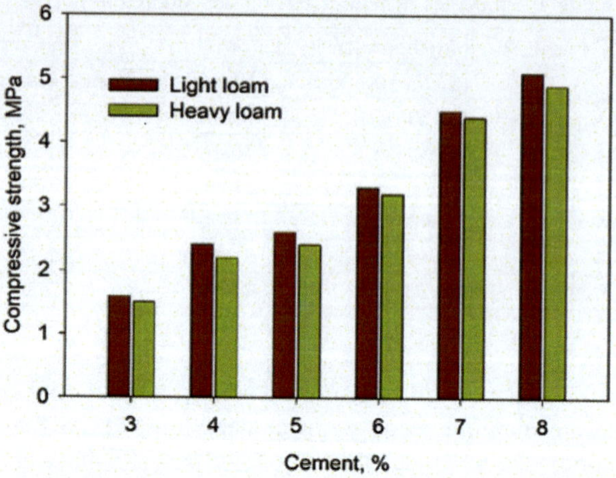

4 Conclusion

The paper shows the influence of physical and chemical indicators of soil quality on its strengthening by cement. Soils with a greater proportion of clay component (heavy loams) require a slightly higher dosage of cement. At the same time, the strengthened system is distinguished by a higher density, which will provide a lower filtering capacity of the consolidated system and predict a longer persistence in the environment.

The efficiency of the calculation method for the selection of the cement dosage is shown, taking into account the composition of the cement, as well as the physicochemical properties of the soil being strengthened (activity and moisture content). The calculated dosage of cement ensures the formation of a consolidated system with a large margin of safety. At the same time, the concentration slightly differs when using light and heavy loam, which is associated with the peculiarities of the dispersed composition of the soils. The convergence of the results of the calculated and experimental methods for determining the required amount of cement, which ensures the formation of a given strength of the reinforced system, has been proved.

Acknowledgements The reported study was funded by RFBR, according to research project No. 19-38-90091. The work was realized using equipment of High Technology Center at BSTU named after V. G. Shukhov.

References

1. Trautvain AI, Yakovlev EA, Gridchin AM (2021) Analysis of the effectiveness of the new generation of additives in improving physical and mechanical properties of highway subgrades. Mater Sci Forum 1017:101–110
2. Trautvain AI, Akimov AE, Grichanikov VA (2020) Stabilization of the clay soil of the increased humidity for road construction. Lect Notes Civil Eng 95:66–72
3. Kalachuk TG, Shirina NV (2020) About irreversible changes in soils strength properties after dynamic loads. IOP Conf Ser Mater Sci Eng 753(2):022051
4. Igosheva LA, Grishina AS (2016) Review of the main methods of strengthening foundation soils. Bull Perm Natl Res Polytech Univ Constr Geotech 7(2):5–21
5. Trautvain AI, Manokhin AV (2020) Influence of baustab stabilizing additive on changes in soil water retention capacity. Bull BSTU Named After V.G. Shukhov 6:8–17
6. Selitskaya NV, Bodyakov AN (2018) The effect of temperature-humidity treatment in terms of soil concrete strength based on technogenic raw materials of ballast stone cleaning. Bull BSTU Named After V.G. Shukhov 11:41–46
7. Mavliev L, Vdovin E, Stroganov V, Konovalov N (2021) Road cement-mineral materials with granulometric and hydrophobic additives. Lect Notes Civil Eng 141:20–28
8. Plyuta K (2019) Determination of the percentage of lime in the strengthening of clay soils using pH. IOP Conf Ser Mater Sci Eng 667(1):012079
9. Vdovin E, Mavliev L, Stroganov V (2020) Interaction of clay soil components with portland cement and complex additive based on octyltriethoxysilane and sodium hydroxide. IOP Conf Ser Mater Sci Eng 012031
10. Du Ch, Yang G, Zhang T, Yang Q (2019) Miltiscale study of the influence of promoters on low-plasticity clay stabilized with cement-based composites. Constr Build Mater 213:537–548
11. Dmitrieva TV, Markova IYu, Strokov VV, Bezrodnykh AA, Kutsyna NP (2020) Efficiency of stabilizers of various composition for strengthening the soil with a mineral binder. Constr Mater Prod 3(1):30–38
12. Mavliev L, Vdovin E, Stroganov V, Konovalov N (2021) Road cement-mineral materials with granulometric and hydrophobic additives. Int Conf Sustain Energy Syst Innov Perspect 141:20–28
13. Zvyagintseva AS (2016) Features of the design and construction of road pavements with foundations of hardened soils. Young Sci 12(116):284–288
14. Berdigulova ZT (2019) Features of soil stabilization in road construction. Format. Eng Technol 2(2):4–11
15. Efimenko VN, Charykov YM (2018) Prospects for the development of electrothermal geotechnology of soil transformation in road construction. Bull Tomsk State Univ Archit Civil Eng 20(1):197–207
16. Lofler M, Slobodchikova NA (2018) Methods for selecting the composition of soils fortified with lime for road construction. Proc Univ Invest Build Prop 8(2)(25):141–147
17. Strokova VV, Lyutenko AO, Lebedev MS (2012) Predicted and experimental methodology for estimation of cement quantity to achieve optimum curing conditions of stabilized soil. VIII Jordana International del Asfalto, pp 1–7

Author Index

S. V. Klyuev (ed.), *Digital Technologies in Construction Engineering*, Lecture Notes in Civil Engineering 173, https://doi.org/10.1007/978-3-030-81289-8